THE LIBRARY
1975

D0975695

MODERN METHODS OF GEOCHEMICAL ANALYSIS

Monographs in Geoscience
General Editor: Rhodes W. Fairbridge
Department of Geology, Columbia University, New York City

B. B. Zvyagin
Electron-Diffraction Analysis of Clay Mineral Structures—1967

E. I. Parkhomenko
Electrical Properties of Rocks—1967

L. M. Lebedev
Metacolloids in Endogenic Deposits—1967

A. I. Perel'man
The Geochemistry of Epigenesis—1967

S. J. Lefond
Handbook of World Salt Resources—1969

A. D. Danilov
Chemistry of the Ionosphere—1970

G. S. Gorshkov
Volcanism and the Upper Mantle: Investigations in the Kurile Island Arc—1970

E. L. Krinitzsky
Radiography in the Earth Sciences and Soil Mechanics—1970

B. Persons
Laterite—Genesis, Location, Use—1970

D. Carroll
Rock Weathering—1970

E. I. Parkhomenko
Electrification Phenomena in Rocks—1971

R. E. Wainerdi and E. A. Uken
Modern Methods of Geochemical Analysis—1971

In preparation:

A. S. Povarennykh
Crystal Chemical Classification of Minerals

MODERN METHODS OF GEOCHEMICAL ANALYSIS

Edited by

Richard E. Wainerdi

Associate Dean of Engineering, and
Head, Activation Analysis Research Laboratory
College of Engineering
Texas A&M University
College Station, Texas

and

Ernst A. Uken

Scientific Advisory Council
Pretoria, South Africa

With a Foreword by
Sir Edward Bullard, F. R. S.

PLENUM PRESS • NEW YORK–LONDON • 1971

Library of Congress Catalog Card Number 75-157148

SBN 306-30474-0

A Division of Plenum Publishing Corporation
227 West 17th Street, New York, N.Y. 10011

United Kingdom edition published by Plenum Press, London
A Division of Plenum Publishing Company, Ltd.
Davis House (4th Floor), 8 Scrubs Lane, Harlesden, NW10 6SE, England

Printed in the United States of Amercia

FOREWORD

The founders of geology at the beginning of the last century were suspicious of laboratories. Hutton's well-known dictum illustrates the point: "There are also superficial reasoning men...they judge of the great operations of the mineral kingdom from having kindled a fire, and looked into the bottom of a little crucible." The idea was not unreasonable; the earth is so large and its changes are so slow and so complicated that laboratory tests and experiments were of little help. The earth had to be studied in its own terms and geology grew up as a separate science and not as a branch of physics or chemistry. Its practitioners were, for the most part, experts in structure, stratigraphy, or paleontology, not in silicate chemistry or mechanics.

The chemists broke into this closed circle before the physicists did. The problems of the classification of rocks, particularly igneous rocks, and of the nature and genesis of ores are obviously chemical and, by the mid-19th century, chemistry was in a state where rocks could be effectively analyzed, and a classification built up depending partly on chemistry and partly on the optical study of thin specimens. Gradually the chemical study of rocks became one of the central themes of earth science. It was, however, a very tedious task; and not many able people were prepared to devote their main efforts to the labyrinthine, and insufferably boring, progress of the wet chemical analysis of a set of rocks where each specimen needs a week of careful work to determine its main constituents and the determination of the many trace elements is hardly to be contemplated. Efforts to get the work done on a large scale, as a business proposition were also not entirely satisfactory owing to the risk of error in a routine operation.

There was clearly an enormous incentive to develop quick and reliable methods. Because the earth is so large, but yet has processes on all scales from the atomic to the global, the need is not for tens or hundreds of analyses but for hundreds of thousands. There are about 90 elements in the rocks, most of them in abundances measured in parts per million or parts per billion. How should we estimate them? How far can we believe the results? These are the themes of this book.

There is a further question: why should we estimate them? What are we trying to do? There are two kinds of answers; some things are important because they are interesting and some are interesting because they are practically important. The possibility of determining not only the main constituents of a rock but also a large number of trace elements and their distribution among the constituent minerals should characterize the rock in such detail as to make it effectively unique. If this can be done it should be possible to be much more certain than we are at present about the displacement of the great transcurrent and transform faults of the continents and the oceans by comparing the displaced rocks on the two sides. In this way we should be able to get additional confidence in the fits of the continents and to work out the history of their movements. We should also be able to assign the continental remnants stranded in midocean to their former positions. Were the Seychelles once alongside Madagascar or were they attached to the mainland of Africa or perhaps to Australia or Antarctica? The success of purely geometrical methods of fitting the continents around the Atlantic should not blind us to the ambiguities in the Indian Ocean and still more to the almost untouched subject of the arrangement of the continents in pre-Mesozoic times. The record is so blurred by time and later movements that we need all the help we can get from geochemistry.

On the practical side, nothing is more certain than that we shall have to mine and use poorer grades of ore and that it is undesirable to waste our efforts in grinding and chemically processing lumps of barren rock. There must be a great future for what a computer engineer would call "real time" chemical analysis; such as the determination of one or a few valuable elements in a lump of rock as it passes on a conveyor belt, for example. Behind this is the larger question of ore genesis; we all know that "ore is where you find it" but this is not a very fruitful rule for prospecting. Can we, "by looking into a little crucible" learn enough about rocks to understand why ore is where it is, and then direct our search to places where the omnipresent trace elements have come together into workable ore? Why should a field in western England, and no field anywhere else, have several feet of pure strontium sulfate just under the surface? How can such a concentration of so rare an element occur and why just in one field? Such questions are practically important and also intellectual problems of the first class; it just happens that they are a little unfashionable at the moment.

By the time this book is published we shall have the first cores from the deep drilling in the oceans by project JOIDES and the first specimens from the moon. No one can tell what enlightenment and what problems these will bring. They cannot fail to tell us much about the history of the earth and of the solar system. The proper chemical analysis of these materials, gathered with so much effort, will be the culmination of two of the great adventures of our age. This book is, indeed, timely.

EDWARD BULLARD, F.R.S.

PREFACE

The field of geochemical analysis is at once an ancient art, and a modern science. It is becoming of importance in every country and, lately, in the areas between countries, and in the heavens above them. The economic pressure behind mineral exploration is only part of the reason geochemical analyses are important. The basic urge to know and understand the processes of the geologic past requires that analytical chemistry, of a most sophisticated sort, be applied to presently available materials in order to attempt to provide basic data for speculation about what went on in past eras.

Analytical geochemistry has been applied to moon samples, to undersea and ocean samples, and to many terrestrial samples, but very few workers have applied more than one of the methods described in this book to the same sample. There are many methods which are mutually complementary, and an important objective of this effort is to acquaint geochemical analysts with recent developments in competitive and complementary methods so that they can utilize them, where needed, to complete the analytical description of a given sample.

The Editors wish to thank all of the contributors to this book for their efforts, and particularly to thank Sir Edward Bullard, F.R.S., for his significant and thoughtful Foreword.

RICHARD E. WAINERDI
ERNST A. UKEN

CONTENTS

Chapter 3

Chemical Analysis and Sample Preparation

by V. C. O. Schüler

Chapter 4

Ion-Exchange Chromatography

by H. F. Walton

Chapter 5

Colorimetry

by Gordon A. Parker, and D. F. Boltz

Chapter 8

Atomic Absorption

by L. R. P. Butler and M. L. Kokot

Chapter 9

X-Ray Techniques

by H. A. Liebhafsky and H. G. Pfeiffer

Chapter 10

Radiometric Technique

by L. Rybach

Chapter 1

INTRODUCTION

R. E. Robinson, W. R. Liebenberg, and S. A. Hiemstra

National Institute for Metallurgy
Johannesburg, South Africa

1. CLASSICAL DEFINITION OF GEOCHEMISTRY

Geochemistry, the study of the chemistry of the earth, is one of the original boundary sciences, relying heavily on the techniques and principles of chemistry and physics to study the distribution of elements in the earth and to explain the processes that give rise to these distribution patterns. Many of those same early scientists that contributed to the knowledge and techniques of chemistry also contributed notably to the knowledge of geochemistry, simply because their subjects for study were natural products of the earth, in many instances minerals and rocks. However, the ideas and objectives of inorganic chemistry crystallized more quickly than those of geochemistry. It was easy to gather data on the composition of natural materials—but how could these be organized and classified into a formal, logical science? What is meant by the term "Geochemistry"?

Among the early attempts to provide a formal definition for geochemistry, the same elements can be noted in the attempts made by W. A. Wernadski, A. E. Fersman, and V. M. Goldschmidt. The classical definition given by Goldschmidt in his papers published in the interval from 1923 to 1937 can still be regarded as the most precise. Rankama and Sahama translate this definition as follows:

(a) To establish the abundance relationships of elements and nuclides in the earth.

(b) To account for the distribution of elements in the geochemical spheres of the earth, e.g., in the minerals and rocks of the lithosphere and in natural products of various kinds.

(c) To detect the laws governing the abundance relationships and the distribution of the elements.

The first stage of the science of geochemistry is analytical—the gathering of data on the distribution of elements. This book on modern methods of geochemical analysis thus deals with the techniques that can be used to realize the first object of geochemistry—that of accumulating data on the composition of natural materials.

This process started in an informal way when Bronze-Age man first realized that certain ores contained the elements necessary to his trade. Early in the 19th century, John Dalton, J. L. Gay-Lussac, and Amedeo Avogadro made the necessary observations that led to the ideas that chemical compounds had fixed compositions and that molecules and atoms exist. The necessary foundations were laid for quantitative analyses to be carried out, using the classical but laborious techniques of wet chemical analysis.

During the 19th century, a wealth of data on the composition of geologic materials, viz., minerals and rocks, were accumulated. The Swedish chemist Jöns Jacob Berzelius made notable contributions both to the science of geochemistry and to the science of chemistry, and is known for his many analyses of minerals and his compilation of an accurate table of atomic weights.

During the 19th century and the first half of the 20th century this process of accumulation of data continued, and, in the period 1908–1924, Frank Wigglesworth Clarke's invaluable compilation *The Data of Geochemistry* saw print.

The technique of wet chemical analysis is still, perhaps, the most valuable technique in the analysis of natural materials, since many of the modern physical methods are dependent on standards analyzed in the classical way. It is a laborious method, however, and for many elements the detection limits are poor. A need for more rapid and more sensitive methods was felt long ago.

In the meantime, the science of optical spectrography based on the classical experiments of Isaac Newton in 1664 and Joseph von Fraunhofer in 1814 was slowly evolving. In 1860 and 1861 Robert Bunsen and Gustav

Kirchhoff discovered caesium and rubidium by means of optical spectrography. In 1922 Assar Hadding demonstrated that optical spectrography is a powerful geochemical tool. From about the third decade of the 20th century it was used increasingly for gathering the facts of geochemistry, particularly the distribution of minor and trace elements in rocks and minerals. Today, optical spectrography is still used extensively for the analysis of large numbers of samples of rocks, minerals, and soil, and the instrumentation varies from the simple and dependable to the huge, sophisticated, direct-reading equipment capable of performing hundreds of analyses a day. In addition to techniques based on optical emission spectrography, infrared spectrophotometry and atomic absorption are being increasingly applied to geochemical problems. Although emission spectrography evolved into a rapid and sensitive tool with low detection limits for many elements, high precision could usually be attained only with difficulty, and the technique had limitations in the analysis of major components.

At the end of the 19th century, two epoch-making scientific discoveries were made within a period of a few months: the detection of X rays by Wilhelm Conrad Röntgen on November 8, 1895, and the discovery of natural radioactivity by A. H. Becquerel in February 1896. These discoveries form the foundation stones of a number of very powerful analytical techniques that became significant in the postwar years of the 20th century, and from about 1960 they have climbed to a position that threatens to overshadow the classical techniques for the analysis of geologic materials.

During 1903 the Russian botanist M. S. Tswett discovered a technique —which later became known as chromatography—for the separation of leaf pigments. In 1931 R. Kuhn and his coworkers demonstrated the efficacy of the method in separating vitamin-A fractions, and since that time the technique of chromatography has grown rapidly to the eminent position it occupies today.

Nowadays the number of instruments and techniques that can be used to analyze geologic materials is increasing rapidly, and they are continuously being refined.

The attainment of speed has always been antithetical to the achievement of high accuracy, and it still is. However, today we are nearer the ideal in both speed and accuracy in the analysis of geologic material than ever before.

While the pages of this book explore in detail the different chemical and physical methods for analyzing materials from the earth, it is also advisable to turn our attention briefly to the use that is being made of these analyses.

2. THE ROLE OF GEOCHEMICAL ANALYSIS

2.1. Geology

In geology very extensive use is made of the techniques and results of geochemical analysis. Broadly speaking, one can distinguish four fields of application: geochemistry, mineralogy, petrology, and exploration.

2.2. Geochemistry

Geochemical analysis provides the fundamental data from which the distribution of every element and nuclide in the main rock types of the earth, the lithosphere, the hydrosphere, the biosphere, and even the atmosphere and outer space, is calculated or estimated. Because of the availability of an increasing number of accurate analyses, old figures for these quantities can be refined and made more and more accurate.

Geochemical analyses are also essential in the study of the various processes in and on the earth that cause elements and nuclides to migrate, disperse, fractionate, or concentrate. Sometimes these processes are well defined and easy to detect with only a few analyses; other times the trends are weak and the concentration values are open to much variation, so that numerous chemical analyses and involved statistical calculations are necessary to detect them.

2.3. Mineralogy

In mineralogy the chemical composition of the minerals studied is of the utmost importance. No description of a mineral is complete without its chemical composition.

Isomorphous series are present in many of the most important rock-forming mineral families. The correlation of the physical properties and the chemical compositions of these minerals form an important branch of mineralogy. The laws that control the admittance, capture, or substitution of minor or trace elements in the structure of these minerals are very important because of the influence on differentiation trends and the enrichment of certain rare elements in the residual products of differentiation.

In the past the study of rare minerals that were available in very small amounts presented serious difficulties. Whereas the techniques for studying the physical properties of minute amounts were available, it was often extremely difficult or impossible to obtain high-quality chemical analyses of these materials. The advent of certain physicoanalytical techniques, notably that of electron-microprobe analysis, can be regarded as a major advance.

2.4. Petrology

In igneous petrology extensive use is being made of chemical analyses in the comparison of rocks from different areas and in the study of differentiation trends and petrographic provinces.

In metamorphic petrology the chemical compositions afford important clues to the original nature of the rocks and the processes responsible for their transformation, and also to whether metasomatic processes were operative or not.

In sedimentary petrology the chemical composition of sediments and their constitutive minerals may provide information on the provenances, the chemical and physical conditions that prevailed during their formation, and the weathering potential of these rocks.

One fundamental problem in petrology is that of obtaining samples that are representative of the parent-rock body. One or a few samples cannot reveal the amount of variation of rock bodies that may cover tens or hundreds of square miles. The availability of methods whereby accurate analyses of many samples can be made in a relatively short time promises to revolutionize the study of petrology.

2.5. Exploration

In the technique known as geochemical prospecting, analyses of samples of eluvial soil or plant material are made in the hope that areas in which there are anomalous concentrations of certain elements will be found. If such areas are found, they are studied in more detail, and this technique sometimes leads to the discovery of ore deposits of a grade and size suitable for exploitation. The technique is critically dependent on analytical methods that are rapid enough to permit thousands of analyses to be carried out in a short time, and sensitive enough to detect low concentrations of the elements being sought.

More indirect techniques include the study of an element known to be associated with the metals sought, for which a better analytical technique is available. One example of this is that a mercury halo is associated with certain ore deposits, and there exist extremely sensitive analytical techniques for the determination of mercury.

Sometimes a geochemical study of certain geologic formations and rock bodies may afford clues to the presence or absence of ore deposits in them.

2.6. Mining

Strictly speaking, the term "geochemical analysis" should not be related to activities other than geochemistry and geochemical prospecting. But it is such rigid adherence to formal and somewhat arbitrary definitions that has been responsible for so much compartmentalization in science, and it is hardly an encouragement for cross fertilization between different disciplines. It will become clear that almost all of the techniques described in this book will have application in many directions other than those conforming to the strict definition of geochemistry.

Probably the first real instance of a geochemical method of analysis being of direct use to the mining engineer was the application of radiometric methods of analysis in the deep-level mines both on the Witwatersrand and in the Orange Free State. The rather crude Geiger counters of the period immediately following World War II proved to be invaluable to the mining engineer by providing a farily good idea of the exact locality of the gold- and uranium-bearing conglomerates and also of their content of uranium. Many of the skills of low-cost, deep-level mining revolve round the ability of the mines to operate on those mine sections that are "payable," and to leave untouched those sections that are "unpayable." Any delays between the sampling of a newly exposed face and the receipt of the analytical results must inevitably bring about delays in mining and a reduced output for the mine. The use of a Geiger counter at rock faces provided a rough but nevertheless valuable assessment of the uranium content, which could be related to both gold and uranium payability.

More recently the demand for rapid on-the-spot analysis in mining has become even more apparent. The development of an efficient and economical rock-cutting unit by the mining research laboratory of the Chamber of Mines of South Africa threatens to revolutionize gold-mining techniques in South Africa. This new technique has made it possible to cut out a slice of rock containing the valuable constituents, rather than to blast it out with explosives; and it promises to greatly diminish the amount of waste rock associated with the valuable constituents, and thus considerably reduce the mining costs and increase the concentration of the valuable constituents in the feed to the metallurgical treatment plant. The maximum benefit to be derived from rock-cutting depends on the ability to decide beforehand where the valuable constituents are situated. In the gold mines in South Africa this does not at the moment represent too serious a problem since the gold and uranium are contained in fairly well-defined reefs that can be observed quite easily. Nevertheless, it has been proved that the gold and uranium are not distributed evenly throughout the reefs, and the possibility of being able to cut out only those portions of the reefs with payable quantities of gold is very attractive.

At the moment no obvious solution to this problem is in sight. What is required is a method of analysis that will give at least a semiquantitative estimate of the gold content, in the region of a few parts per million, in a relatively short time, and will be capable of being used underground at the rock face. It appears that the techniques of neutron activation or x-ray fluorescence cannot at present achieve the sensitivity required for them to be of much value; but further developments in the direction of stronger activation sources could well bring this very demanding problem into the region of being solvable.

Another area in mining where geochemical methods of analysis could play an important role in the future is in that of opencast mining, particularly for base metals such as copper and zinc. Many of the recently discovered, large ore bodies of base metals not only are of low grade but also are very complex in structure.

A typical case is the complex carbonatite ore body at Phalaborwa, in the Transvaal, which is being mined and processed by the Palabora Mining Company. This body consists of many zones of material—high-titanium magnetite, low-titanium magnetite, phoscorite-rich areas, carbonatite-rich areas—all of which contain copper in varying amounts. Certain areas of the deposit can be considered to contain payable quantities of copper, others not; and in the planning of the mining strategy it has been necessary to obtain a fairly accurate three-dimensional model of the whole body. The samples were obtained by drilling on a preestablished grid pattern, but the examination of all the samples by chemical analysis proved to be tedious and expensive. Even with the accurate three-dimensional model, the mining engineer would find it invaluable to have a simple instrumental method of analysis that could be used for checking the grade of each section of the pit before blasting, so that its destiny—whether the treatment plant or waste heap—can be established as early as possible. This concept is not far-fetched science fiction. The latest instruments using radioactive materials as sources of both γ rays and neutrons (e.g., californium 252) are coming close to achieving just this.

2.7. Ore Dressing

Where mining stops and ore dressing commences is perhaps a debatable point, but, if one accepts that the ore-dressing stage of the operation begins with the mechanical crushing of the ore to be treated, it is fairly obvious that there is scope for, and benefits to be derived from, rapid methods of analysis. Crushing, grinding, and the treatment of waste material represent inefficiency and wastage of money. The greater the degree of selectivity between waste and valuable constituents that can be exercised at each ore-dressing stage, the more efficient will be the operation.

The method that has been used the most successfully is that of radiometric analysis, which is now accepted as a standard procedure.

This method of analysis has been engineered to a point where either completely full trucks of ore can be "analyzed" or individual rocks on a moving belt can be separated on the basis of their content of radioactive materials.

Obviously this method is limited to the ores that contain naturally radioactive elements (which means essentially uranium and thorium). However, the problems in imparting a short-lived radioactivity to other ores, so as to enable separations to be carried out, are not really insurmountable.

Over the years the ore dresser has made use of just about every physical property of mineral particles to concentrate valuable constituents from waste materials. The use of properties such as specific gravity, size, and shape are well known and have been used since ancient times. Some of the more recent developments are of greater interest. Instruments have now been developed that measure the light reflected from individual mineral particles and, depending on the color, provide a signal that will allow one colored mineral to be separated from another. These instruments work at very high speed. If one considers the basic features, one realizes that this "color sorter" is only another form of activation analysis—that is to say, the measurement of photons of a certain wavelength produced by the excitation of the mineral by photons of a range of wavelengths.

When it is appreciated that most of mineral dressing deals with major concentrations of constituents and not with trace elements, it does not seem unrealistic to imagine that all the other "activation" techniques of analysis can be seriously considered as being sufficiently rapid to make mineral separation feasible.

2.8. Chemical Processing

The role of chemical and instrumental analysis in chemical processing is a major one. In the modern metallurgical processing plant, particular attention is being paid to accurate metallurgical balances, to the control and automation of all stages of the process, and to the specification analysis of the final product.

The specifications, imposed in the first instance by the nuclear-power and atomic energy programs for metals of extremely high purity, have caused repercussions in many other metallurgical practices.

The optical spectrograph is now becoming a standard instrument of metallurgical plant control, and the degree of expense and sophistication introduced into the latest automatic spectrographs is in itself an indication of the importance of this type of analysis to the metallurgist.

Very recently the mass spectrograph was applied to the control and final-product specification analysis of the metallurgical plant. The production of titanium dioxide pigment from ilmenite is a case where a rigid control over impurities such as iron, vanadium, and chromium is necessary at all stages of the operation, and where the mass spectrometer is apparently doing a very successful job.

3. SUMMARY

One can expect that in future years analytical methods will show many profound improvements and will have greatly increased applications.

Geochemical requirements are among the most challenging. Speed and sensitivity are probably the primary requisites, and for this reason the methods developed will probably find equally valuable applications in other areas, such as the mining, dressing, and chemical processing of minerals.

Thus, this book, which presents those methods of analysis of interest to geochemists, must not be read and studied with a restrictive attitude. Most of the methods described are standard techniques employed in a host of analytical laboratories throughout the world. The wider the audience that this book attracts, and the greater the degree of cross fertilization between the many analysts, the more all will benefit.

Chapter 2

STATISTICS

A. B. Calder

Newcastle upon Tyne Polytechnic
Newcastle, England

1. INTRODUCTION

1.1. The Nature and Scope of Statistics

Statistical methods are sometimes described as merely being techniques for treating numerical data. It is necessary, however, to restrict both the nature of the data and the reasons for studying them, before such methods can rightfully be called statistical. Statistics is concerned essentially with data that have been obtained by taking a set of observations, in the form of measurements or counts, from a source of such observations. The ultimate purpose of statistics is to infer, from the set of observations made, something about the source of the observations from which the set was taken and, in particular, to make estimates of certain numerical characteristics or parameters of the source of the observations.

Consider a granitic outcrop approximately 50 ft in diameter, of fairly uniform appearance, structure, and composition. Now consider the case of a chemist who is confronted with the task of assessing the cobalt content of this particular outcrop. Clearly the whole outcrop cannot be removed, crushed, and analyzed. Instead the chemist takes a number of samples, say 20 from different spots in the area. Each sample is crushed, thoroughly mixed, and a suitable portion of, say, 2.0 g may be weighed out for spectrochemical analysis. The results obtained may be as follows:

Co (in ppm):

0.25 0.20 0.25 0.22 0.24 0.23 0.23 0.28 0.26 0.23

0.24 0.24 0.25 0.22 0.23 0.21 0.23 0.22 0.21 0.20

Having obtained these values, the chemist will report that the average cobalt content for the granitic outcrop considered is (0.23 ± 0.005) ppm, where 0.005 is the standard error, discussed later. The statistical nature of the chemist's results is evident. One may, of course, regard the entire outcrop as a source of many millions of units or aliquots, each of 2.0 g. The average cobalt content of the set of 20 granite samples is of no particular interest in itself, but the 20 results may be used to make some inference regarding the average cobalt content for the entire outcrop, i.e., in estimating a numerical characteristic, the average.

The variation between the 20 results suggests that a similar variation probably exists for the outcrop as a whole, so that if a further set of 20 specimens were taken, a different average would probably be obtained. One may therefore ask whether such a determination i.e., the average of a set, can provide reliable information about the true average for the entire outcrop. It will be demonstrated that by statistical methods one can, in fact, ascertain the limits between which the true average must lie.

1.2. The Importance of a Statistical Approach in Analytical Chemistry

Each observation in analytical chemistry, no less than in any other branch of scientific investigation, is inaccurate in some degree. While the "accurate" value for the concentration of some particular constituent in the analysis material cannot be determined, it is reasonable to assume that the "accurate" value exists and it is important to estimate the limits within which this value lies.

The description of an analytical procedure is never complete without reference to the error involved in the process. It is therefore imperative that the chemist should be familiar with the basics of statistical methods, not only from the point of view of consistency in the general presentation of analytical results, but also in order to derive reliable estimates from the observed data. This eliminates the risk of misinterpretation and permits the interested analyst to assess correctly the application of a particular analytical technique within his own field of study. For example, where large sampling errors arise it is not enough to suggest "replication at the sampling stage" without further qualification. A quantitative estimate of the sampling variation will avoid the effects of overreplication with possible sacrifice of speed, or underreplication with loss in precision. It is also important to bear in mind that sampling errors may vary for different types of the same material. An example may be found in the variation in trace-element content of granites of different grain size. Again when the analytical treatment involves several stages (multistage analysis) it is desirable to indicate how the total error of determination is distributed. Such presentation may indicate how the time of operation can be cut down in any particular stage without significantly affecting the precision of the final estimate. Such considerations, of course, are not meant to imply that the analytical chemist is not interested in the general "accuracy" of his results. Many a chemist does, however, tend to interpret these results intuitively and, therefore, often rather inefficiently.

It should be emphasized that the statistical approach is concerned with the appraisal of experimental design and data only, whereas the analytical approach is often confined to the analytical process. Statistical techniques can neither detect nor evaluate constant errors or bias; the detection and elimination of inaccuracies are analytical problems. Nevertheless, statistical techniques can assist considerably in determining whether or not inaccuracies exist, and also whether procedural modifications have effectively reduced them. It would be wrong for the research worker to therefore assume that mathematical calculations by themselves can obviate the need for common sense and sound analytical technique in the chemical laboratory. To quote from Chambers[1]: "...statistical methods are merely tools for a research worker. They enable him to describe, relate and assess the value of his observations. They cannot make amends for incorrect observation nor can they of themselves provide a single fact of psychology, biology or any other subject of research".

It is therefore appropriate to briefly outline the mathematical treatment required for evaluating analytical data. This treatment is by no means exhaustive, but it is hoped that it will already reveal to the reader the statistical nature of certain problems encountered in analytical chemistry. Statistical concepts are more fully discussed in a monograph by Pantony.[2] An introduction to actual statistical calculations is provided by Chambers.[1] For further background reading in the application of statistical methods to chemical analyses Saunders and Fleming[3] and Youden[7] may be consulted.

2. BASIC CONCEPTS

2.1. Variation

All analysts must have observed a certain amount of variability or *variation* in their results (see Section 1.1). The science of statistics is to a very large extent concerned with the study of variation, and statistical methods are therefore applicable to all types of work where this phenomenon is encountered. There remains, unfortunately, a tendency on the part of the analyst to ascertain only average values and often the important characteristic of variation is completely ignored. In this connection such loose statements as "good duplicates" and "reasonably reliable data" are qualitative statements only. They serve no useful purpose and are even liable to misinterpretation. In chemical analyses the proper evaluation of data is an intrinsic part of the procedure and calls for correct presentation. It will be shown that variation is an important characteristic of a series of quantitative data and that its assessment is essential for arriving at the objective type of conclusion that each analyst should aim to achieve.

2.2. Populations and Samples: Parameters and Statistics

A set of data that is taken from some source of observations for the purpose of obtaining information about the source is called a *sample*, whereas the source of these observations is called a *population*. Statistical methods are concerned with the study of populations and their application permits one to infer from observations made on a sample something about the population from which the sample has been drawn. The distribution of the total population can usually be expressed in a mathematical form by using a small number of constants or *parameters*. These estimates are termed *statistics*. Measures of parameters are often called *descriptive statistics*, since they yield in a condensed form a description of a whole series of observations.

2.3. Types of Population and Variable

Previously (Section 1.1), a population (the granitic outcrop) was described, with a variable or variate (the cobalt content) which may take any value between certain limits. Here *real* (material) population and a *continuous* variable were considered. In statistical work other types of population and variable may be dealt with. The population could be a *hypothetical* one, for example, the possible results of an experiment; it could be a *finite* population, for example, the number of granitic outcrops in the United

Kingdom; or it could be an *infinite* population, for example, the possible yields of a chemical reaction. The variable may be continuous, as above, or *discontinuous*, for example, the number of defective items in a batch. In this chapter the variables will be denoted by X.

2.4. Probability and Random Selection

In the throwing of an ordinary, symmetrical, cubical dice the total number of possible events is six. A particular event, such as throwing a four, constitutes one out of these six, and therefore we say that the *probability* of throwing a four is $\frac{1}{6}$. Similarly the probability of throwing an even number is $\frac{3}{6}$, i.e., $\frac{1}{2}$. We assume that all the events are likely or equally probable.

In expressing limits of error, the probability P may be defined as the probability that a quantity will lie between certain limits. In tests of significance (see Section 6) the probability P' is the probability that a quantity will lie outside the limits. For given limits the sum of the probabilities that the quantity will be either within or outside these limits is 1 (i.e., the one or the other of these hypotheses will certainly be correct):

$$P + P' = 1 \qquad \text{or} \qquad P' = 1 - P$$

The term *random* is used in connection with the selection or allocation of one or more objects from a group. In the statistical sense a selection is said to be random when it is made in such a way that every object has the same probability of being chosen. Random sampling, as we shall see later, is often difficult to achieve. The human mind has a tendency to exercise a preference or bias, according to some physical characteristic which influences the senses. To overcome this type of influence, complete mechanization of the sampling procedure is desirable. Sometimes it may be necessary to select a number of items, in which case an individual may have a personal bias for particular numbers. This may be overcome by consulting a table of random numbers prepared by mathematicians for this purpose. If one does not have a set of random numbers available it may be more convenient to draw cards from a pack.

3. MEASURES OF LOCATION AND DISPERSION

If a whole population is examined for a particular property, a plot of values of the measurement of this property vs the number of times a value occurs (i.e., its frequency) is termed the *frequency distribution* of the variable

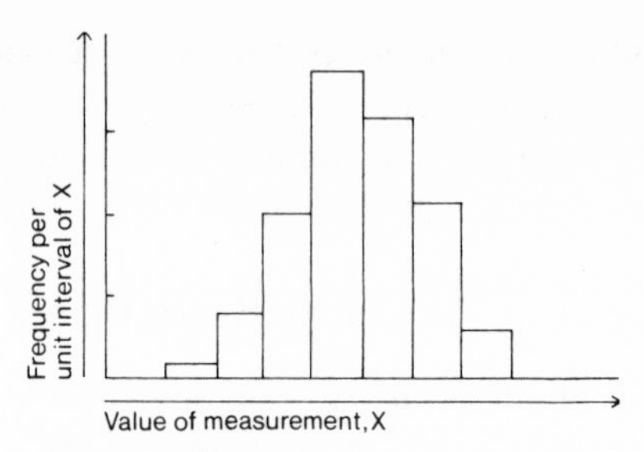

Fig. 1. Histogram of measurements.

in question. It is usually impossible to draw a continuous curve through the points of a frequency distribution because of the relatively small number of measurements made. In such cases the values are grouped into increments of equal size (i.e., class intervals) and the distribution then takes the form of a *histogram* shown in Fig. 1. If, however, a very large number of observations are made enabling a smaller class interval to be used, i.e., if the points along the base line are much more numerous, then the boundary of the histogram becomes a smooth continuous line corresponding in shape to the curve shown in Fig. 2.

The concept of a frequency distribution is basic to all statistical work and we must therefore now consider the parameters defining a frequency distribution and their numerical expression. The measures of primary interest are the location of the distribution or *average* value of the variable, and the spread of the distribution, i.e., the *variability*.

3.1. Average: Arithmetic Mean

The average value may be expressed in a number of forms, such as, the arithmetic mean, the median, or the mode. The best known and most useful form of average is the arithmetic mean, usually referred to simply

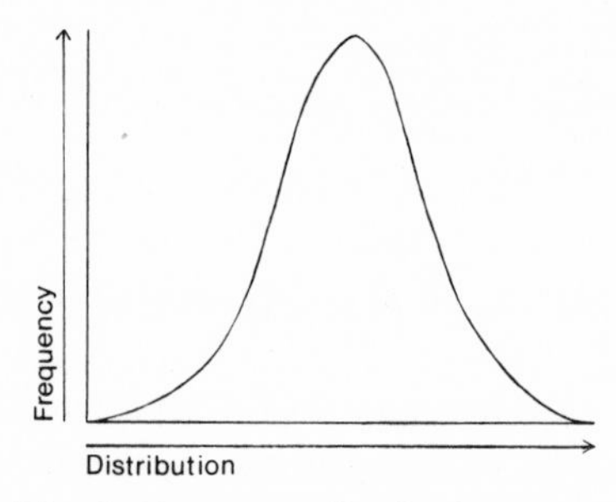

Fig. 2. Frequency–distribution curve.

as the "average" or "mean." For a series of n observations $X_1, X_2, \ldots, X_n$, the mean, written as $\bar{X}$, is given by

$$\bar{X} = 1/n(X_1 + X_2 + \ldots + X_n) = \Sigma(X)/n$$

where Σ represents the summation. Our purpose here is not merely to calculate the mean of a series of sample values, but to estimate the mean in the population from which the sample was drawn. The corresponding population parameter is written as μ and it can be shown that the sample mean, $\bar{X}$, is in fact a satisfactory estimate of μ.

3.2. Variability: Standard Deviation and Variance

The only useful measure of scatter or dispersion suitable for critical work is the *standard deviation*. For a series of n observations $X_1, X_2, \ldots, X_n$ the standard deviation is given by $s = [\Sigma(X - \mu)^2/n]^{1/2}$; in practice μ is not generally known, and we rely on the sample estimate $\bar{X}$. The quantity $s_1 = [\Sigma(X - \bar{X})^2/n]^{1/2}$ is not, however, an unbiased estimate of the corresponding population parameter σ and this effect is counterbalanced by replacing n by $n - 1$ in the denominator. The general rule is to use Roman letters for sample estimates (e.g., $\bar{X}$ for mean; s for standard deviation) and Greek letters for population parameters (i.e., μ for mean; σ for standard deviation). Hence, the standard deviation is given by $s = [\Sigma(X - \bar{X})^2/n - 1]^{1/2}$ with the usual notation. The square of the standard deviation, i.e., s^2, is known as the mean square or *variance* v of the population. The interesting property of variance is that it is linearly additive provided that the sources of the variances are independent.

3.3. Degrees of Freedom

A concept of great importance in statistical analysis is that of *degrees of freedom*. Consider again the series of n observations $X_1, X_2, \ldots, X_n$ and their sum $\Sigma_1(X)$. If the average (or sum) is fixed and n is also fixed, then we can give $n - 1$ of the X's any values we please but the nth will be fixed. Thus, for $n = 6$ and $\Sigma(X) = 25$ if we give to X_1 up to X_5 the values 7, -2, 6, 8, -1 summing to 25, then X_6 must be 7. We say there are five degrees of freedom given by $n(= 6)$, minus the number of constraints $(= 1)$ that have been placed on the system by fixing the average (or the sum). A small sample (in the statistical sense) of observations tends to underestimate the variance, and a better estimate is obtained by dividing the sum of squares $\Sigma(X - \bar{X})^2$ by the number of degrees of freedom $n - 1$, whence v(variance, unbiased estimate) $= [\Sigma(X - \bar{X})^2/n - 1]$ (see above).

It must be pointed out that the number of degrees of freedom will not always be simply $n - 1$, since in certain more advanced calculations there may be r constraints $r > 1$ and therefore only $n - r$ degrees of freedom.

3.4. Calculation of the Standard Deviation

The calculation of the standard deviation is not a difficult matter when the observations consist of small integral values and $\bar{X}$ is also integral. This is seldom the case in practice and to facilitate the arithmetic the following procedure is adopted for obtaining the sum of squares. Subtracting some quantity M from the X's and working with the quantities $X - M$ we have

$$\Sigma(X - M)/n = \Sigma(X)/n - M = \bar{X} - M$$

and, therefore,

$$\bar{X} = \Sigma(X - M)/n + M$$

Now

$$\begin{aligned}
\Sigma(X - M)^2 &= \Sigma[(X - \bar{X}) + (\bar{X} - M)]^2 \\
&= \Sigma(X - \bar{X})^2 + 2\Sigma[(X - \bar{X})(\bar{X} - M)] + \Sigma(\bar{X} - M)^2 \\
&= \Sigma(X - \bar{X})^2 + 2(\bar{X} - M)\,\Sigma(X - \bar{X}) + n\left[\frac{\Sigma(X - M)}{n}\right]^2 \\
&= \Sigma(X - \bar{X})^2 + \frac{[\Sigma(X - M)]^2}{n}
\end{aligned}$$

since

$$\Sigma(X - \bar{X}) = 0$$

Thus

$$\Sigma(X - \bar{X})^2 = \Sigma(X - M)^2 - \frac{[\Sigma(X - M)]^2}{n}$$

If a calculating machine is available we simply put $M = 0$, giving

$$\Sigma(X - \bar{X})^2 = \Sigma(X^2) - [\Sigma(X)]^2/n$$

3.5. Coefficient of Variation

An error or variation is expressed as a coefficient of variation. A percentage coefficient of variation is derived from the standard deviation (Section 3.2) according to the relation

$$\text{coefficient of variation } (V) = \frac{100s}{\bar{X}}$$

4. FREQUENCY DISTRIBUTION

4.1. Normal Distribution

The majority of frequency distributions encountered in analytical practice are of much the same shape as that shown in Fig. 2, which depicts the shape of a normal or Gaussian distribution. Such curves are virtually, though not exactly, symmetrical. They may be made nearly symmetrical by a suitable transformation of the variable, such as expressing the variable in logarithms for example. Many of the methods of statistics depend on the assumption that the variables under consideration are normally distributed, and most of the methods cannot strictly be applied unless this assumption is justified. Biological variables are, for example, often distributed in such a manner, but it is not safe to assume that any particular distribution is normal without prior examination.

4.2. Equation and Properties of the Normal Distribution

The normal curve has an exact mathematical formula and its equation can be given in terms of the standard deviation:

$$\log_e\left(\frac{y}{y_0}\right) = -\frac{1}{2}\left(\frac{X-\mu}{\sigma}\right)^2$$

where y is the height of the curve above the horizontal axis and y_0 is the value of the ordinate corresponding to $X = \mu$. The equation therefore defines y for any value of X between $\pm\infty$.

This curve has important properties. It is a continuous curve and applies to continuous variables, where the difference between one value of the variable and the next can be infinitesimally small. It is symmetrical about the mean μ and falls off on either side, tailing off asymptotically to the X axis in both directions. Only the portion drawn in Fig. 2 is of practical importance. The points of inflexion occur at $X = \mu \pm \sigma$. The area under any part of the curve is proportional to the fraction of the population lying between the corresponding values of X. Thus, if a single observation is drawn at random from this population the probability that it will lie between $X = p$ and $X = q$ is proportional to the area between these limits. The probability P' of main practical interest here is that of an observation lying outside the limits $\mu \pm d$. The value will, of course, depend on σ, and to calculate it we form first what is known as the *normal deviate*:

$$u = \frac{X-\mu}{\sigma} = \frac{d}{\sigma}$$

The probability P corresponding to any value of $u(= d/\sigma)$ may be derived mathematically. Some numerical values of P' for given values of u and vice versa are given in Table I:

Table I

Probability of a Random Observation Lying outside the Limits $\mu \pm d$

$u(= d/\sigma)$	$P'(= 1 - P)$	$P'(= 1 - P)$	$u(= d/\sigma)$
1	0.3173	0.10	1.64
2	0.0455	0.05	1.96
3	0.0027	0.01	2.58

These figures mean that if a value X of the variable is selected at random from a normal distribution, there is a 0.3173 (or 31.7%) probability that its normal deviate will lie outside the limits ± 1, a 0.0455 (or 4.5%) probability that it will exceed ± 2, and so on. These are often referred to as the 68 and 95% confidence limits of a series of results. Such probabilities refer to the plus and minus range, and are called double tailed; those referring to the plus or minus range alone are called single tailed, for which the probability is exactly one half of that of the double tailed data. Alternatively, referring again to Table I and selecting round-figure probabilities, there is a 0.10 (10%) probability that the normal deviate will exceed ± 1.64 and a 0.01 (1%) probability that it will exceed ± 2.58.

4.3. Justification for the Assumption of Normality in Practice

The majority of continuous variables encountered in analytical chemical practice are distributed in approximately normal fashion. A moderate displacement from normality seldom gives rise to appreciable errors, since (a) it can be shown that the distribution of the mean tends to normality with increase in sample size, irrespective of the form of the parent distribution (central limit theorem), and the measures of probability in significance tests are computed with reference to the sample average, and (b) marked deviations from normality can be greatly reduced by transformation of the data onto another scale, such as the logarithmic scale, for example.

4.4. Binomial and Poisson Distributions

In some cases the significance of observed results may be tested by the use of the *binomial distribution.* This distribution consists of the expansion of the expression

$$N(p+q)^n = N\left(p^n + np^{n-1}q + \cdots + \frac{n(n-1)\cdots(n-r+1)}{r!}p^{n-r}q^r + \cdots + q^n\right)$$

Usually in probability problems, p is the probability of an event occurring (i.e., a success) and q is the probability of an event not occurring (i.e., a failure). If both are expressed as fractions, then $p + q = 1$. N is the number of *trials* made and n is the number of *events* in each trial. It can be shown that the mean of a binomial distribution $(p + q)^n$ is np and that the standard deviation is $(npq)^{1/2}$. The form of the binomial distribution varies considerably depending on the values of p and n. When p is not small but n is large enough, it can be verified that the binomial distribution approximates to the normal error curve.

One important practical case is when p is very small, that is, the probability of the event happening is very small, but n is large, so large, in fact, that np is not insignificant. It can be shown that if p is small while n is large, such that $np = m$, the binomial expansion of $(p + q)^n$ approximates closely to the series

$$e^{-m}(1 + m/1! + m^2/2! + \ldots + m^n/n!)$$

This is known as the Poisson series and any distribution which corresponds to the successive terms of this series is called a *Poisson distribution.* The mean of the distribution is m and the standard deviation is $m^{1/2}$. Data conforming approximately to Poisson distributions occur widely in science. They also arise in many industrial problems.

These distributions can be applied to the study of counting techniques such as those used in dust-particle analysis and the quantitative estimation of radioactivity. In these measurements only a small proportion of the total number of particles present are counted by the device used and so the rare event is that a particle happens to be counted. Compared with the application of the Gaussian expression, however, their use is limited.

5. STANDARD ERROR

5.1. Sampling Distributions

In most instances we are more interested in the reliability of the sample mean $\bar{X}$ than with that of an individual observation. When systematic errors are eliminated the standard deviation of a set of repeated measurements may be used to give an assessment of the random error of the mean. More definite conclusions about the reliability of a mean can be obtained by considering the scatter of a series of means, each obtained from a number n of repeated measurements, about the true value of the measurement. Consider a population of single results divided into groups or samples, each containing a set of n values. The distribution of the $\bar{X}$'s is known as a *sampling distribution* and may be treated in exactly the same way as the parent distribution.

5.2. Standard Error

Obviously, the mean of a sample will generally not be the same as the mean of the whole population. If n is large then two means may not differ very much, but if n is small they may. It is possible to express mathematically the manner in which the means of different samples of given size are distributed. It can be shown that this distribution is itself normal and such that its mean equals the true mean of the whole population, while its standard deviation equals $\sigma/\sqrt{n}$, where σ is the standard deviation of the whole population. In Fig. 3 it is illustrated how the distribution of the means of samples retains its normal form, but decreases in dispersion with increase in sample size. The standard deviation of the sampling distribution of the mean has been given a special term, the *standard error* and may be used for probability calculations concerning the mean in exactly the same way as the standard deviation is used in connection with single observations. Since

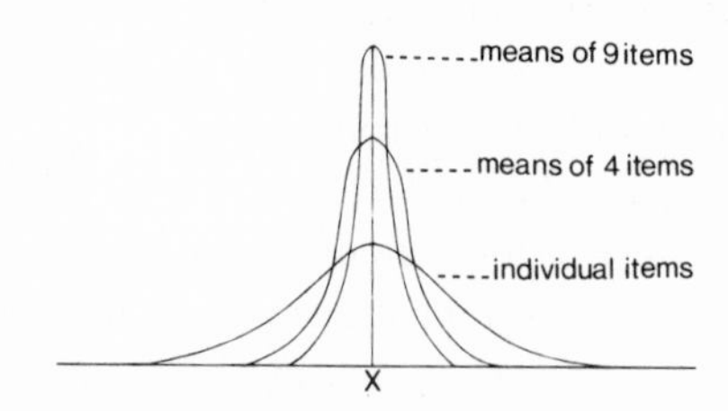

Fig. 3. Distribution of the means of samples. X is the grand average for entire population.

an experiment or analysis usually consists of a single sample only (using sample in the statistical sense) the standard error cannot be estimated directly and recourse is made instead to the relationship already referred to above, namely, $\sigma_{\bar{X}} = \sigma/\sqrt{n}$. In practice ($\sigma$ being generally unknown), an unbiased estimate of $\sigma_{\bar{X}}$ is obtained by calculating $s_{\bar{X}} = s/\sqrt{n}$, where $s_{\bar{X}}$ is an estimate of the standard deviation of the series of means that would be obtained if further samples were drawn.

5.3. Theoretical Derivations

5.3.1. Variance of a Function of Variates

Let A be a function of X and Y, i.e., $A = f(X, Y)$. If $\bar{A}$, $\bar{X}$, and $\bar{Y}$ are the mean values of A, X, and Y, respectively, then $\bar{A} = f(\bar{X}, \bar{Y})$. Consider two values of X and Y which differ from their means by the small amounts ΔX and ΔY, respectively. The deviation ΔA of the corresponding value of A from its mean is given by

$$\Delta A \simeq \frac{\partial A}{\partial X}\,\Delta X + \frac{\partial A}{\partial Y}\,\Delta Y$$

If a large number n of pairs of values of X and Y are considered, the variance of the resulting values of A about their mean will be $\Sigma(\Delta A)^2/n$. But

$$(\Delta A)^2 = \left(\frac{\partial A}{\partial X}\right)^2 (\Delta X)^2 + 2\left(\frac{\partial A}{\partial X}\right)\left(\frac{\partial A}{\partial Y}\right)(\Delta X)(\Delta Y) + \left(\frac{\partial A}{\partial Y}\right)^2 (\Delta Y)^2$$

Therefore

$$\sigma_A{}^2 = \left(\frac{\partial A}{\partial X}\right)^2 \Sigma(\Delta X)^2/n + 2\left(\frac{\partial A}{\partial X}\right)\left(\frac{\partial A}{\partial Y}\right)\Sigma[(\Delta X)(\Delta Y)]/n + \\ + \left(\frac{\partial A}{\partial Y}\right)^2 \Sigma(\Delta Y)^2/n$$

The term $[\Sigma(\Delta X)(\Delta Y)]/n$ represents the *covariance* of X and Y and is defined as the average value of the product $(X - \mu_X)(Y - \mu_Y)$. But if the X's and Y's are independent, i.e., if the probability that any one of them will assume a certain value does not depend on the value assumed by any other, then $\Sigma[(\Delta X)(\Delta Y)] = 0$. Now, $\Sigma(\Delta X)^2/n = \sigma_X{}^2$ and $\Sigma(\Delta Y)^2/n = \sigma_Y{}^2$. Therefore,

$$\sigma_A{}^2 = \left(\frac{\partial A}{\partial X}\right)^2 \sigma_X{}^2 + \left(\frac{\partial A}{\partial Y}\right)^2 \sigma_Y{}^2$$

5.3.2. *Variance of a Sample Mean*

Consider a large distribution of values of the variable X. If $\bar{X}$ is the mean of a sample of n of these values, then

$$\bar{X} = (X_1 + X_2 + \ldots + X_n)/n$$

As $X_1, X_2, \ldots, X_n$ are drawn at random from the distribution, they can be regarded as independent of one another and, therefore,

$$\bar{X} = f(X_1, X_2, \ldots, X_n)$$

From Section 5.3.1 we can deduce $\sigma_{\bar{X}}^2$ thus:

$$\sigma_{\bar{X}}^2 = \sigma_{X_1}^2/n^2 + \sigma_{X_2}^2/n^2 + \ldots$$

Now $\sigma_{X_1}^2, \sigma_{X_2}^2, \ldots$ are each equal to the distribution variance σ^2, and as there are n terms in the expression for $\sigma_{\bar{X}}^2$, it follows that

$$\sigma_{\bar{X}}^2 = n \cdot \sigma^2/n^2 = \sigma^2/n$$

As $\sigma_{\bar{X}}^2$ is the variance of the sample mean about the population mean μ, $\sigma_{\bar{X}}$ is the standard deviation of $\bar{X}$ about μ, i.e., the standard error of $\bar{X}$; thus,

$$\sigma_{\bar{X}} = \sigma/\sqrt{n}$$

5.4. The *t* distribution

For a variable X distributed normally about a mean μ with standard deviation σ, we can calculate the probability that a random observation will lie within or outside certain limits by obtaining the normal deviate (Section 4.2) as $u = (X - \mu)/\sigma$, and entering the appropriate probability table. Similar calculations may be made in connection with a sample mean $\bar{X}$ by taking $u = (\bar{X} - \mu)/\sigma_{\bar{X}} = (\bar{X} - \mu)\sqrt{n}/\sigma$. In practice σ will generally not be known and we have to depend on the estimate s. We then have corresponding to the normal deviate the quantity t given by $t = (\bar{X} - \mu)/s_{\bar{X}} = (\bar{X} - \mu)\sqrt{n}/s$. For large values of n the distribution is nearly the same as that of the normal deviate, but for small values the discrepancy becomes more marked. The values of t vary with the number of degrees of freedom in the calculation of s and with the probability level chosen for the limits. A table of values of t for the 0.95 probability level is given in Section 10.1.

6. TESTS OF SIGNIFICANCE

6.1. Introductory Note

If two mean results differ from one another, two hypotheses can be advanced to explain the discrepancy; either it is due to chance experimental error, or it is due to a *significant difference*. The first hypothesis is known as the *null hypothesis*, and the level of significance of this hypothesis can be estimated by using *tests of significance*.

6.2. Normal Deviate Test

This test is used to determine whether a sample mean $\bar{X}$ differs significantly from a known population mean μ, the population standard deviation σ being known. If the difference between $\bar{X}$ and μ is due only to chance, i.e., the null hypothesis is assumed, this means that $\bar{X}$ is a sample mean drawn from a normal distribution of such means about μ. The probability that a sample mean (chosen at random) will lie outside the limits given by $\pm u$ is $P'(= 1 - P)$. If P' is small, the probability that a chosen sample mean will have as large a value as u will be small. The probability, then, that the sample mean has been drawn from the normal distribution is small and the null hypothesis is likely to be incorrect. P' can therefore be regarded as a measure of the *level of significance* of the null hypothesis. The exact level of significance selected for acceptance or rejection of the null hypothesis varies with the type of problem considered. Conventionally, but arbitrarily, an effect is considered to be significant if the probability P' is less than 0.05 and highly significant if it is less than 0.01. In other words, if $P' > 0.05$ the null hypothesis is accepted, and the difference $\bar{X} - \mu$ is ascribed to chance. If $P' < 0.05$, the null hypothesis is rejected, and the difference $\bar{X} - \mu$ is considered to be significant. Referring to the table in Section 4.2 we find that the value of u corresponding to $P' = 0.05$ is ± 1.96. If u lies outside these limits, $\bar{X} - \mu$ is significant, whereas if u lies within these limits, $\bar{X} - \mu$ is not significant.

6.3. *t* Test

The t test is used to decide whether the difference between two sample means is significant when the population mean μ and standard deviation σ are unknown. The quantity t which has a theoretical limiting value for a given probability level, is calculated from two sets of experimental data.

As in the normal deviate test, if the calculated value exceeds the theoretical value at $P' = 0.05$, the null hypothesis is likely to be incorrect, and there is a significant difference between the two sets of results (see Section 10.1).

6.4. Variance Ratio (F) Test

Another method by which two samples of values of a quantity can be compared with one another is the "variance ratio" test. Here we have to introduce the F distribution, F being the ratio of the two variances (always the larger variance divided by the smaller variance). If we have two independent estimates of variance from the same normal population, s_1^2 (d.f. $= n_1$) and s_2^2 (d.f. $= n_2$), the probability that the ratio $F = s_1^2/s_2^2$ exceeds a certain value is a function of n_1 and n_2 only, and this function can be referred to tables for various levels of probability (see Section 10.2).

7. REGRESSION

7.1. Introduction

The scatter diagram obtained when values of Y, a quantity liable to random error, are plotted against X, a quantity with negligible error, is shown in Fig. 4. The line of best fit through these points can be drawn by eye but is dependent upon the subjective judgment of the person who draws it. Consequently the position of the line will differ slightly from person to person. A line of best fit independent of individual judgment must be drawn mathematically. Such a line is called a *regression line*. Regression provides the best-fitting relation between two quantities, one of which is subject to chance error. For example, in instrumental analysis the calibration involves plotting the instrumental "response" (Y, e.g., "wave height" in polarography, "drum reading" in absorptiometric analysis) against concentration

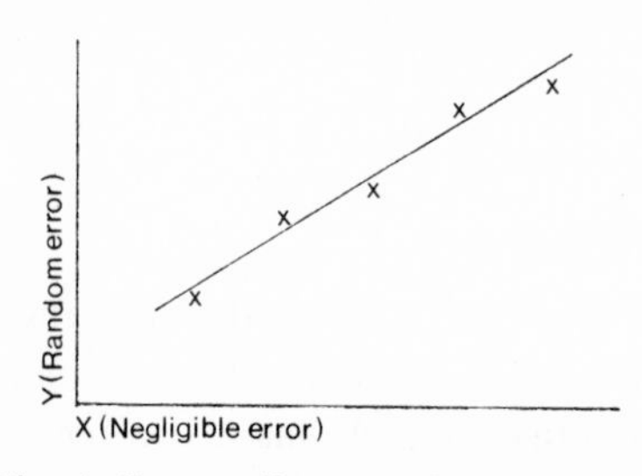

Fig. 4. Scatter diagram of Y against X.

of standard (X). The experimental error in X is negligible but an appreciable error may arise in Y. The simplest case, and the one we shall deal with here, is when the relation between Y and X should theoretically be a straight line.

7.2. Method of Least Squares

When drawing a line of best fit, an attempt is made to minimize the total divergence of the points from the line. In computing the line mathematically, the same idea is followed, but it has been found that the best line (the regression line) is one which minimises the sum of the squares of the vertical deviations from the straight line. This approach is known as the *method of least squares.*

If the equation of the regression line is

$$y = mx + a$$

and if the experimental points are $(X_1, Y_1), (X_2, Y_2), \ldots, (X_n, Y_n)$ and y_1 is the value of Y calculated from the equation of the regression line at $x = X_1$, the vertical deviation Δ_1 of the point (X_1, Y_1) from the line is given by

$$\Delta_1 = Y_1 - y_1 = Y_1 - (mX_1 + a)$$

and

$$\begin{aligned}(\Delta_1)^2 &= (Y_1 - mX_1 - a)^2 \\ &= Y_1{}^2 + m^2X_1{}^2 - 2mX_1Y_1 - 2aY_1 + 2amX_1 + a^2\end{aligned}$$

If z represents the sum of the squares for all similar squared deviations, then

$$z = \Sigma(Y^2) + m^2\Sigma(X^2) - 2m\Sigma(XY) - 2a\Sigma(Y) + 2am\Sigma(X) + na^2$$

where n is the number of points. For a given set of points all the sums $\Sigma(\;)$ in the above equation will be constants, and the value of z will vary with the values given to m and a. The conditions for z to be a minimum are

$$\left(\frac{\partial z}{\partial a}\right)_m = 0$$

$$\left(\frac{\partial z}{\partial m}\right)_a = 0$$

Now

$$\left(\frac{\partial z}{\partial a}\right)_m = -2\Sigma(Y) + 2m\Sigma(X) + 2na = 0$$

i.e.,

$$\frac{\Sigma(Y)}{n} = \frac{m\Sigma(X)}{n} + a$$

or

$$\bar{Y} = m\bar{X} + a$$

The regression line therefore passes through the point $(\bar{X}, \bar{Y})$. Again

$$\left(\frac{\partial z}{\partial m}\right)_a = 2m\Sigma(X^2) - 2\Sigma(XY) + 2a\Sigma(X) = 0$$

Therefore,

$$\Sigma(XY) = m\Sigma(X^2) + a\Sigma(X)$$

But $a = \bar{Y} - m\bar{X}$, so that

$$\Sigma(XY) = m\Sigma(X^2) + (\bar{Y} - m\bar{X})\Sigma(X)$$

Writing $n\bar{X}$ for $\Sigma(X)$ gives

$$m = \frac{\Sigma(XY) - n\bar{X}\bar{Y}}{\Sigma(X^2) - n\bar{X}^2}$$

and therefore

$$m = \frac{\Sigma(XY) - \Sigma(X)\Sigma(Y)/n}{\Sigma(X^2) - [\Sigma(X)]^2/n}$$

The regression line is then given by $y - \bar{Y} = m(x - \bar{X})$, a being eliminated by the result that when $y = \bar{Y}$, then $x = \bar{X}$.

7.3. Variance about Regression

It can be shown that the variance of a set of points about their regression line[3] is given by

$$s_Y{}^2 = \frac{\Sigma(Y^2) - n\bar{Y}^2 - m^2[\Sigma(X^2) - n\bar{X}^2]}{n - 2}$$

7.4. Error of Estimation of *X* from Regression Line

In instrumental methods of analysis, the instrument is calibrated with substances of known concentration. A regression line is used as the calibration curve for determining the concentration of an unknown substance from the instrumental "response." The error involved in predicting x values from this curve can be computed by recognized statistical methods[3] as $\pm s_X t$, where t has $(n - 2)$ degrees of freedom, and

$$s_X{}^2 = (s_Y{}^2/m^2)[(n + r)/nr + (Y - \bar{Y})^2/m^2[\Sigma(X - \bar{X})^2]],$$

where $s_Y{}^2$ represents the variance about regression (see Section 7.3), and r denotes the appropriate replication.

8. ANALYSIS OF VARIANCE

The technique of analysis of variance is a powerful tool for determining the separate effects of different sources of variation in experimental data. To illustrate the principles involved we shall examine its usefulness for interpreting the results of biological assays.[15]

In bioassay processes, doses or diets of increasing magnitude are given to groups of animals and the biological responses observed. The response measured may be, for example, the gain in weight by a whole animal during a period on a specific diet (vitamin assay) or the gain in weight of a particular organ (hormone assay), and so on. In an ideal situation the response of each animal to a given dose would always be the same. In addition, the responses at the different dose levels would be a linear function of the logarithm of the dose. This situation is virtually never found in practice and the object of the statistical technique is to compare the variation in response due to increasing dosage with that due to other causes, particularly individual animal variation within a dose group. It can then be assessed whether the variation due to increasing dosage is significant when viewed against the background of animal variability. If the latter contributes significantly to the total variation the selected bioassay process is unsatisfactory and the responses do not truly represent the effect of increasing dosage. It is also possible from the statistical analysis to decide whether the response–log dose relationship is linear within the limits imposed by animal variation.

The calculations required for an analysis of variance in a problem of this type with a total of n observations arranged in c columns each containing n/c observations are shown in Table II. Y represents an individual response and X is the logarithm of the dose for each group, i.e., the quantity which is varied from one column to another, and m is the average response for a dose group, i.e., the column mean.

The results may be interpreted as follows. The mean square (variance) due to dose difference M_2 is compared with that due to animal variability

Table II

Analysis of Variance[a]

Source of variation	Sum of squares	d.f.	Mean square
Total	$S_1 = \Sigma Y^2 - (\Sigma Y)^2/n$	$n-1$	$M_1 = S_1/(n-1)$
Between columns	$S_2 = \frac{n}{c}(\Sigma m^2) - (\Sigma Y)^2/n$	$c-1$	$M_2 = S_2/(c-1)$
Within columns	$S_3 = S_1 - S_2$	$n-c$	$M_3 = S_3/(n-c)$
About regression	$S_4 = S_2 - S_5$	$c-2$	$M_4 = S_4/(c-2)$
Due to regression	$S_5 = \frac{n}{c}\left(\frac{(\Sigma Xm - \Sigma X \Sigma m/c)^2}{\Sigma X^2 - (\Sigma X)^2/c}\right)$	1	$M_5 = S_5$

[a] Courtesy *The Statistician.*[15]

M_3. If M_2 is very much greater than M_3 the response gives a reasonable measure of dose level or concentration and can be used for assay purposes; i.e., the bioassay process is sensitive. To confirm that the two variances are significantly different, the ratio (larger mean square)/(smaller mean square) is calculated and compared with the theoretical value of the variance ratio F for $P' = 0.05$ and the appropriate degrees of freedom.

Additional information regarding the suitability of the process for purposes of biological assaying is obtained by carrying out a regression analysis on the data, assuming that the average group response m is a linear function of log dose X. The sum of squares due to doses S_2 is subdivided into two components: (a) S_4, the sum of the squared deviations of the (X, m) points from the regression line, and (b) S_5, the sum of the squared deviations from the mean of the theoretical responses calculated from this line. S_4, the sum of squares about regression, measures the scatter of the (X, m) points about the line. S_5, the sum of squares due to regression, measures the squared deviations from the grand mean due to the fact that there is an m/X regression line with finite slope.

Although M_4 may be greater than M_3, it is reasonable to assume, providing M_4/M_3 (the mean-square ratio about regression) $< F$ for $P = 0.95$, that the difference between variances "within columns" and "about regression" is not significant, or that they are sample variances from the same distribution. Therefore the scatter of the (X, m) points about their regression is only what might be expected from animal variation and it is thus reasonable to assume that there is a linear relationship between response and log dose. If M_4/M_3 is greater than F, the scatter about regression is greater than can be ascribed to animal variability thereby suggesting that the true relationship is not linear.

In instrumental analytical techniques the treatment of known standard dilutions (in solid or solution form) is replicated according to some recognized instrumental procedure and the instrumental response or reading is noted. In such techniques the "within-columns" variation is always negligible compared to the "between-columns" variation and this is fully demonstrated in typical calibrations. In geochemical assaying procedures, it may be quite considerable since, in general, the reproducibility of measurements on geologic systems in usually poor. Nevertheless, in instrumental techniques the analyst often assumes the within-columns variation to be less than it actually is. For example, in the colorimetric determination of manganese in a specific rock sample to be analyzed, the pretreatment involves oxidation of the manganese to permanganate with subsequent measurement of the optical density of the colored solution. Usually known standard solutions of $KMnO_4$ are made up covering the required concentration range and the colorimetric measurements are carried out in duplicate. The closeness of the duplicates (or even replicates) merely indicates the precision of the colorimetric procedure and yields a false comparison with the "between-concentration" (i.e., "between columns") variation for the analytical process as a whole. A more correct procedure of determining the precision of the method, would involve preparing a known set of standards conforming both in general composition and physical form to the test samples. The required concentration range with regard to manganese should also be covered, of course. The standards should then be treated in every respect in the same way as the test samples. A simplified treatment of variance and its evaluation may be found elsewhere.[4]

9. SAMPLING METHODS

9.1. General Comments

An analytical determination, whether carried out in the laboratory or on factory scale, is only as useful as the sample on which the measurement is made. No amount of replication at the analytical stage will increase the accuracy of our information, or make up for deficiencies in the sampling technique during the collection of the analysis material. If the analysis material is homogeneous in character, i.e., of a single phase of solid or liquid such as a bar of metal alloy or a tank of homogeneous liquid, for example, little attention need be paid to the sampling process, since any part of the material will be properly representative of the whole. In the laboratory it is generally not particularly difficult to obtain a sample which is truly representative of the whole, although errors may arise due to faulty mixing in routine-control operations (Section 11). In dealing with heterogeneous material, involving a mixture of two or more phases, such as in sludges or minerals, a small sample taken from any one part of the aggregate is liable to give a false picture of the aggregate as a whole. Furthermore, with heterogeneous material no sampling process can ever give an exact representation of the aggregate and a sampling error will therefore occur. This error will depend upon the size of the sample, the variability of the material, and the sampling scheme adopted. The sampling error must be sufficiently low to meet the requirements for which the sampling is being made. The requirements are, then, to obtain a sample that will provide an unbiased estimate of the corresponding aggregate parameter, and the sampling error showed be reduced to some adequately low level. An important objective of statistical examination here should be to minimize expenditure of time and effort on the sampling and analytical procedures, in order to obtain the degree of precision which the situation demands.

9.2. Remarks on Analytical Procedure

The standard deviation has the same dimensions as the variable to which it refers: but the variance s^2 has the advantage of additivity. This implies that if there are a number of independent causes of variation operating simultaneously on a system with variances $s_1^2, s_2^2, \ldots$, the resulting total variance s_T^2 is given by

$$s_T^2 = s_1^2 + s_2^2 + s_3^2 + \ldots$$

Since the coefficient of variation V is by definition related directly to the standard deviation s, $V^2 \propto s^2$ and, hence,

$$V_T^2 = V_1^2 + V_2^2 + V_3^2 + \ldots$$

In the analysis for a particular element in the sample to be analyzed, there are two main factors which may contribute to the normal variation V_T. The first is the variation or error involved in the laboratory V_L, i.e., the error inherent in the analytical technique, and the second, the variation or error involved in the collection of the analysis sample V_S, i.e., the sampling error. From the above equation relating component variances to total variance, it follows that

$$V_T^2 = V_L^2 + V_S^2$$

If there was no error in sampling (i.e., $V_S = 0$), the total variation would depend entirely on the analytical technique (i.e., $V_T = V_L$) and could be reduced by replication of analysis.

In multistage analytical techniques that often involve some form of chemical pretreatment prior to an instrumental assay, it is desirable to consider how the error is distributed within the analytical system. Such considerations eliminate the possibility of confusion arising in the discussion of the method and may suggest how the operational time may be reduced. Suitable presentation of the error distribution also permits the interested analyst to assess the application of that particular analytical procedure within his own field of work. We can therefore write

$$V_T^2 = V_{\text{instr.}}^2 + V_{\text{chem. pr.}}^2 + V_S^2$$

For example, in the colorimetric determination of copper in ore samples, the analytical method involves a chemical pretreatment to obtain the sample in solution before the instrumental measurement may be made.

9.3. Computation of Error Distribution

How do we determine the magnitudes of these quantities for, say, a granitic outcrop? V_T^2 may be obtained by calculating s_T^2 from the analytical data for n rock samples, taken at random from the particular area concerned (uniform within itself with regard to rock type). V_L^2 is obtained by eliminating the effect of V_S^2 and computing s_L^2 from the data for several similar portions of the same granite sample; i.e., the *complete analysis* is performed on each of n similar portions of a carefully mixed ground-rock sample. Similarly, $V_{\text{instr.}}^2$ is derived via $s_{\text{instr.}}^2$ which is calculated from the analytical results obtained for n aliquots of one particular extract.

In the above example the total error may be reduced in practice, either by (a) increasing the number of samples taken and analyzing each sample individually, or (b) if V_S is large in comparison with V_L, by making a composite sample of n subsamples and analyzing the composite sample. In (a) the error of the mean is given by

$$V_1^2 = (V_T^2)/n = (V_L^2 + V_S^2)/n$$

where n represents the number of samples taken. In (b) the error of estimation V_2 is given by the relationship

$$V_2^2 = V_L^2 + (V_S^2)/n$$

Any replication contemplated in the sampling or analysis must necessarily involve the following considerations: (a) the permissible error to be attached to the final result, (b) the relative magnitudes of the analytical and sampling errors, (c) the obvious necessity of keeping the sampling within practicable limits of expenditure of time and labor, and (d) the futility, in many cases, of replicating beyond the point where the major components have equal magnitude in the final result.

In the above examples V_1 and V_2 may be referred to as the errors of determination of the method as opposed to the analytical error V_L, and that fraction of the error contributed by the analytical method is given by $V_L^2/(nV_1^2)$ and V_L^2/V_2^2, respectively. A useful account of sampling methods is given by Laitinen.[5]

10. STATISTICAL TABLES

10.1. Values of t

Table III

Degrees of freedom	Value of t, $P' = 0.05$	Degrees of freedom	Value of t, $P' = 0.05$
1	12.71	15	2.13
2	4.30	20	2.09
3	3.18	30	2.04
4	2.78	40	2.02
5	2.57	60	2.00
6	2.45	120	1.98
7	2.36	∞	1.96
8	2.31		
9	2.26		
10	2.23		

10.2. Values of F ($P' = 0.05$)

Table IV[a]

n_2	n_1[b] 1	2	3	4	5	10	20	∞
1	161	200	216	225	230	242	248	254
2	18.51	19.00	19.16	19.25	19.30	19.39	19.44	19.50
3	10.13	9.55	9.28	9.12	9.01	8.78	8.66	8.53
4	7.71	6.94	6.59	6.39	6.26	5.96	5.80	5.63
5	6.61	5.79	5.41	5.19	5.05	4.74	4.56	4.36
10	4.96	4.10	3.71	3.48	3.33	2.97	2.77	2.54
20	4.35	3.49	3.10	2.87	2.71	2.35	2.12	1.84
∞	3.84	2.99	2.60	2.37	2.21	1.83	1.57	1.00

[a] Extracts from tables in Fisher and Yates: *Statistical Tables for Biological, Agricultural, and Medical Research*, reproduced by kind permission of Oliver and Boyd, Ltd., Edinburgh.
[b] n_1 is the number of degrees of freedom for the *greater* estimate of variance.

11. APPLICATIONS TO ANALYTICAL METHODS

11.1. Introduction

The errors which affect an experimental result may be divided into two types: (a) determinate, i.e., constant or systematic errors, and (b) indeterminate errors.

Determinate errors (a) are those which can be avoided or whose magnitude can be determined and the measurements corrected. A determinate error may have the same value throughout a series of observations and may remain constant under a variety of conditions (a *constant* error) or it may vary in magnitude and sign with the conditions (a *systematic* error). Determinate errors are numerous and include instrumental errors and those arising from reagents, errors associated with the manipulations of an analysis, personal errors, and errors which have their origin in the chemical or physicochemical properties of the system.

Indeterminate errors (b) are revealed by the slight variations that occur in measurements made by the same observer with the greatest care under as nearly identical conditions as possible. They may be described as the errors which remain even when all effects due to determinate errors have been eliminated.

Important factors in the choice of an analytical method include sensitivity, accuracy, precision, selectivity, concentration range, availability of standards, sampling requirements, instrumental and equipment costs, and time of analysis. Sensitivity, accuracy, and precision are of primary importance.

The *precision* of a result is its reproducibility; the *accuracy* is its nearness or closeness to the truth. A determinate error causes a loss of accuracy which may or may not affect the precision depending on whether the error is constant or variable. Indeterminate errors lower reproducibility, but by making sufficient observations it is possible to reduce the scatter to within certain limits so that the accuracy may not necessarily be affected. *Sensitivity* in chemical analysis concerns the ability to detect the difference between very small amounts of the analysis element, and is generally used to indicate the smallest quantity of element that can be detected or what is better defined as the *lower limit of detection.* In analysis we obtain a reading from an instrument (the instrumental response) which, after correction for background effects (i.e., noise level, blank, etc.), is taken as a measure of the amount of analysis-element present. It is thus really a difference in two responses which determines the concentration of element present, and at the lower detection limit these responses are similar in magnitude. Difficulties of interpretation must obviously arise and it is therefore necessary to have available an objective method for assessing when a *real* difference in response exists. The requirement of objectivity involves the application of statistical method. Accordingly, Kaiser and Specker[9] have defined the minimum detectable difference in response as $k\sqrt{2}s_b$, where s_b is the standard deviation of the blank reading and the values of k are those associated with the probability levels of the normal distribution, i.e., $k = 1.00$, 1.96, and 3.00 for 68, 95, and 99.7% confidence levels, respectively.

11.2. Reduction in Operational Time

In Sections 9.2 and 9.3 the distribution of errors is discussed with special reference to spectrophotometric analytical data. Two examples based on actual spectrographic results are given in Table V.

It is evident from Table V that for potassium the error inherent in the spectrographic technique accounts for 25% of the total analytical error, and about 2.5% of the error of determination. For manganese, the spectrographic error accounts for some 50% of the total analytical error and about 2% of the error of determination.

Table V

Distribution of Errors[a]

Variate	Analytical, V_L^2			Sampling, V_S^2	Total, $V_L^2 + V_S^2$	Error of determination, %
	Spectrographic	Chemical pretreatment	Total analytical			
K (major)	9	27	36	364	400	20
Mn (trace)	25	24	49	1176	1225	35

[a] Variance as V^2, based on single observations.

The spectrographic errors are therefore negligible and the analytical errors relatively unimportant, or in other words, the normal variation is almost wholly accounted for by the sampling variation. It is obvious that effective reduction of the errors of determination can only be achieved by replication at the sampling stage. Assuming a 25× replication in the sampling technique, the *errors of determination* for K and Mn[6] would become $\sqrt{36 + 364/25} = 7.1\%$ and $\sqrt{49 + 1176/25} = 9.8\%$, respectively. Now, in the spectrographic method employed, for example, it has often been recommended that duplicate spectra should be recorded for each sample to be examined. The idea behind this is to increase the reproducibility of the instrumental performance. From the purely analytical point of view, there is really no justification for such replication when we view the process of determination as a whole. In Table V the *spectrographic variances* for K and Mn are, respectively, 9 and 25, and these refer to results evaluated from duplicate spectra. Suppose we were to record only one spectrum per sample to be examined. This would be equivalent to doubling the spectrographic variance. Hence, if we evaluate our final result from a single spectrum only, but still using a 25× replication at the sampling stage, the total variance for K would increase from $(9 + 27 + 364/25)$ to $[(9 \times 2) + 27 + 364/25] = 59.6$. For Mn, the total variance would increase from $(25 + 24 + 1176/25)$ to $[(25 \times 2) + 24 + 1176/25] = 121$. The errors of determination would therefore increase from 7.1 to 7.7% for K, and from 9.8 to 11.0% for Mn, respectively.

The differences in precision as between single and duplicate instrumental analyses are therefore not sufficiently large to justify a preference for either alternative. The operational time is, however, reduced to approximately one-half when only single spectra are taken.

Let us now investigate more closely the instrumental stage of the analysis. In flame-spectrographic methods we plot θ_E/θ_B against concentration, where θ_E is the galvanometer deflection (i.e., the "blackening") for the analysis line; θ_B is the corresponding deflection for the background (i.e., the background spectrum of the carbon monoxide flame from an air–acetylene source) measured at a convenient specified point in the immediate neighborhood of the analysis line concerned. The background varies in density throughout the length of the spectrum, but at any particular wavelength it should be reproducible for a series of spectra. Its introduction therefore compensates for fluctuations in the light source or small variations in total emission from one spectrum to the next. It serves, in fact, as an internal standard. Each photographic emulsion carries in addition to the sample spectra, a number of standard spectra from which the θ_E/θ_B *vs* concentration relationship is derived. In determining the elements K and Mn it is necessary to obtain four readings: θ_B(K), θ_E(K), θ_B(Mn), and θ_E(Mn). Since the analysis lines for K and Mn, and hence their respective backgrounds, lie very close together in the spectrum, the question arises whether one background reading could serve for both lines? For example, in the spectrographic determination of potassium and manganese, using the K(4044 Å) and Mn(4130 Å) lines, respectively, the background readings shown in Table VI were obtained,[8] from an emulsion carrying 30 spectra.

Table VI

Background Reading, θ_B (cm)[a]

θ_B(K)	θ_B(Mn)	θ_B(K)	θ_B(Mn)	θ_B(K)	θ_B(Mn)
26.5	26.8	24.8	24.7	25.9	26.3
26.4	26.9	24.7	24.9	26.5	27.3
27.2	26.6	25.8	25.9	26.5	26.4
27.2	27.4	26.2	26.0	26.8	26.7
27.2	27.3	26.9	25.8	26.6	26.5
26.6	25.9	26.2	26.1	27.0	27.3
27.4	26.0	26.0	25.8	26.2	27.7
27.3	27.3	26.3	26.7	26.2	26.2
26.8	26.4	26.6	26.2	26.4	26.7
26.4	26.1	26.0	25.8	26.1	26.8

[a] Courtesy, *J. Appl. Statistics.*

The differences are small but not consistent; if we regard them as a sample from the hypothetical population formed by the differences from an infinite number of future assays, it may be determined whether the observed values could have been drawn from a population where the mean difference is zero. With the usual notation and writing the differences in background as X, we have

$$n = 30 \qquad \Sigma(X^2) = 8.92 \qquad s^2 = 0.30754$$

$$\Sigma(X) = 0.2 \qquad (\Sigma X)^2/n = 0.00133 \qquad s = 0.554$$

$$\bar{X} = 0.0066 \qquad \Sigma(X - \bar{X})^2 = 8.91867 \qquad s_{\bar{X}} = 0.1011$$

The statistic t is calculated from

$$t = (\bar{X} - \mu)/s_{\bar{X}}$$

where μ is the hypothetical population mean. On the null hypothesis, $\mu = 0$, and

$$t = (0.0066 - 0)/0.1011 = 0.065$$

Entering the t table with d.f. $= 29$, we have $t = 2.045$ ($P' = 0.05$). The mean is therefore not significantly different from zero. Further tests confirmed these results and on the basis of these the null hypothesis was accepted. Hence, there is no need to read both backgrounds and in this case the background for the manganese line was recommended for use from the point of view of operational suitability.

Let us now examine the combined effect of these procedural modifications at the instrumental stage of the analysis. Instead of recording two spectra per sample, leading to four subsequent measurements per spectrum, the proposed modifications only require the measurement of one spectrum per unknown sample, and only three subsequent photometric observations per spectrogram. The total exposure time has therefore been reduced by one-half, and the time associated with photometric evaluation has been reduced by five-eighths. Most important, is, however, the fact that no significant sacrifice has been made in reproducibility. In actual practice, making allowance for photographic processing, emulsion change over and so on, the saving in operational time in the complete analytical process has been found to be about 40%. If one is dealing with several hundred samples per week, this saving in time would be invaluable. Further examples of how simple statistical methods can be applied to the problem of standardization in spectrochemical analysis, are given elsewhere.[8]

11.3. Sampling Errors

If analytical results display a below-normal precision, then the cause may lie in faulty sampling. The latter may be due to insufficient grinding and/or mixing of the sample material.

In Table VII some analytical results are given for Fe_2O_3, as obtained colorimetrically from samples that were mixed for different periods of time.

It is apparent from Table VII that there is a substantial variation in Fe content within each set, although the individual duplicates are in good agreement with each other. These results, therefore, suggest a relatively large sampling error, which may be due to the lack of homogeneity in the

Table VII

Analytical Results from Samples That Were Mixed for Different Periods of Time

Time of mixing	Set 1, 1 min		Set 2, $1\frac{1}{2}$ min		Set 3, 3 min		Set 4, 5 min	
	mg Fe_2O_3		mg Fe_2O_3		mg Fe_2O_3		mg Fe_2O_3	
	0.180 0.180	0.1800	0.193 0.195	0.1940	0.183 0.190	0.1865	0.170 0.173	0.1715
	0.187 0.187	0.1870	0.193 0.187	0.1900	0.197 0.187	0.1920	0.173 0.170	0.1715
	0.183 0.182	0.1825	0.193 0.190	0.1915	0.170 0.173	0.1715	0.173 0.173	0.1730
	0.201 0.206	0.2035	0.183 0.180	0.1815	0.173 0.194	0.1835	0.173 0.170	0.1715
	0.167 0.162	0.1645	0.185 0.182	0.1835	0.173 0.182	0.1775	0.173 0.173	0.1730
	0.197 0.200	0.1985	0.170 0.167	0.1685	0.181 0.177	0.1790	0.170 0.173	0.1715
	0.195 0.197	0.1960	0.170 0.170	0.1700	0.195 0.177	0.1860	0.175 0.170	0.1725
	0.198 0.202	0.2000	0.183 0.170	0.1765	0.183 0.178	0.1805	0.172 0.173	0.1725
Standard deviation	0.01308		0.01062		0.00632		0.00069	
Coefficient of variation	6.9		5.7		3.5		0.4	

prepared sample. The calculated standard deviation and coefficient of variation clearly demonstrate the fact that longer periods of mixing systematically improve the precision of the results.

11.3.1. Experimental Procedure

(a) Known solutions of purified 6*N* HCl containing Al and (varying) Fe were treated to yield mixed 8-quinolinolate precipitates.[12] The precipitates were filtered, washed, dried in an oven, and finally ignited at 450°C. The mixed-oxide concentrates thereby obtained were ground in an agate mortar, each one for a specified interval of time (in this case, four in all, at 1, $1\frac{1}{2}$, 3, and 5 min, respectively). Then from each concentrate a set of eight 5-mg samples were weighed out and analyzed for Fe_2O_3 by the colorimetric method prescribed by Scott.[13] The only modification was introduced at the colorimetric stage proper, when two aliquots of the final analysis solution were colored up and measured and the average Fe_2O_3 content evaluated by reference to a calibration curve.

(b) Ten 20-ml aliquots of a standard Fe solution (falling within the concentration range of the prescribed method) were colored up in the usual manner and the results noted.

11.3.2. Results

The results for (a) and (b) above are shown in full in Tables VIII and IX. These results have been subjected to a statistical analysis and Table X records the standard deviation and coefficient of variation for each set. It should be noted that in the normal recognized method there is no duplication at the colorimetric stage, although the effect of such duplication on the coefficient of variation is negligible.

11.3.3. Discussion

It is immediately apparent from the results in Table VIII that there is a substantial variation in Fe content within each set, although the duplicates are extremely close (cf. Table IX), suggesting a relatively large sampling error (i.e., lack of homogeneity in the prepared concentrate). It is also noticeable, as would be expected, that the variation within a set decreases with the time of mixing and approximates eventually to that associated with the precision of the colorimetric procedure.

The above results are specially significant in works-routine practice where "unqualified" junior assistants may be given a set of instructions in

Table VIII

Analytical Data for Sets

Set 1, 1 min mixing		Set 2, 1½ min mixing		Set 3, 3 min mixing		Set 4, 5 min mixing	
mg Fe_2O_3		mg Fe_2O_3		mg Fe_2O_3		mg Fe_2O_3	
0.180 0.180	0.1800	0.193 0.195	0.1940	0.183 0.190	0.1865	0.170 0.173	0.1715
0.187 0.187	0.1870	0.193 0.187	0.1900	0.197 0.187	0.1920	0.173 0.170	0.1715
0.183 0.182	0.1825	0.193 0.190	0.1915	0.170 0.173	0.1715	0.173 0.173	0.1730
0.201 0.206	0.2035	0.183 0.180	0.1815	0.173 0.194	0.1835	0.173 0.170	0.1715
0.167 0.162	0.1645	0.185 0.182	0.1835	0.173 0.182	0.1775	0.173 0.173	0.1730
0.197 0.200	0.1985	0.170 0.167	0.1685	0.181 0.177	0.1790	0.170 0.173	0.1715
0.195 0.197	0.1960	0.170 0.170	0.1700	0.195 0.177	0.1860	0.175 0.170	0.1725
0.198 0.202	0.2000	0.183 0.170	0.1765	0.183 0.178	0.1805	0.172 0.173	0.1725

Table IX

Reproducibility Data for Colorimetric Procedure

mg Fe_2O_3	(20 ml aliquots)	mg Fe_2O_3
0.201		0.200
0.201		0.201
0.200		0.200
0.200		0.202
0.201		0.200

Table X

Statistical Analysis of Experimental Data

Set	Standard deviation	Coefficient of variation
1	0.01308	6.9
2	0.01062	5.7
3	0.00632	3.5
4	0.000694	0.4
Colorimeter	0.000494	0.2

which it is required to "thoroughly grind and mix the concentrate." In such cases it may be assumed that 2, 3, or 4 min "thorough mixing" is all that is required. Further, after mixing some dozen concentrates, the procedure becomes tedious for junior (and senior?) analysts and this fact must be accepted *realistically*. It would, of course, be wrong to stipulate any particular time of mixing, as the actual process of grinding and mixing will always depend on the efficiency of the analyst concerned: *at the same time*, assuming a high degree of efficiency in the above experiments, it would appear that *at least* 5 min grinding and mixing is required for this particular stage in the analytical process.*

11.4. Calibration

In Section 7 we discussed the best-fitting relation between two quantities, one of which is subject to chance error. In experimental work, the values for that quantity Y are usually measured for given values of the other (X). Thus, in gravimetric analysis, for example, Y would represent the weight of precipitate produced under certain specified conditions by a concentration of the element represented by X. Let us, by way of illustration, consider data obtained during the development of a gravimetric method of assay for the estimation of aluminium (as oxinate). A set of typical results ($n = 6$) is shown in Table XI.

Referring to Section 7.2 and using the results in Table XI, we derive

* This section is reproduced in its entirety by kind permission of the Council of the Institute of Chemistry of Ireland.[14]

Table XI

Gravimetric Assay for Al[a]

Al (in mg)	Al oxinate (in mg)
3.5	59.5
7.0	118.3
14.0	232.0
17.5	298.4
21.0	352.8
24.5	416.7

[a] Courtesy, *Anal. Chem.*[10]

the equation relating weight of precipitate to concentration as

$$x = (y + 1.03)/16.96$$

and we can compute the error involved in predicting x values from this expression by statistical methods referred to in Section 7.4.

Now, in teaching practice, for example, it is usual to draw up a series of gravimetric exercises and ask the student to carry out each in duplicate. It is doubtful if this procedure provides the student with anything other than practice in manipulation, and certainly the information is extremely limited. A more correct procedure would be to record additional data for each exercise in the manner described above at the expense of the number of exercises to be performed and to use the technique of regression for dealing with the data in each case. These recommendations would, of course, apply equally well to the analyst who is developing a new method of gravimetric assay. The advantages in adopting such a procedure are as follows: (a) more information is immediately available from the results. A correct measure of the precision of the analysis is obtainable for the full range of concentration investigated. If we put the regression (calibration) line equation in the form $y = mx + a$ (see Section 7.2), the error in $|a|$ can be estimated and it can be determined whether $|a|$ is significant, both *statistically* and, therefore, *chemically*, in which case a "blank" effect (constant error) is present. The standard deviation of an intercept is given by $s_a = s_Y\{\Sigma X^2/[n\Sigma X^2 - (\Sigma X)^2]\}^{1/2}$ and the value for $t = |a|/s_a$ is examined by means of the t table with $n - 2$ degrees of freedom. A value of t in excess of the selected critical value is evidence that the data do not support the expectation that the calibration graph passes through the origin. This should lead to further work in the laboratory to find the cause for this departure. A positive blank here could indicate the presence of Al impurity in the reagents and/or solutions employed. A negative blank, although unusual, could indicate the presence of an impurity (e.g., citric acid) which complexes with part of the aluminium, thereby rendering it innocuous. (b) It is possible to determine if a significant difference exists between the *theoretical* and *observed* values of m, the theoretical value in this example being given by $[Al(C_9H_6ON)_3] \div [Al]$. The difference between m(theoretical) and m(observed) is tested by pooling the variance estimates for each [zero for m(theoretical)]. Thus, the variance of the difference is given by $s_Y^2[1/(\Sigma X^2 - \bar{X}\Sigma X)]$ and

$$t = \frac{|\, m(\text{theor.}) - m(\text{obs.}) \,|}{\sqrt{s_Y^2 \left\{ \dfrac{1}{\Sigma X^2 - \bar{X}\Sigma X} \right\}}}$$

with $n - 2$ degrees of freedom. A significant difference implies that the reaction is not fully stoichiometric, but it should be noted that this by itself does not justify rejection of the method. It is also important to note that the above information would not be derived from an examination of the data of replicate analyses carried out for only one point on the concentration scale. In other words, if we wish to examine the reliability of a gravimetric (or, for that matter, instrumental) method of assay, it is essential to obtain estimates of m and $|\, a \,|$. The extra time involved is completely justified in view of the information to be obtained.

(c) The treatment outlined disposes with the need for graphical computation and eliminates errors associated with reading graphs.

11.5. Blank Variation

In determining the concentration of a particular element X in the analysis material, it is essential during any particular run of analyses to make a blank determination, i.e., to obtain a measure of the amount of X in the reagents and solutions employed. Any error arising due to the presence of X in the solutions, if it exists, will obviously be constant from one standard dilution to the next; and when we balance test and reference solutions in a photometer, we are, in effect, carrying out a simple subtraction, viz., X(analysis material + blank) − X(blank), which gives the value of X referred to the analysis material. The same argument applies when we are dealing with the addition of solid reagents (e.g., in fusions) in our analytical scheme, provided that there is no variation in the *X* content of the reagent. If the solid reagent is not homogeneous it would be futile to carry out routine analyses using specific aliquots of reagent and to base the estimate of the blank on the *X* content of only one such aliquot, since the error concerned will add to the existing analytical errors. The possibility of such errors arising is best eliminated by thorough mixing of the reagent(s) at the commencement of the analysis. Reagent purification only adds to the time of determination and is not always feasible or even necessary, provided the average *X* content of the reagent is not sufficiently high to cause a reduction in photometer sensitivity.

Such effects have been encountered for various batches of laboratory-grade reagents. For example, the variation in Fe content of a particular batch of $KHSO_4$ was found to be of the order of 20% ($P = 0.95$). Obviously the use of such heterogeneous material for the fusion of analysis samples in trace-Fe determinations could give rise to serious errors, depending on the quantities of analysis and fusion materials used and the range of concentration being investigated.

The calibration curves shown in Fig. 5 demonstrate the effect. In the first calibration 75 g $KHSO_4$ (in lots) were fused in a silica crucible, the melts dissolved in 25% citric acid (200 ml altogether, to render Al innocuous) and the extracts combined and made up with distilled water to 500 ml (A). A standard Fe solution (B) was then prepared in which 1 ml (B) = 30 μg Fe. Standard dilutions were obtained by adding to 50-ml aliquots of (A) varying amounts of Fe, i.e., 2, 4, 6, 8, and 10 ml (B). To each standard dilution in a 100-ml flask was added 1 ml of 25% thioglycollic acid followed by concentrated ammonia drop by drop until a red coloration just appeared due to the formation of the intensely colored ferrous thioglycollate ion $[Fe(S \cdot CH_2 \cdot COO)]_2^-$, and a further two drops in excess. The solution was cooled and made up to the mark with distilled water. The optical density

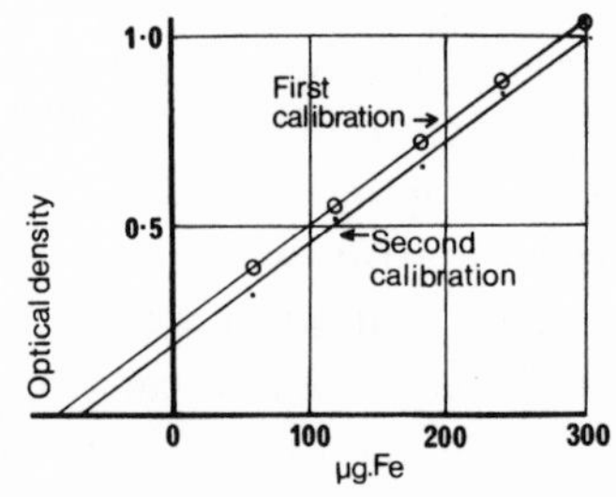

Fig. 5. Blank variation due to iron in $KHSO_4$.
Courtesy of *Spectrovision*.

was measured (against water) on a Unicam SP 500 Spectrophotometer at $\lambda_{max} = 530$ mμ using 1-cm cells. In the second calibration the same general procedure was adopted, except that the base solution (A) was not prepared in bulk; each base solution aliquot was obtained by fusing a 7.5-g portion of $KHSO_4$ selected from the reagent bottle, extracting the melt in 20 ml of 25% citric acid and making up to 50 ml. The relatively large scatter of points about the second calibration (regression) curve is due to variation in the Fe content of the batch of $KHSO_4$ used; the error attaching to the blank estimate may be evaluated from the regression analysis (see Section 7).

In certain cases it may not be possible to eliminate the effect of variation in the blank. All that can then be done is to determine the analytical precision with the blank effect present; provided the error falls within the desired limits, there is no need to remove the interfering element, bearing in mind also the consequent increase in analysis time and analytical manipulation, if the latter procedure is adopted. For example, small amounts of Al in TiO_2 (0.05–0.35%) can be determined by fusing a specific aliquot of the titania with NaOH, and extracting the melt with water. To a specified aliquot of the extract (containing 10–75 μg Al) in a degreased separating funnel are added 45 ml of NH_4OH–NH_4Cl buffer (to yield pH 9.0) and 10 ml of 1% w/v oxine in chloroform. The mixture is shaken for 3 min and allowed to settle; the organic layer is then run off into a 1-cm cell and its optical density measured (against a reagent blank) at $\lambda_{max} = 390$ mμ. A series of standard dilutions of Al_2O_3 in solution form (a) in extracts of melts of specified aliquots of pure TiO_2, and (b) without the addition of TiO_2, gave the calibration (regression) curves shown in Fig. 6a and b, respectively. From the relatively large scatter associated with the former curve and practically absent from the latter, the blank obtained in Fig. 6a suggested possible interference due to the presence of Ti caused by incomplete and nonreproducible hydrolysis of the titanate. This was con-

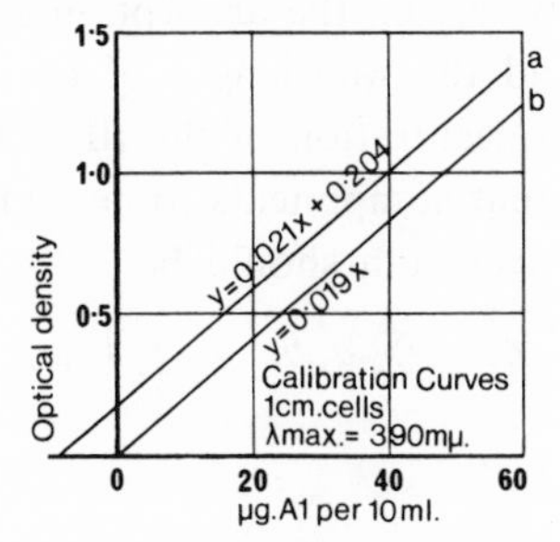

Fig. 6. Blank variation due to aluminum in TiO_2. Courtesy of *Spectrovision.*

firmed by deriving the absorption spectrum of titanyl oxinate in chloroform which indicated λ_{max}(Ti) = 380 mμ and partial absorption at λ_{max}(Al) = 390 mμ. Future results may be read from the Al calibration curve and the precision evaluated as in ordinary regression analysis. In the event of the precision being found not to be acceptable, the Ti can be removed by an initial precipitation with cupferron prior to extraction and color development.*

11.6. Multicomponent Methods of Analyses

In spectrophotometric analysis we measure the amount of light absorbed A at a specific wavelength λ by a known length of solution b and relate this to the concentration c of some particular constituent in the solution. The fundamental relation between these quantities is given by

$$\log (I_0/I) = A = abc$$

where I_0 is the intensity of the incident radiation, I is the intensity of the transmitted radiation, A equals $\log (I_0/I)$ which equals the absorbance, and a is a constant characteristic of the absorbing material and wavelength concerned.

In spectrophotometric methods for the quantitative analysis of mixtures, the analysis depends upon the fact that the absorption spectrum of a mixture is a linear superposition of the absorption spectra of the individual components. The analysis therefore involves the setting up of a series of simultaneous linear equations.[15]

The instrumental response (the absorbance in this case) due to the jth component at the ith wavelength will be

$$A_{ij} = a_{ij}bc_j$$

where a_{ij} is the absorptivity, i.e., the absorption constant characteristic of the absorbing species and the wavelength of the radiation; b is the cell thickness and c_j is the concentration of the jth component. Assuming the absorbances due to different components to be additive, the total observed absorbance at the ith wavelength should be

$$A_i = \sum_j a_{ij}bc_j = \sum_j k_{ij}c_j$$

where $k_{ij} = a_{ij}b$.

* This section is reproduced in its entirety by kind permission of Pye Unicam Ltd., Cambridge.[11]

For example, in the analysis of neodymium and praseodymium in acid solution, the absorbance A_1, measured at $\lambda = 575$ mμ, is given by

$$A_1 = k_{(575\ \mathrm{m}\mu)(\mathrm{Nd})} \cdot c_{\mathrm{Nd}} + k_{(575\ \mathrm{m}\mu)(\mathrm{Pr})} \cdot c_{\mathrm{Pr}}$$

and the absorbance A_2, measured at $\lambda = 590$ mμ, is given by

$$A_2 = k_{(590\ \mathrm{m}\mu)(\mathrm{Nd})} \cdot c_{\mathrm{Nd}} + k_{(590\ \mathrm{m}\mu)(\mathrm{Pr})} \cdot c_{\mathrm{Pr}}$$

Writing the linear equations in matrix form, we have

$$\begin{bmatrix} A_{(575\ \mathrm{m}\mu)} \\ A_{(590\ \mathrm{m}\mu)} \end{bmatrix} = \begin{bmatrix} k_{(575\mathrm{m}\mu)(\mathrm{Nd})} k_{(575\ \mathrm{m}\mu)(\mathrm{Pr})} \\ k_{(590\mathrm{m}\mu)(\mathrm{Nd})} k_{(590\ \mathrm{m}\mu)(\mathrm{Pr})} \end{bmatrix} \begin{bmatrix} c_{\mathrm{Nd}} \\ c_{\mathrm{Pr}} \end{bmatrix}$$

or

$$[A_i] = [k_{ij}]\,[c_j]$$
$$A = KC$$

Similarly in mass-spectrographic methods for the quantitative analysis of complex petroleum fractions, the analysis depends upon the fact that the mass spectrum of a mixture is a linear superposition of the mass spectra of the individual components. To determine the contribution of each component at a given mass, it is necessary to measure the mass spectrum of each component in order to find the peak height for a given partial pressure of the component. Since the peak heights are directly proportional to the partial pressure of the gas, the coefficients determined for the pure component can then be substituted into the series of simultaneous equations. Thus, the peak height due to the jth component measured at mass m will be

$$H_{mj} = x_{mj} p_j$$

where x_{mj} is the peak height at mass m with unit pressure of the jth component in the inlet sample system; p_j is the partial pressure of the jth component. Writing the equations in matrix form, we have

$$H = XP.$$

Let us return to our system of linear equations of the form $A = KC$. Assuming the A_i and k_{ij} have been measured, the problem is to determine the best value of the c_j. If the number of elements in the matrix A is just equal to the number in the matrix C, the equations have a unique solution. Error in one of the A_i values produces an error in the C matrix. When the problem is overdetermined (by measuring more A_i than the number of

c_j to be determined) the equations have no unique solution but a "best" solution which is found from the least-squares criterion. It is often desirable in multicomponent analysis to overdetermine the concentrations, thereby minimizing the effects of random experimental errors.

Let $\Delta = A - KC$ be the matrix of errors. Differentiating the product $\Delta^*\Delta$ with regard to C to find the best value of C gives[†]

$$K^*A = K^*KC$$

Putting $K^*A = A'$ and $K^*K = K'$, the equation becomes

$$A' = K'C$$

A' and C have the same dimension which is the order of the square matrix, K'. Since the matrix K' is symmetric, evaluation of the determinant or inverse of K' requires less effort than that required for a nonsymmetric matrix of the same size.

The above methods have been applied with considerable advantage in the evaluation of mass-spectrographic and absorption-spectroscopic analytical data, as, for example, in the simultaneous uv spectroscopic analysis of molybdenum, titanium, and vanadium in solution. In mass-spectrographic work the number of equations to be solved may be in the region of seven or eight, but the analysis may be simplified initially by first fractionating the mixture into several components or component mixtures by distillation.

11.7. Particle Counting

The Poisson distribution was discussed in Section 4.4. One of the best instances of large n and small p is afforded by radioactivity. A small mass of radium contains many millions of atoms. In a specified interval of time a moderate number of atoms, constituting a very small proportion of the number in the mass, will change with the emission of α or β particles. If we treat the occurrence as one of pure chance, the Poisson distribution provides a mathematical model, to which the variation in the number of particles emitted in the specific time interval may be expected to conform. For example, Rutherford and Geiger using the scintillation method counted the number of α particles emitted by polonium per unit of time. Their results are given below; f is the number of times n α particles were observed.

[†] Δ^* represents transpose of Δ, i.e., the matrix whose rows are the columns of Δ.

n	0	1	2	3	4	5	6	7	8	9	10	11	12	13
f	57	203	383	525	532	408	273	139	45	27	10	4	1	1

We shall find the mean number of α particles emitted and the Poisson distribution corresponding to this mean.

The mean number of particles is found to be 10105/2608 = 3.87. The terms of the Poisson series with $m = 3.87$ are

$$e^{-3.87}(1 + 3.87 + 3.87^2/2 + 3.87^3/6 + \ldots)$$

that is, 0.021 + 0.081 + 0.156 + 0.202 + 0.195 + 0.151 + 0.100 + 0.054 + 0.026 + 0.011 + 0.004 + 0.002 + 0.001 + 0.000.

On multiplying by 2608 the successive terms become 55, 211, 407, 527, 509, 417, 261, 141, 68, 29, 10, 5, 3, 0, which are in good agreement with the values of the frequency given above.

REFERENCES

1. Chambers, E. G., *Statistical Calculation for Beginners*, University Press, Cambridge, 1958.
2. Pantony, D. A., *A Chemist's Introduction to Statistics, Theory of Error and Design of Experiment*, Roy. Inst. Chem. London, Lecture Ser. 1961, No. 2, 1961.
3. Saunders, L., and Fleming, R., *Mathematics and Statistics for Use in Pharmacy, Biology and Chemistry*, Pharmaceutical Press, London, 1966.
4. Calder, A. B., *The British Chemist* **34**, 60 (1967).
5. Laitinen, H. A., *Chemical Analysis*, McGraw-Hill, New York, 1960.
6. Calder, A. B., *Evaluation and Presentation of Spectro-Analytical Results*, Hilger & Watts, London, 1959.
7. Youden, W. J., *Statistical Methods for Chemists*, John Wiley & Sons, New York, 1951.
8. Calder, A. B., *J. Appl. Stat.* **9**, 170 (1961).
9. Kaiser, H., and Specker, H., *Z. Anal. Chem.* **149**, 46 (1955).
10. Calder, A. B., *Anal. Chem.* **36**, 27A (1964).
11. Calder, A. B., *Spectrovision* No. 13, 8 (1965).
12. Mitchell, R. L., *The Spectrochemical Analysis of Soils, Plants and Related Materials*, Commonwealth Bureau of Soil Science, Tech. Comm. No. 44A, England, 1964.
13. Scott, R. O., *Analyst*, 66, 142 (1941).
14. Calder, A. B., *Operational Statistics in the Analytical Laboratory*, Lecture delivered to Conference on Statistics in Research and Quality Control (Institute of Chemistry of Ireland) held in Dublin, April (1967).
15. Calder, A. B., *The Statistician* **16**, 203 (1966).

Chapter 3

CHEMICAL ANALYSIS AND SAMPLE PREPARATION

V. C. O. Schüler

Anglo American Corporation of South Africa, Ltd.
Johannesburg, South Africa

1. INTRODUCTION

Instrumental techniques usually require careful sample preparation. In many cases some preliminary chemical treatment has to be performed before the instrumental measurements can be made. It is therefore essential that the user of modern analytical instruments has a knowledge of sample-preparation techniques, as well as an insight into some of the wet chemical procedures that are applicable to the analyses of geologic samples.

2. SAMPLE PREPARATION

Any analytical result can only be as accurate as the sample that was used for the determination of its constituents. It is therefore of paramount importance that the portion of material used is completely representative of the bulk sample submitted for analysis. The sample received by an analytical laboratory must be assumed to be representative of the material from which it was taken. The analyst often has little or no control over the collection of samples. The importance of representative sampling, especially in the field, cannot be overemphasized (see Chap. 2, Section 11.3).

The sampling procedure actually constitutes part of the analytical process, and as such it will affect both the precision and the accuracy of the results. Precision may be defined as the concordance of a series of measurements of the same quantity, while accuracy expresses the correctness of the measurement. Good precision, therefore, does not necessarily imply good accuracy.

2.1. Sampling Procedures

Geochemical samples submitted for an analysis could be in the liquid, solid, or, possibly, gaseous phase.

(a) *Solutions.* If a sample of well or river water, for example, is submitted to the laboratory as a clear solution, it is comparatively easy to extract a representative portion for analysis. A suitable portion may be taken after thorough mixing. The latter may be achieved by inverting or shaking in closed containers. It should be remembered that liquid samples must be stored in closed containers in order to reduce concentration changes due to evaporation of the solvent.

(b) *Solids.* In the case of ores and rocks the samples are usually ground in stages. Successively smaller portions are taken for further grinding, until a small amount finally remains. This portion is ground fine enough so that a representative portion may be withdrawn after mixing. The minimum quantity of sample that may be taken at the various stages of grinding will depend on the nature of the material and its particle size. These minimum quantities have been determined empirically.[1] Sample crushing can be done manually by pounding on a steel plate with a hammerlike muller. The sample may then be ground further in a pestle and mortar. Powered mechanical devices are, however, normally used—especially if large numbers of samples have to be dealt with. Coarse samples are first broken down in a "jaw crusher" which consists of two steel plates. Large sample lumps are broken up by the hammering motion of one plate against the other. Crushing may be followed by milling or pulverizing in laboratory machines which may be of a disc-, ball-, rod-, or swing-mill type.

During these processes intermittent screening is often employed. The oversize particles which do not pass through the particular sieve are reground until the whole sample passes through it. This is followed by thorough mixing so that the sample may be divided into smaller portions.

2.2. Sample Dividing

The bulk sample must be reduced in such a manner that each portion remains representative of the original. A simple means of achieving this without the use of special equipment is by dumping the sample on a clean, flat surface to form an inverted cone. It is then flattened and the resulting circular layer of sample is divided into four equal segments. Two opposite segments are discarded and the remainder again "coned and quartered," until the sample bulk has been reduced sufficiently for the analyst to proceed with his determination.

The sample-splitting procedure may be carried out more readily by mechanical devices. One of the better known sample dividers is the "riffle" or the "Jones splitter." It consists of a hopper, the base of which is fitted with a series of adjoining chutes arranged in such a manner that a sample passing through the device will be directed in two directions by alternate chutes and collected in one of two troughs standing on either side of the splitter.

Another sampling device consists of a disc mounted horizontally on a rotatable spindle. The sample is fed into a hopper above the disc. As the disc is rotated the sample is deposited onto it in the form of a ring. The hopper is mounted on a screwlike spindle which also rotates, lifting the hopper as sample distribution on the disc proceeds. By means of an adjustable sector any desired sample volume may be removed from the stationary disc.

The weight of the final portion to be retained for chemical analysis will depend on the method to be employed, but a sample weighing between 50 and 100 g is generally sufficient. It is of utmost importance that no sample contamination should occur during these operations. In order to eliminate abrasion of the grinding surfaces of the mills they must be made of materials that are considerably harder than the sample to be ground. For most rock crushers special steels are used in the manufacture of the grinding surfaces. Agate vessels should be used if contamination by the metal crushers is to be eliminated completely. It is important that the mechanical crushing and milling devices should be designed in such a way that they retain a minimum amount of material after the sample has passed through. This will simplify cleaning procedures and the risk of contaminating subsequent samples. In addition use may be made of inert materials such as quartz, with which the crushers may be flushed, to avoid intersample contamination.

The fineness to which a sample has to be ground may be determined by replicate analyses of the ground material. The "spread" of the analytical results will give an indication of the homogeneity of the material, although these results will also indicate errors due to limitations of the analytical procedure itself.

Rule-of-thumb sample-preparation procedures have in most cases been developed for each particular type of geologic material. When preparing material for chemical analyses it is not only essential to retain a homogeneous sample from which representative portions may be taken, but the material must also be of sufficient fineness. The finer the material is ground, the more readily it will succumb to chemical attack, such as acid digestion or fusion with the various fluxes. As a general rule, geologic material which is amenable to acid digestion should be ground fine enough to pass through a 200-mesh sieve. Samples which are to be decomposed by fusion with fluxes should, on the other hand, at least pass through a 100-mesh sieve, provided that the material is sufficiently uniform to yield representative samples at this particle-size level.

2.3. Homogenizing

The final portion from which amounts will be taken for the analysis must be as homogeneous as possible, since "dip" samples are usually taken at this stage. After grinding the sample to a suitable fineness it is intimately mixed. This can be done manually by rolling the sample on a clean sheet of paper, plastic, or rubberized cloth. Mechanical devices for homogenizing samples or "blenders" are available commercially. A popular type consists of a hollow cube rotated about an offset axis. Another device consists of two cylinders assembled in the form of a V. These are rotated about an axis such that the apex describes a circle in the vertical plane.

2.4. Sample Storage

The finely crushed and homogenized samples should be carefully stored in clean, dry containers with positive closures such as screw or snap-on caps. All containers should be immediately labelled to avoid confusion and to facilitate identification.

3. DISSOLUTION OF GEOLOGIC SAMPLES

Once a geologic sample has been reduced to sufficient fineness, it has to be brought into solution before a wet chemical analysis can be performed. There is no universal method for obtaining a solution of the sample, since the procedure to be adopted will depend on the nature of the material to be analyzed. In the following sections the more general methods of obtaining solutions of solid materials are outlined. It is advisable to test the solubility of a small portion of the sample before subjecting the entire sample to a specific treatment. If a particular sample is soluble in water, then full use should be made of this fact. Very few geochemical samples are, however, water soluble excepting some of the salts such as rock salt (sodium chloride) and nitre (potassium nitrate).

3.1. Acid Digestion

Many geologic materials are soluble in acids or in mixtures thereof. Oxidizing or, in some cases, reducing agents may be added to effect solution. In many cases the sample will not dissolve completely, but the constituent to be determined may dissolve out readily. For example, acid-soluble oxides or sulfides may occur with insoluble silicates.

The following acids are frequently used to obtain solutions of the elements or minerals indicated. If the solubility of an unknown sample is to be determined, the various acids should be tested in the sequence outlined below.

3.1.1. Hydrochloric Acid

Carbonate and oxide ores of Ba, Ca, Fe, and Mg are soluble in hydrochloric acid.

Hydrochloric acid plus reducing agent. Hydrochloric acid plus reducing agents such as stannous chloride ($SnCl_2$) or hydrazine hydrochloride will more readily dissolve Mn, U, Cd, and Zn oxide ores, than will hydrochloric acid by itself.

Hydrochloric acid plus oxidizing agent. Hydrochloric acid plus oxidizing agents such as nitric acid, bromine, potassium chlorate, and perchloric acid may be used for dissolving oxide ores such as those of Cu, Ce, Pb, Mo, U, and Zn.

3.1.2. Nitric Acid

HNO_3 in the dilute or concentrated form may be used to dissolve ores of Cd, Co, Cu, Pb, Mn, and Ni. For sulfide ores it is often beneficial to make the solvent more oxidizing by the addition of bromine. With the addition of hydrochloric acid the strongly oxidizing solvent aqua regia is obtained. It consists of three parts hydrochloric acid to one part nitric acid. Aqua regia is used for sulfide ores and may be made more oxidizing by the addition of potassium chlorate or bromine.

3.1.3. Sulfuric Acid

H_2SO_4 may be used for ores of Al, Be, Mn, Th, Ti, U, and Pb. In the latter case lead sulfate is formed which is comparatively insoluble in dilute sulfuric acid and may be separated as such after dilution. It may then be dissolved in a solvent such as a solution of ammonium acetate.

3.1.4. Hydrofluoric Acid

HF is an excellent solvent for many silicates forming volatile silicon tetrafluoride, which may be completely expelled from the sample solution by heating in the presence of a high-boiling-point acid such as sulfuric acid. Since hydrofluoric acid attacks silicates, glass vessels cannot be used, and platinum or plastic ware such as Teflon is a requirement.

3.1.5. Perchloric Acid

$HClO_3$ is a powerful oxidizing agent with a relatively high boiling point. It is an excellent solvent for certain refractory materials such as chromite. Perchloric acid should not be heated in contact with organic matter as it is strongly oxidizing and may react with explosive violence.

Acid digestions are in most cases the preferred means of bringing geologic materials into solution as they introduce a minimum of salts and cations into the final sample solution. In cases where anionic constituents are to be determined, care must be taken that the corresponding acids are less volatile than the solvent. For instance, if fluorspar is heated with sulfuric acid the hydrofluoric acid formed would be volatilized and lost when heating the solution. A subsequent fluorine determination would therefore be invalidated.

3.1.6. Pressure Leaching

The decomposition of geologic samples, especially silicates, may be facilitated by conducting the dissolution under pressure. For this purpose "bombs" constructed from, or lined with material inert to the decomposing solution are used. Teflon, which is inert to hydrofluoric and other acids, is suitable for lining such vessels. Bernas[2] has described such a Teflon-lined pressure-reaction vessel and a procedure for decomposing silicates with aqua regia and hydrofluoric acid. The time required to decompose a 50–300-mg sample portion appears to be 30–40 min. After cooling and diluting, boric acid is added to dissolve precipitated metal fluorides.

3.2. Sample Fusions

Geologic materials that do not dissolve in acids may be brought into solution by fusing the sample with a suitable flux.

3.2.1. Types of Fluxes

Broadly speaking, there are four types of fluxes:

Alkaline fluxes. These include sodium carbonate, potassium carbonate, potassium hydroxide, sodium hydroxide, and sodium peroxide. Sodium peroxide also has oxidizing properties. The alkaline fluxes may be used individually or as mixtures. Mixtures usually have a melting point lower than the individual components, thus facilitating lower fusion temperatures.

Acid fluxes. The better known acid fluxes are potassium and sodium bisulfate, and their dried counterparts, the pyrosulfates.

Oxidizing fluxes. These contain sodium peroxide, potassium nitrate, or sodium nitrate, usually in admixture with alkaline components.

Neutral fluxes. Examples of neutral fluxes include sodium fluoride, borax, and lithium fluoride.

3.2.2. Choice of Crucible

The type of crucible to be used for fusions must be chosen such that the attack on it is negligible. The material from which the crucible is made should not contain any of the elements for which the sample is being analyzed.

Platinum crucibles. These may be used for fusions with potassium and sodium carbonate or with potassium pyrosulfate. However, the samples should not contain metals which alloy with platinum such as lead, zinc, tin, bismuth, silver, gold, or copper. Mixtures from which these elements are reduced to metals from their components can also not be tolerated. In addition, sulfur, selenium, tellurium, phosphorus, arsenic, and antimony all combine readily with platinum on heating, and they should therefore be absent.

Due to the high cost of platinumware, samples of unknown composition should not be fused in platinumware, but should rather first be tested in crucibles made of some other material.

Molten sodium peroxide attacks platinum metal and should not be used, but alkaline fusions with sodium and potassium carbonate, alone or admixed, can safely be undertaken. Fusions with bi- and pyrosulfates of sodium and potassium, alone or mixed with sodium fluoride or lithium fluoride, may be done in platinumware.

Iron, nickel, silver, or zirconium crucibles. These may be used for alkaline and oxidizing fusions. In each case the crucibles are attacked to some extent, and a crucible must therefore be chosen that is made of a metal not sought in the sample. The crucible material should be such that it can easily be separated in the subsequent analytical procedure, if necessary. Zirconium crucibles, although expensive, are the most suitable for most alkaline fusions, since they are attacked only slightly by alkaline fluxes.

Silica crucibles. Silica crucibles are excellent for fusions with acid fluxes, such as sodium and potassium bi- or pyrosulfates, which are commonly used for samples low in silicates.

3.2.3. Choice of Flux

For samples of unknown composition, pilot fusions should first be undertaken. If the sample is thought to consist mainly of refractory oxides with a low silica content, fusion with potassium pyrosulfate in a silica crucible is suggested. For siliceous material an alkaline fusion with a mixture of equimolar amounts of the carbonates of sodium and potassium in platinum crucibles may be attempted. The fusion may with advantage first be tested in a nickel or iron crucible. In this case sodium peroxide may be included in the flux to reduce the fusion temperature.

Silica-rich samples may also be fused in platinumware with a flux consisting of potassium pyrosulfate and lithium or sodium fluoride to attack the silica or silicates. In this case the cooled fusion mass may be digested with sulfuric acid in the platinum vessel and heated to expel silicon as the tetrafluoride. The procedure is not applicable where silicon has to be estimated.

3.2.4. Fusion Procedure

Once the type of flux to be used has been decided upon, the most convenient type of crucible is chosen. The sample and the flux are placed into the crucible, mixed well by stirring, and gently heated. The ratio of sample weight to flux weight should be in the range of 1:10–1:20. Once the mass starts to melt the crucible is gently swirled over the heat source, until a clear melt is obtained. In the case of sodium peroxide–sodium carbonate flux fusions this will require about 5–10 min, depending on the burner temperature. Fusions with sodium carbonate, potassium carbonate, and potassium pyrosulfate may require as long as 30 min, even at high temperature, for satisfactory decomposition of the sample. Once the molten fusion becomes clear the crucible and contents are cooled and the melt is dissolved in water or acids, depending on the particular analysis.

Melts from potassium pyrosulfate fusions are usually dissolved in dilute acids. Alkali-fusion melts are normally first treated with water, giving rise to two phases—a solution and a hydroxide precipitate. The whole mixture may be acidified to obtain complete solution of the sample. The precipitate may also be filtered or centrifuged off and examined separately. For example if chromite ore is fused with a mixture of sodium carbonate and sodium peroxide, and on cooling leached with water, the solution will contain the water-soluble sodium chromate and aluminate formed during fusion. The precipitate, on the other hand, will contain ferric hydroxide and other material insoluble in sodium hydroxide–sodium carbonate solutions, such as nickel hydroxide which will result if a nickel crucible is used for the fusion. By filtering the alkaline solution one thus achieves a separation of the iron and chromium present in the sample, facilitating their estimation.

4. METHODS OF SEPARATION

Once the sample to be analyzed has been obtained in solution, the constituent to be determined must often first be separated from the other components. These may interfere with the estimation depending on what particular technique is to be employed.

Only an outline of the more common separations and techniques can be given in the scope of this chapter.

4.1. Precipitation Methods

4.1.1. Group Precipitation

The classical group-precipitation scheme of qualitative chemical analysis consists essentially of separating the cations into groups of elements, according to the solubilities of their salts.

The sample solutions are treated successively with reagents which form insoluble components with certain cations. Usually each reagent forms insoluble components with several elements or groups of elements. The precipitated elements are then isolated by filtering or by centrifuging. They are then redissolved and separated further by the addition of more specific reactants. Details of such procedures may be found in texts on qualitative inorganic analysis.[3,4]

Many of these reactions find application in quantitative analyses, where a constituent may have to be isolated before it can be determined. In such cases it is of importance that the precipitation reaction should be quantitative, and preferably rapid and specific.

4.1.2. Controlled pH Precipitations

In addition to the above scheme, a number of separations may be achieved by the precipitation of various hydroxides at controlled pH values. Many cations are precipitated as hydroxides, provided the alkalinity or the pH is above a certain value. The latter is characteristic of the metal itself, as well as the valency state in which it is present in solution.

The pH value may be controlled by the addition of solutions of ammonia or sodium hydroxide, together with suitable buffering agents. If the pH values at which two or more elements hydrolyze are sufficiently different, a separation may be achieved. Either ammonium or sodium hydroxide may be used to adjust the pH value of the solution from which the cations are to be hydrolyzed. Both of these may, however, be used in certain instances to achieve greater selectivity.

In the case of ammonia, some metals like copper, nickel, and cobalt, for example, form soluble ammine complexes in the presence of an excess of ammonium radicals. Thus, if ammonium hydroxide solution is added to an acid solution of cupric ions, a hydroxide precipitate will appear at a pH value of approximately 5.4. If an excess of ammonium hydroxide or an ammonium salt is added the cupric ammine complex $Cu(NH_3)_4^{2+}$ will be formed, and the precipitate will redissolve. This property may be put to use in separating the elements that form soluble ammine complexes from those which do not, such as Fe(III), aluminium, titanium, etc.

Similarly, if sodium hydroxide is used to increase the pH value of a solution, the cations will be hydrolyzed at those particular pH values at which precipitation of the cations occur. Advantage may frequently be taken of the amphoteric nature of some elements, in that they form soluble sodium salts in the presence of excess sodium hydroxide. For example,

$$Al^{3+} + 3OH^- \rightarrow \underset{\text{(precipitate)}}{Al(OH)_3}$$

$$Al(OH)_3 + NaOH \rightarrow \underset{\text{(soluble)}}{NaAlO_2} + 2H_2O$$

Thus, metals such as Al, Be, Zn, Sn, Pb, As, Sb, V, and Mo, which are amphoteric in nature and therefore soluble in an excess of sodium hydroxide, may be separated from Fe, Ti, Zr, Th, Se, Y, and the rare-earth elements. The hydroxides of the latter are insoluble in sodium hydroxide.

4.2. Electrolytic Separations

4.2.1. Electrodeposition

This electrochemical procedure as applied to chemical analysis, is usually confined to the isolation of metals from aqueous solutions of their salts. When the potential of a direct current applied to two inert electrodes immersed in an electrolyte of the metal salt solution is increased from zero upwards, a potential will be reached where electrolysis commences. At this stage the metal ions will migrate towards the negative electrode, to be plated onto the cathode. This potential is known as the decomposition potential and is characteristic of the electrolyte in question. If the electrolyte consists of salts of more than one metal, it is possible, by controlling the potential, to separate these metals by plating them out one at a time. Their decomposition potentials must, however, differ by at least 0.25 V if a virtually complete separation is to be achieved.

4.2.2. Mercury Cathode Cell

A number of useful electrolytic separations can be achieved by means of this apparatus. It consists essentially of a glass vessel containing a mercury pool, which is made the cathode of an electrolysis system. The solution to be electrolyzed is poured into the vessel and an inert platinum electrode is inserted into the solution to form the anode. A direct current having a potential higher than the metals to be separated is applied to the electrodes. The metals will be electrodeposited into the mercury to form an amalgam. In this way metals such as Fe, Cu, Co, Ni, Zn, Cr, Cd, Mn, Pb, Bi, and Sn may be separated from a solution containing elements that cannot normally be plated from the solution, such as Al, Ti, U, the alkaline earths and rare earths. Dilute sulfuric acid solutions are usually employed for this type of separation. An advantage of this separation procedure is that no reagents have to be added, leaving the electrolyzed solution free of possible contaminants and high salt concentrations.

4.3. Solvent Extraction

If two immiscible solvents are intimately contacted by mechanical mixing, any substance soluble in both will distribute itself between the two solvents in a definite proportion. This ratio is termed the distribution coefficient for the solute between the two solvents.

In analytical chemistry this principle of solvent extraction is frequently applied.[5,6] Conditions have to be correctly controlled to transfer certain substances more or less quantitatively from an aqueous solution to an organic solvent. The aqueous solution is shaken together with the organic solvent in a stoppered separatory funnel and then allowed to stand to allow the two phases to separate out. They may then be drawn off one at a time.

Common metal salts are not normally soluble in organic solvents. This is to be expected because of their ionic nature and the tendency for the metal ions to be solvated. Before such metals can be extracted into organic solvents it appears that uncharged species must first be formed. All the water molecules coordinated to the metal ion must also be displaced. However, these requirements make it possible to achieve selective separations by solvent extraction. By forming extractable species only of the metal to be separated, it may be isolated by solvent extraction from the other metals in solution, provided these remain in the ionic state.

A neutral ion may be formed by coordination of anions with the metal cation, thus excluding solvated water. Some charged coordination complexes can also associate to form neutral extractable ions. Chelating agents may be used to form neutral ions.

For solvent extraction to be successful the operating conditions have to be closely controlled. Sufficient complexing agent must, for instance, be present to ensure that all the metal ions are complexed. An excess of complexing agent is therefore essential, since reversible reactions are usually involved. In some cases the valency state of the metal ion influences the complexing reaction. For example, Fe(III) forms complexes more readily than Fe(II). The addition of a salt containing the same anion as the species to be extracted may effectively increase the distribution coefficient. The extraction of uranium, for example, as the nitrate complex from a solution of nitric acid with ether is enhanced by the addition of either ammonium nitrate or aluminium nitrate. This effect may be due to the higher concentration of complexing ions as well as the solvation tendencies of the added salts to help remove the water of hydration from the extractable species.

In some cases the extraction of metals can be rendered more selective by the use of so-called masking agents. These prevent selected metals from being extracted into the organic phase. Two types of masking agents are generally recognized: (a) those which form charged-ion complexes and thus prevent extraction, and (b) masking agents which form complexes that are stronger than those which are obtained with the complexing agent itself. Thus, aluminium can be extracted in the presence of Fe(II) as the 8-hydroxyquinolate by the prior addition of an alkaline cyanide. The stable ferrocyanide ion will be formed, which does not react with 8-hydroxyquinoline.

5. METHODS OF DETERMINATION

In wet chemical analyses the final concentration of the element of interest is usually determined by gravimetric or volumetric procedures.

5.1. Gravimetric Analysis

In gravimetric analyses the element to be determined must be isolated and weighed in its pure form, either as the element itself or as a stoichiometric compound. For this purpose the separation procedures already described may be used once a solution of the sample has been obtained. If the other constituents of the sample are known, simplified separation procedures may often be applied. Where possible, a specific reagent should be used that will yield an insoluble compound of only that element which is of interest. Various techniques may be used to make a reagent specific for certain applications. For instance, it is sometimes possible to complex the other ions present so that they do not react with the precipitant. Many organic reagents can be used in this way. For example, nickel reacts with dimethylglyoxime in an ammoniacal solution to give a quantitative precipitate. However, in an ammoniacal solution, iron, if present, will also precipitate out. This may be prevented by forming a soluble iron citrate or tartrate complex by the addition of these reactants to the acidic solution.

Once the precipitate has been formed quantitatively it is usually separated by filtering or centrifuging. The precipitate is then washed, dried, and finally weighed. In the case of organic precipitants the precipitate may be dried at about 110°C and weighed as such, provided that a stoichiometric precipitate is formed. Otherwise the precipitate may be gently ignited in air to form the oxide, which is then weighed. It is, of course, essential that all transfer operations are done without any loss of the precipitate, as this would obviously negate the results.

In some cases neat gravimetric procedures based on electrolytic separations may be employed, provided the decomposition potentials of the salts of any other metals in the solution differ sufficiently from that of the element sought. For instance, copper may be plated quantitatively from sulfuric acid solution onto preweighed platinum electrodes. Reweighing the electrode after the copper has been deposited onto it affords a convenient means of estimating the amount of copper present in the sample.

5.2. Fire Assay

In fire- or dry-assaying techniques, the metal or metals sought are separated from the other components of the ore by heating the sample. Suitable reagents, referred to as fluxes, are usually added and the separated metal is then weighed directly. The method is used mainly for the estimation of gold-, silver-, and platinum-group metals.[7,8]

A weighed amount of ore, say 30–150 g, is mixed with a flux consisting usually of sodium carbonate, lead oxide (litharge), and borax or sodium fluoride. A reducing agent such as powdered charcoal is also added. The mixture is fused in a fire-clay crucible at red heat. The lead oxide is reduced by the charcoal to metal droplets, which will alloy the precious metals present in the sample. The silica, alumina, and other sample constituents form a slag with the remaining components of the flux. The molten mass is poured into a conical steel mold, in which the lead containing the precious metals will settle to form a "button" at the bottom. After cooling, the mold is emptied and the lead button is broken away from the slag. The lead button is now placed onto a flat roasting dish or cupel which is made of bone ash, magnesium oxide, or other porous material. With further heating under oxidizing conditions, the lead is oxidized and partly volatilized. It is also partly absorbed by the cupel, leaving a bead of precious metal which can then be weighed after cooling. If more than one precious metal is present in the bead, chemical separations are employed to separate the metals by selective dissolution.

The fire-assay method is exceedingly sensitive, and by weighing the separated precious metal on a suitable microbalance, ores containing less than 1 ppm of gold, say, may be evaluated quantitatively.

5.3. Volumetric Analysis

Volumetric or titrimetric analyses consist of the determination of that volume of a standard solution, or reagent of known concentration, which is required to react quantitatively with the substance to be determined. In titrimetric analysis the reactions must go to completion rapidly and stoichiometrically in dilute solutions. At the equivalence point or endpoint, there must be a marked change in some detectable property of the test solution so that the completeness of the reaction may be easily recognized. Broadly, titrimetric procedures[9] fall into the following categories:

(a) Acid–base reactions, which consist of the combination of hydrogen and hydroxyl ions to form water.

(b) Oxidation–reduction reactions involve a valency change of the reactants.

(c) Precipitation reactions are accompanied by the formation of an insoluble precipitate.

(d) Complex-formation reactions cause a negligibly dissociated ion or compound to be formed. The equivalence point of the reaction may be determined either visually or by instrumental aids.

5.3.1. Visual Indicators

In certain cases where a highly colored solution constitutes one of the reactants, the sudden disappearance of its color may indicate the equivalence point. If the color of the reactant which is added is discharged by the reaction, then the equivalence point will be at that value, beyond which further reagent additions begin to color the test solution. In such cases a correction will have to be made for the slight reagent excess that is required to produce the color. A well-known example of this type of titration is the reaction of reduced ions in acid solution, with a standard potassium permanganate solution. A large number of organic substances is available which will intensify and therefore assist the detection of endpoints.

Acid–base indicators have the characteristic of changing color at different pH values or hydrogen ion concentrations. This color change may either be associated with ionized and unionized forms of the substance or it may be due to inherent structural changes which are pH sensitive.

Reduction–oxidation indicators or "redox" indicators are organic substances, such as dyestuffs, which change color at definite oxidation–reduction potentials. These may therefore be used to follow titrations which involve a change in oxidation potential. Some substances may be used as indicators because they form a colored compound with a metal ion, provided the latter is in a certain state of oxidation. Thiocyanate ions, for example, react with ferric, but not with ferrous, ions to form a blood-red compound. This phenomenon may be utilized for the determination of titanium. Ferric ions are added to a solution of titanium initially present in the reduced state. In the presence of thiocyanate, the endpoint will be indicated by a pronounced red color, which will result from the presence of excess ferric ions.

Several visual methods are available for the determination of endpoints in precipitation and complex-formation reactions. They lend themselves, however, to specific rather than general applications. For example, in the titration of chloride ions with silver ions, sodium chromate may be used as an indicator. When all the chloride ions have reacted with the silver which is being added as titrant, any excess silver ions will react to form a red silver chromate precipitate, thus indicating the endpoint.

Metal ion indicators are used extensively for the complexometric titrations of metals.[10] These indicators are dyestuffs which react with metal ions to form strongly colored complexes. The indicator must form a fairly stable colored compound with the metal to be titrated. The complex which the metal forms with the titrant must, however, be more stable than the indicator complex. After the addition of the indicator, a highly colored complex will be formed. During the titration, the titrant will react with the metal to form a more stable complex. At the endpoint the highly colored metal-indicator complex will be destroyed due to the removal of the metal by the titrant. At this point the color will suddenly disappear, indicating that the equivalence point has been reached.

5.3.2. Instrumental Indicators

Most of the conventional volumetric methods rely on visual endpoint detection, but instrumental procedures have by now also been well established. Acid–base titrations are easily followed potentiometrically by taking successive pH measurements. Oxidation–reduction titrations may also be followed potentiometrically with the use of suitable electrodes. Titrations that are accompanied by changes in the conductivity of the solution may have their equivalence point determined by conductance measurements. Precipitation reactions are an example of this application. An alternating current may be used to reduce the polarization and other electrolytic effects at the electrodes.

6. ADVANTAGES, DISADVANTAGES, AND LIMITATIONS OF WET CHEMICAL ANALYSES

6.1. General Considerations

The development of modern instrumentation has made the rapid, virtually complete instrumental analysis of large numbers of samples possible. In most cases the conventional wet chemical procedures are more time consuming. Further, except for the simple routine analyses, experienced analysts are required to conduct delicate chemical manipulations. However, many instrumental methods require some chemical treatment before the samples can be presented for instrumental measurement. For example, many instrumental techniques require the sample to be in the form of a solution, and in some cases interfering elements must first be removed. It is thus to the operator's advantage to have a good knowledge of the principles of wet chemical analysis.

The comparatively modest capital outlay to equip a wet-chemical-analysis laboratory may be considered an advantage, although this may only be significant if small numbers of samples have to be analyzed.

Wet chemical methods are, in general, less suited for the estimation of trace constituents than are many instrumental procedures, although these may have only specific applications. For instance, where concentrations of less than 10 ppm of a particular element have to be determined, very few gravimetric procedures are satisfactory. A well-known exception to this is the fire-assay procedure for precious metals which, however, is a dry-assay technique.

A basic advantage of the so-called classical methods is that they offer a direct means of analysis. The element sought may be isolated as the element, or as one of its compounds, and subsequently weighed as such. Alternatively, the concentration of a particular element may be deduced from some well-established stoichiometric relationship. For these reasons wet chemical methods are still used as reference methods in cases where different instrumental techniques render different results. The major role which wet chemical procedures play in the field of standardization warrants a more detailed discussion.

6.2. Standardization

Most instrumental methods are comparative techniques and independent standardization procedures are necessary. In many cases instruments can be calibrated with so-called pure substances, such as reagent grade chemicals or their highly refined counterparts. With many instrumental procedures, especially if no prior chemical treatment is included, the presence of elements or compounds other than those sought may have significant effects on the results obtained. To compensate for such matrix effects the instruments are calibrated with standard samples which are similar in composition to those to be analyzed. Reference samples are carefully prepared and analyzed for the constituents for which they will serve. These analyses are usually performed by a number of different procedures. Preferably several analysts from different laboratories should be involved. Wet chemical methods are used extensively for this purpose, and there are no indications that they will, in the foreseeable future, be replaced by other procedures.

A number of commercial firms and bureaus market such standard, reference, or comparative samples, and for many applications such samples are adequate. For the analysis of geologic samples it is, however, often difficult to obtain samples similar to those to be analyzed. It then becomes necessary for the laboratory to prepare and analyze its own reference samples. A series of standards covering the whole range of concentrations likely to be encountered should, if possible, be prepared so that they can be used for the compilation of instrument calibration curves. Once these curves have been established it is often only necessary to see whether two standards yield results that still lie on the calibration curve. This procedure will verify that no shift in instrument calibration has occurred since the original calibration.

The analysis of such reference samples is usually carried out by wet chemical procedures, where the criterion is accuracy rather than speed. Once all the analytical results have been collected they should be analyzed statistically to determine the validity of the results (see Chap. 2), and, if possible, any analytical bias should be removed. Results that suggest an analytical bias are obviously erroneous and should be discarded. The weighted average of all acceptable values should be calculated and used to allocate a standard value to the standard sample.

REFERENCES

1. Taggart, A. F., *Handbook of Mineral Dressing*, John Wiley & Sons, New York, 1953.
2. Bernas, B., *Anal. Chem.* **40**, 1682 (1968).
3. Vogel, A. I., *A Textbook of Macro and Semimicro Qualitative Inorganic Analysis*, Longmans, Green and Co., London, 1954.
4. Feigl, F., *Qualitative Analysis by Spot Tests*, Elsevier, New York, 1947.

5. Morrison, G. H., and Freiser, H., *Solvent Extraction in Analytical Chemistry*, John Wiley & Sons, New York, 1957.
6. Stary, J., *The Solvent Extraction of Metal Chelates*, H. Irving, ed., Pergamon Press, London, 1964.
7. Beamish, F., *The Analytical Chemistry of the Noble Metals*, Pergamon Press, Oxford, 1966.
8. Dillon, V. S., *Assay Practice on the Witwatersrand*, Cape Times, Cape Town, 1955.
9. Vogel, A. I., *A Textbook of Quantitative Inorganic Analysis*, 3rd ed., Longman, London, 1962.
10. Schwarzenbach, G., *Die Komplexometrische Titration*, Ferdinand Enke, Stuttgart, 1957.

Chapter 4

ION-EXCHANGE CHROMATOGRAPHY

H. F. Walton

University of Colorado
Boulder, Colorado

1. INTRODUCTION

Ion-exchange chromatography is a versatile method of separation and concentration which is applicable to many kinds of analysis, inorganic, organic, and biochemical. The experimental technique is simple and in most cases fairly rapid. In the great majority of inorganic applications, all that is needed is a tube about 25 cm long and 1 cm i.d., constricted at one end, with a glass wool plug or sintered disc to support a column of exchanger some 15 cm high. A typical ion-exchange column used for analytical purposes is shown in Fig. 1. Obviously, the column can be scaled up or down according to sample size. A common mistake is to use a column that is too large. Gravity flow is generally sufficient with a convenient linear flow rate of 1–2 cm/min. There are a few simple details that should be noted for efficient column operation. One of these is that the ion-exchanging material should be packed as uniformly as possible, and that the column should not be allowed to drain or to accumulate air bubbles during use. The column should always be kept filled with liquid. If the granules of exchanger are sufficiently fine (100–200 mesh or finer) this presents no problem, as surface tension prevents the level of water from dropping more than a couple of millimeters below the top of the bed. (The solvent may, of course, not be water, as mentioned in Section 5.)

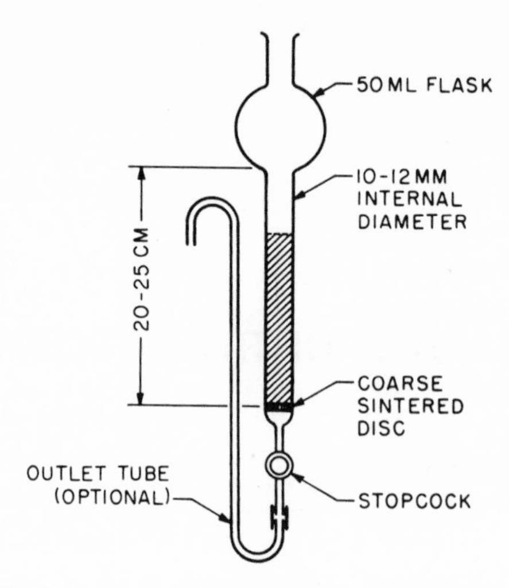

Fig. 1. Ion-exchange column for analytical use.

2. ION-EXCHANGING MATERIALS

There are many kinds of solid ion-exchanging materials, organic and inorganic. All are insoluble polyelectrolytes, that is, they consist chemically of a continuous matrix of very high molecular weight which carries fixed ionic groups. The charge on these groups may be negative or positive, depending on their chemical nature. Associated with these, and having opposite charges, are the small mobile "counterions" which can change places with other small ions of similar charge in the surrounding solution. One more condition is necessary. The structure of the matrix must be such that it permits easy movement of counterions in and out.

2.1. Ion-Exchange Resins

Ion-exchange resins are used in at least $\frac{9}{10}$ of the analytical applications, and the ion-exchange resins most commonly used have a matrix of cross-linked polystyrene. Styrene is mixed with divinylbenzene in liquid form and polymerized in droplets to give solid spherical beads whose size can be controlled. The proportion of divinylbenzene determines the crosslinking or tightness of the matrix. High crosslinking means a high density of ionic charges, but at the same time it means slow diffusion and reduced efficiency in column operation. For most analytical purposes a resin of 8% cross-linking is preferred, that is to say, a resin prepared from 8% divinylbenzene and 92% styrene.

Ionic groups are introduced into the polymer beads by appropriate chemical means. Treatment with fuming sulfuric acid introduces sulfonic acid groups, $-SO_3^-H^+$, giving a cation-exchange resin. By a more complicated reaction the groups $-CH_2N(CH_3)_3^+Cl^-$ may be introduced, thus forming an anion-exchange resin. These two types of resin are by far the most common. Another functional group that may be introduced into polystyrene is the iminodiacetate group, $-CH_2N(CH_2COOH)_2$. This forms chelated compounds with metal ions and binds doubly and triply charged ions very strongly, giving interesting selectivity sequences.[1]

Resins containing the carboxyl group —COOH are also available. The polymer matrix in these resins is not polystyrene, but polyacrylic acid. A great many polymers and condensation products have been made with special functional groups, and a few are made commercially, such as a phosphonic resin and one containing guanidine groups. The latter is highly selective for the platinum metals and gold.[2] An anion-exchange resin selective for borate has sorbitol molecules grafted on to a weak-base derivative of polystyrene.[3] This resin is also available commercially.

2.2. Inorganic Exchangers

Inorganic exchangers of many types have been prepared. The first ion-exchanging materials ever to be made artificially were aluminosilicates. These are little used today because of their sensitivity to acids and alkalis, but the crystalline aluminosilicates called "molecular sieves" may find some use in ion-exchange chromatography. More important are the amorphous or microcrystalline zirconium phosphate, zirconium tungstate, molybdate, and related compounds. Ammonium molybdophosphate (AMP) is a representative of the heteropoly acid group. Other types are silver ferrocyanide[4] and potassium hexacyanocobalt (II) ferrate (II).[5] All these compounds are cation exchangers. Hydrous zirconium oxide is a cation exchanger in alkaline solution and an anion exchanger in acid solution.

The distinctive feature of inorganic exchangers is their very high selectivity for certain ions. Every one of the cation exchangers mentioned is selective for cesium, and absorbs Cs^+ much more strongly than Rb^+ or K^+. They are also selective for radium over barium, and permit very effective chromatographic separations of the heavier alkali metal ions and alkaline earth metal ions. These cation exchangers also allow separation of the alkaline earth ions as a group from the alkali metal ions. The great impetus to their development has been the need to separate cesium from radioactive wastes. Cation exchangers have been used to concentrate traces of cesium in geochemical analysis. The basis of this extraordinary selectivity is not well understood, and it is worth noting that the selectivity is greatest at low pH values (where the exchange capacity is low). The selectivity is much less in well-crystallized materials than in amorphous or microcrystalline solids.[6] Zirconium oxide as an anion exchanger is particularly selective for fluoride, phosphate, and polyvalent anions in general.

2.3. Ion Exchangers Based on Cellulose

Ion exchangers based on cellulose are useful in biochemistry, where large ions and molecules must be absorbed, but they have little application to geochemistry. Phosphorylated cellulose has, however, been used to absorb traces of uranium from natural waters.[7] Carboxymethyl-cellulose combined with dithizone absorbs the same metals that dithizone does, and takes up zinc, lead, copper, and silver from sea water.[8]

3. ION-EXCHANGE SELECTIVITY

3.1. Inherent Selectivity

We have noted that certain ion-exchanging materials show specially strong absorption for certain ions. In general, however, this is not the case. Ion exchange is a competitive process, with one kind of ion displacing another to give an equilibrium which, to a first approximation, follows the law of mass action. Thus, the exchange of sodium and potassium ions between an aqueous solution and a cation-exchange resin, expressed as

$$K^+ + Na_{Res} \rightleftharpoons K_{Res} + Na^+$$

may be represented fairly accurately by an equilibrium constant:

$$E_{Na}^{K} = \frac{[K_{Res}][Na^+]}{[Na_{Res}][K^+]} \tag{1}$$

where the brackets indicate molar concentrations. For exchanges of ions of unequal charge, for example, the sodium–calcium exchange, the concentrations must be raised to the appropriate power:

$$E_{Na}^{Ca} = \frac{[Ca_{Res}][Na^+]^2}{[Na_{Res}]^2[Ca^{2+}]} \tag{2}$$

It will readily be seen that in the example given in Eq. (2), dilution of the solution will force the ion of higher charge (Ca^{2+}) into the exchanger. This effect is put to good use in chromatographic separations. It should also be noted that the quotients E are not true constants, because neither the exchanger nor the solution behaves ideally.

In a sulfonated polystyrene cation-exchange resin with 8% crosslinking the values of E for the alkali metals increase by a factor of 4 as one goes from Li^+ to Cs^+. For the Li^+ to Ag^+ exchange E is about 10. The differences in selectivity are thus not great. In anion exchange the range of E values is larger. In going from F^- to I^-, the E values increase by a factor of nearly 100. The effect is great enough that simple, practical procedures have been worked out to separate fluoride from the products of alkali fusion of rocks by retaining all other ions on a anion-exchange resin.[9,10] For the separation of metal ions, however, the selectivity of the usual cation-exchange resins is not enough, and it is essential to supplement it by another selective process, the formation of complex ions in solution, as outlined in Section 3.2.

Before discussing this topic, however, we should note the "nonchromatographic" uses of ion-exchange columns, the most common of which is the absorption of all cations in a solution and their replacement by an equivalent amount of hydrogen ions. This is done by passing the solution through a column of strongly acidic cation-exchange resin, such as sulfonated polystyrene, in its hydrogen form. The hydrogen ions that are released may be titrated to find the total salt concentration of the original solution. This is a common technique in water analysis, and is far simpler and more informative than the old method of evaporating and weighing the residue. Another use for the complete exchange of dissolved cations for hydrogen ions is to permit the accurate determination of the anions that pass through the column and emerge as their corresponding acids. Well-known and established methods exist for the determination of sulfur in iron pyrites by oxidation to sulfuric acid, and for phosphate in phosphate rock by conversion to phosphoric acid. In each case the solution is freed from interfering metal ions by passing through a cation-exchange resin; then the acid of interest is determined by titration. Recently, this same procedure has been used in conjunction with activation analysis to determine phosphorus in silicate rocks.[11]

Though separations like these are not properly "chromatographic," the question of selectivity still exists, for it is important to keep the acidity of the influent solutions as low as possible. Ion exchange is a competitive process, and if the initial concentration of hydrogen ions is too great, it will prevent complete absorption of the metal ions by the column of cation-exchange resin. Generally, an unnecessary excess of acid or salt should be avoided in the preparation of samples for ion-exchange processing.

3.2. Complex Ion Formation

Complex ion formation is very widely used for separating metal ions from each other. The classic case is the separation of the rare earths or lanthanides by chromatography on columns of cation-exchange resin, using as the eluent a solution of ammonium citrate of controlled pH. The element forming the most stable citrate complex (lutecium) is eluted first, while the one forming the least stable complex (lanthanum) comes out last. Selectivity of the resin itself has little or nothing to do with this separation. The stabilities of the dissolved complexes control the separation, and even though the differences in these stabilities are small, effective separations are possible by using columns that are sufficiently long and have sufficiently high "plate numbers" The plate number is an experimental parameter which expresses the resolving power of a chromatographic column or other multistage separator. Recent publications on the ion-exchange separation of lanthanides cite the use of α-hydroxyisobutyric acid, rather than citric acid, and quite rapid separations for techniques such as activation analysis have been made with this reagent.[12]

Magnesium, calcium, strontium, and barium ions are eluted in this order from a cation-exchange resin by an ammonium lactate solution of continuously increasing pH. This technique is called "gradient elution" and the most stable complex is eluted first. The concentrations in the effluent have been measured by flame photometry.[13] This carefully executed research illustrates how rapidly analytical processes become obsolete. Soon after it was published, the atomic absorption technique came into general use. Interelement effects are much less important in atomic absorption spectroscopy than in flame emission, and the need for separating these ions has now become much less. It is nevertheless quite easy to separate barium from the other alkaline earths by cation exchange, since barium is held very tightly by sulfonic acid resins.

Citrate, lactate, α-hydroxyisobutyrate, and EDTA complexes tend to be more stable, especially for metal ions of high charge. Their complexes are usually uncharged or negatively charged, and therefore not bound by a cation-exchange resin. Metals can thus be separated on the basis of ionic charge. A good illustration is the separation of radioactive Y-90 (charge of $+3$) from Sr-90 (charge of $+2$) on a cation-exchange resin column. Yttrium is eluted by citrate or EDTA while strontium remains behind. The gamma emission of Y-90 is easily counted and serves to measure the amount of the parent Sr-90.[14]

Chloride, sulfate, and nitrate ions form neutral or negatively charged complexes with nearly all metals if the concentrations are high enough. These complexes may therefore be used to pull metal ions off a cation-exchange resin, one at a time, giving elegant separations.[15] This same effect, the formation of negatively charged complex ions, will put metal ions on to an anion-exchange resin. The absorption of metal ions from hydrochloric acid solutions by strong-base anion-exchange resins is so important that it merits a separate discussion.

4. ANION-EXCHANGE SEPARATIONS OF METALS

A dilute solution of Co(II) chloride is pink, due to the hydrated cation $Co(OH_2)_6^{2+}$. If hydrochloric acid is added to this solution to make it 4*M* or more, in acid the color changes to blue, indicating the presence of the complex anion $CoCl_4^{2-}$. If some strong-base anion-exchange resin is mixed with the blue solution, the resin will acquire a blue color as it absorbs this complex anion. The absorption increases with increasing hydrochloric acid concentration up to a maximum value of about 100 ml/g, that is, the weight of cobalt in 1 ml of solution is about 100 times that in 1 g of resin. This maximum absorption is reached at roughly 9*M* HCl. At higher concentrations the distribution ratio *D* drops slightly, as shown in Fig. 2.

Each element has its characteristic graph of distribution ratio against hydrochloric acid concentration. Graphs for most of the metals in the

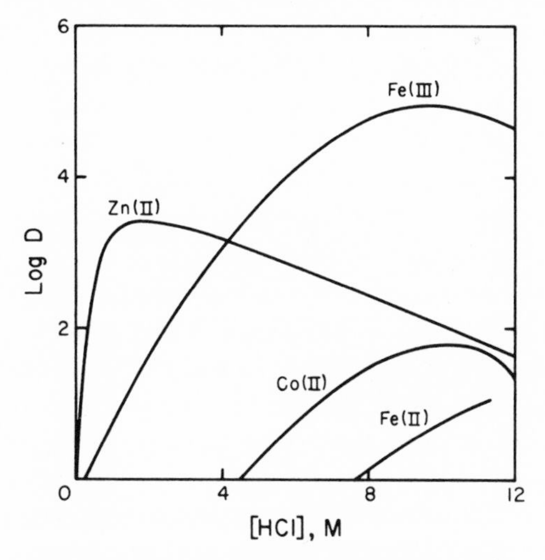

Fig. 2. Adsorption of metals from hydrochloric acid solutions. *D* is the distribution ratio in ml/g.

Table I

Anion Exchange of Metals in Hydrochloric Acid[a]

Element	Oxidation state	$[HCl]_{nax}$, M	log D_{max}
Ag	I	<1	3
As	III	10	1
Au	III	<1	7
Bi	III	<1	4.5
Cd	II[b]	2	3.5
Co	II	9	1.7
Cr	III	sl. ads.	
	VI[c]	str. ads.	
Cu	I	<2	2
	II	4	2
Fe	II	12	1
	III	10	4.5
Ga	III	7	5
Ge	IV	12	2
Hf	IV	12	6
Hg	II[c]	<1	5
In	III	3	1
Ir	III	<1	2
	IV	<1	4
Mn	II	11	0.3
Mo	VI	4	2.5
Nb	V[d]	8	3
Ni	II	no ads.	
Os	III	<1	4
Pb	II	1	1.5
Pd	I;	<1	3
Po	IV	str. ads.	
Pt	IV	<1	3.5
Rh	III	<1	1.5
	IV	str. ads.	
Ru	IV	<2	3
Sb	III	2	3
	V[c]	10	5.5
Sc	III	sl. ads.	
Se	IV[d]	>6	>1
Sn	II	<1	3
	IV	6	4

Table I (*Continued*)

Element	Oxidation state	$[HCl]_{max}$, M	$\log D_{max}$
Ta	V[d]	12	2.5
Tc	VII	4	2.5
Te	IV	str. ads.	
Th	IV	no ads.	
Ti	III	sl. ads.	
	IV	12	1.2
Tl	III	<1	1.5
U	IV	12	2.5
	VI	12	3
V	IV	sl. ads.	
	V	12	3
W	VI	9	1.5
Y	III	no ads.	
Zn	II[b]	2	3.2
Zr	IV	12	3

[a] From L. Meites, *Handbook of Analytical Chemistry*, McGraw-Hill, New York, 1963. Used by permission.
[b] Values of D for Cd and Zn differ significantly below $2M$; in $0.1M$ HCl, $\log D = 1.0$ for Zn, 2.2 for Cd.
[c] Cr(VI), Hg(II), and Sb(V) attack the resin.
[d] Values for Nb(V), Se(IV), and Ta(V) are erratic because of hydrolysis.

periodic table are given in the original papers of K. A. Kraus,[16] and have been reproduced in numerous publications. Representative curves are shown in Fig. 2, and Table I summarizes the data. The distribution ratio, measured by shaking the resin and the solution together until equilibrium is reached, is related directly to the elution volume in a chromatographic column. It is therefore easy to devise an almost unlimited number of separation schemes. The separation of Fe(III), cobalt, nickel, and zinc will serve as an example:

Make sample solution $9M$ in HCl; place on anion-exchange resin column previously washed with $9M$ HCl.

Pass more $9M$ HCl — Ni is eluted.

Pass $4M$ HCl — Co is eluted.

Pass $1M$ HCl — Fe is eluted.

Pass $0.1M$ HCl or water — Zn is eluted.

The analytical literature is full of separations of this kind, and applications to geochemistry are described in Section 7. Anion exchange in hydrochloric acid is a convenient way to separate minor elements as a group from large amounts of common elements, for the alkalis, alkaline earth metals, and aluminum are not absorbed.

Hydrochloric acid is a convenient reagent for such separations, for it can be obtained very pure with minimum trace element contamination. Excess acid can easily be removed from the separated solutions by evaporation. As a practical caution it should be noted that it is imperative for the anion-exchange resin to be pure. Most commercially purchased resins contain not only metallic impurities, but organic impurities and byproducts from the manufacturing processes, that often ruin the separations. Resins must be carefully washed before use. The experimenter may save himself a great deal of trouble by buying specially purified grades of resin.

What can be done with hydrochloric acid can also be done with other acids and salts, whose anions complex metal ions. Studies have been made of anion-exchange separations in hydrobromic acid,[17] hydrofluoric acid,[18] nitric acid,[19] sulfuric acid,[20] and thiocyanate solutions.[21] Separations in hydrofluoric acid are of interest, because the ions of titanium, zirconium, hafnium, niobium, and tantalum hydrolyze strongly in water and are best manipulated in fluoride solutions. Anion-exchange chromatography in hydrofluoric acid is perhaps the best way to separate niobium and tantalum,[22] and is used for this purpose in geochemical analysis.[23] Zirconium and hafnium are jointly isolated by anion exchange in hydrofluoric acid. They may be separated from each other by solvent extraction with thenoyltrifluoroacetone.[24]

The strength of absorption of a metal complex by an anion-exchange resin is determined by two factors that are independent of each other. The first is the stability of the negatively charged complex in solution, which can be measured by the usual techniques of coordination chemistry. The second factor is the strength of binding of the complex by the resin. The complex ion $FeCl_4^-$ has a very low stability in water. There is, in fact, some doubt whether it is formed at all, except in very concentrated hydrochloric acid. It is, however, absorbed very strongly indeed by the resin. Singly charged complexes, such as $FeCl_4^-$, $AuCl_4^-$, and $GaCl_4^-$, appear to be especially strongly absorbed. Theoretical grounds have been advanced to explain this behavior.

5. MIXED SOLVENTS IN ION-EXCHANGE SEPARATIONS

Ion-exchange selectivity can be modified radically by substituting another solvent for water. This was shown by Fritz[25,26] who found that zinc and cadmium ions could be stripped from a cation-exchange resin by hydrochloric acid much more effectively if ethyl alcohol or acetone were added to the water used as solvent. Lowering the dielectric constant of the solvent evidently stabilizes the complexes by increasing the electrostatic attractions between the metal ions and Cl^-. Absorption of many metal ions by anion-exchange resins is increased for the same reason.[27] In some cases, however, the nonaqueous solvent decreases the absorption. Fe(III) is not absorbed at all by an anion-exchange resin from 1*M* HCl in 80–90% acetone.[28] The reason for this apparently anomalous behavior is that the low dielectric constant stabilizes not only the complex ion $FeCl_4^-$, but also the ion pair $H^+FeCl_4^-$. The latter, being uncharged, is not absorbed by the resin. The iron–cobalt–nickel separation cited in Section 4 may now be modified as follows:[29]

> To sample solution in 6*M* aq. HCl, add 4 vols. of acetone. Place on anion-exchange resin column previously washed by 1 vol. 6*M* aq. HCl: 4 vols. acetone. Wash with more of this solvent—Fe is eluted.
>
> Pass 1 vol. 6*M* aq. HCl: 4 vols. acetone: 1 vol. water — Ni is eluted.
>
> Pass 1*M* aq. HCl or water — Co is eluted.

The proportions of water and acetone are not critical, and in practice one simply adds more water to get the next element out. This procedure has certain advantages over that with aqueous hydrochloric acid. Firstly, iron is in large excess in most analytical samples. As a rule small amounts of cobalt and nickel must be determined in the presence of large amounts of iron. If aqueous acetone is used, a small column will suffice to concentrate the minor constituents. Secondly, the hydrochloric acid concentrations are much lower, and thirdly, the viscosity of the solvent is lower and therefore allows faster flow rates than with water.

This separation will also occur on a cation-exchange resin column.[29] Again, iron, which forms an uncharged species, is not absorbed from 80% acetone or 80% tetrahydrofuran–20% 3*M* HCl. Cobalt is eluted before nickel in this case, as it forms a stronger anionic complex. Aluminum, if present, is absorbed by the cation exchanger, since its chloride complex is very weak.[30] Mixed-solvent ion-exchange chromatography has been explored thoroughly by Korkisch and coworkers. They call their method "combined ion exchange and solvent extraction.[29]" Fe(III) is one of relatively few metals that forms uncharged ion pairs under these conditions. Gallium, indium, and bismuth form stable uncharged ion pairs with chloride ions and can be separated from aluminum and from divalent ions.[29,31] Another metal that forms uncharged ion pairs is U(VI). It forms a nitrate complex, apparently $SH^+UO_2(NO_3)_3^-$, where S is an oxygen-containing solvent, tetrahydrofuran or (less effectively) acetone. Thus, other elements are absorbed on a resin column from 90% tetrahydrofuran–10% 6M HNO_3 while uranium passes through.[32] In sulfate or chloride solutions, however, U(VI) forms an anionic species which is strongly absorbed. Traces of uranium are recovered from sea water and marine sediments by making a solution in 90% methyl glycol–10% 6M HCl, and passing this through an anion-exchange resin. Uranium is strongly and selectively absorbed. After washing out accompanying elements with 80% methyl glycol–20% 6M HCl, the uranium is eluted with 1M aqueous HCl and determined fluorimetrically.

A cursory review of the available literature reveals that the mixed-solvent ion-exchange technique has not been adequately exploited in geochemical analysis.

6. SPECIAL ION-EXCHANGE TECHNIQUES

In this section a number of special ways of using ion-exchanging materials in geochemical analysis are reviewed.

6.1. Ligand Exchange

This is primarily a technique for separating organic compounds, but a geochemical application that has been explored is the concentration of amino acids from sea water.[34] The absorbent is a chelating resin loaded with cupric ions. In such an absorbent the cupric ion is coordinated with a nitrogen atom and two oxygen atoms in the functional group of the exchanger, and its fourth coordination site is free to accept a water molecule, an ammonia molecule, or another electron-donating ligand.[35] When sea water is passed through a column of this resin the amino acids become absorbed, displacing the coordinated water molecules. They are later displaced from the column in concentrated form by passing aqueous ammonia.

Ligand-exchange chromatography is used primarily for compounds having basic nitrogen atoms, but it can also separate amino acids and polyhydroxy compounds.

6.2. Isotopic Ion Exchange

In activation analysis it often happens that the activity of a minor component is masked by an enormously greater activity of a major component or matrix element. To measure the minor component it is not always necessary to remove the matrix element. It is sufficient to replace the radioactive matrix by a nonradioactive isotope of the same element. This may be achieved by a simple ion-exchange procedure.[36] For example, large amounts of radioactive Ba-133 are separated from trace amounts of radioactive sodium, silver, zinc, and other elements by passing a solution of the chlorides in 0.1*M* HCl through a column of sulfonated polystyrene cation-exchange resin, saturated with inactive barium ions. This resin has a high affinity for barium ions, and other divalent ions are only slightly absorbed and retarded as the solution flows through the column. Radioactive barium ions, however, exchange easily with isotopic nonradioactive barium ions in the resin and are retained. The column is rinsed with a small volume of dilute, inactive, barium chloride solution, to wash out the weakly absorbed ions of the minor constituents.

A variation of this method is used to separate strongly absorbed minor constituents from a weakly absorbed matrix element. Again a column of cation-exchange resin is used which is loaded with strongly absorbed cations, for example, Ba^{2+}. The solution containing the weakly absorbed matrix ions—say Na-22—and strongly absorbed trace activities is passed. The matrix passes on, and is washed out with, say, inactive NaCl solution, while nearly all the trace elements are retained. These may be counted on the column or eluted by appropriate means.

"Isotopic ion exchange" has been adapted to thin-layer chromatography. A mixture of Sr-90 and Y-90 is placed on a thin-layer plate coated with strontium sulfate and developed with dilute sulfuric acid. The yttrium moves ahead while the radioactive strontium exchanges with the absorbent and stays behind.[37]

6.3. Precipitation Ion Exchange

This is another method for concentrating the trace elements present in a matrix of another element.[38] The salt mixture to be analyzed is dissolved in a minimum amount of water and absorbed in a small excess of cation-exchange resin. This is rinsed with a very little water, followed by dioxane. The effluent, which contains some of the trace elements, is reserved. The resin is now dried, and the resulting mixture of resin and precipitated salt placed on top of a new column of resin previously washed with 70% dioxane–30% 12*M* HCl. The whole column is now washed with more of this solvent. A small volume of wash liquid suffices to bring out nearly all the trace elements, leaving most of the matrix behind. The method works for elements whose chlorides are sparingly soluble in 12*M* HCl, that is, Na, K, Ba, Sr, and also Ca, Mg, Li. By modifying the solvent other matrix elements can be precipitated, including Ni, Cr, Mn, Pb, and Al.[39]

6.4. Difference Chromatography

This is a method for accurately measuring small differences in composition among very similar solutions. It was designed for studying variations in the composition of sea water.[40] A "standard" sample of sea water was passed through a long column of cation-exchange resin until equilibrium was established between the resin and the standard solution. Then a quantity of sea water of slightly different ionic composition was injected into the top of the column. Each ionic constituent traveled down the column at its own rate. This depends on the slope of its absorption isotherm, giving concentration steps or fronts which may be detected by appropriate means at the column outlet. Several very sensitive differential techniques are now available for measuring such fronts. They include potentiometric, conductometric, refractometric, and thermometric (heat of absorption) techniques. The latter has been developed since the paper of Mangelsdorf was published. Possibly "difference chromatography" will find applications to industry as well as oceanography.

6.5. Ion-Exchange Papers

Papers are now made from cellulose fibers and finely ground ion-exchange resins in about equal proportions by weight. Four types of resin are used: strong acid (sulfonic resin), weak acid (carboxylic resin), strong base (quaternary ammonium resin), and weak base (polyamine resin with secondary and tertiary amine groups). Another type of ion-exchanging paper has the ionic functional groups attached to the cellulose itself. They have been used analytically in two ways—for paper chromatography of inorganic and organic ions, and as a "filter" to collect strongly absorbed ions from a large volume of solution in preparation for their measurement by such means as x-ray fluorescence or radioactivity.

Significant literature exists describing the use of resin-impregnated papers for the chromatography of inorganic ions.[41,42] Broadly, the paper-chromatography separations parallel the column separations discussed earlier. In paper chromatography much smaller amounts of material are handled than in columns, but the former are more rapid and give sharper separations. The "number of theoretical plates" in a 20-cm strip of resin-impregnated paper is much more than in a column 20 or 30 cm long. In the references cited, Sherma shows that aqueous–nonaqueous solvent mixtures can be used with both cation- and anion-exchanging papers to give selective migration of certain ions. For example, using a cation-exchange resin paper and $0.6N$ nitric acid in 90% tetrahydrofuran, gold moved with the solvent front, U(VI) migrated at an appreciable rate ($R_F = 0.13$), while other elements stayed close to the origin. Tests such as this will surely find use in geochemical analysis.

One important practical detail must be noted by potential users of this technique in the laboratory.[43,44] The solvent is allowed to move a little way up the paper strip, and then the sample spot is placed 1–2 cm *behind* the solvent front. The paper is then replaced in the solvent chamber and the solvent left to move up the paper. If the sample spot is placed on the dry paper and the solvent allowed to overtake it in the usual way, severe streaking may result. The reason is that the resin particles undergo a drastic change in swelling and in ionic content when the solvent meets them.

Chromatography on resin-impregnated paper can be used as a quick exploratory tool to predict the results of column operations, but it must be used with caution as the cellulose matrix does on occasion influence the migration of the ions. It is a nice way to measure the stability of complex ions,[44] especially complexes with low formation constants. Modified cellulose papers, those in which the ionic groups are built into the cellulose molecules, have also been used widely for inorganic paper chromatography.[45]

Turning to the use of ion-exchanging paper as a filter for strongly absorbed ions, the technique is to clamp the paper in a circular aperture through which the solution is made to flow. Usually two or three paper discs are used together. The paper circles loaded with metal ions are generally dried and analyzed by x-ray fluorescence or other means without further treatment.[46]

7. APPLICATIONS TO GEOCHEMICAL ANALYSIS

In this section we shall note a few representative applications of ion-exchange techniques to geochemical problems.

7.1. Concentration of Traces of Metals

Sea water has a high concentration of sodium and chloride ions, and therefore an absorption process of high selectivity is needed to separate and concentrate trace elements. Some ion-exchange procedures that are in use have already been cited.[8,33] Resins with special functional groups, such as carboxylate or iminodiacetate (chelating), are very selective for "heavy metals" and have a relatively low affinity for sodium ions. Thus, traces of copper, zinc, and bismuth have been absorbed from sea water by passing it through a sodium-form carboxylic resin.[47] These elements are later stripped off the exchanger by passing dilute hydrochloric acid through the column and a spectrographic determination may subsequently be performed. Chelating resins have been used in a similar way with industrial waste waters.[48] Strong-base quaternary anion-exchange resins can absorb those metals that form strongly absorbed chloride complexes. These are primarily the trivalent metals that give singly charged complexes such as MCl_4^-. Because sea water is already $0.5M$ in chloride ions it is not necessary to add much hydrochloric acid. Thus, bismuth, whose concentration in sea water is about 2 parts in 10^{11}, is removed by adding 90 ml of concentrated hydrochloric acid to 10 liters of sea water and passing through a small column of quaternary-base resin.[49] It is eluted with $1M$ nitric acid.

Cesium is concentrated from sea water, river water, and marine sediments by using selective inorganic exchangers such as ammonium molybdophosphate.[50,51] In river waters the cesium concentration is about 2 parts in 10^{11}. Radioactive Cs-137 has been collected from sea water, milk, and other environmental samples,[52] with potassium hexacyanatocobalt(II) ferrate(II). Radium has been recovered from sea water by coprecipitation with calcium carbonate followed by separation on a cation-exchange resin of the sulfonated polystyrene type.[53]

Schemes have been proposed for the separation and determination of major cations in sea water,[54–56] that is, Na^+, K^+, Mg^{2+}, Ca^{2+}, Sr^{2+}, and Ba^{2+}. The techniques depend on the relative strengths of absorption by a sulfonated polystyrene resin, as well as other selective reactions, such as the precipitation of K^+ by sodium tetraphenylboron.

The analyst should never restrict himself to one method of separation, and the method that most often complements ion exchange for the concentration and separation of metals is solvent extraction. Brooks[57,58] has used the Craig countercurrent extraction technique to separate trace metals in geochemical analysis, and in one of his methods,[57] sea water is passed through a succession of extraction tubes containing dithizone in carbon tetrachloride. The elements Tl, Au, Cu, Pd, and Pt are retained in that order. Another reagent used in the same way is 8-hydroxyquinoline in chloroform. Here, gold is the first element to be retained, followed by tin. Eight liters of sea water gave 20 mg of combined solid extract. After removing the solvent and ashing, a concentration ratio of 400,000 was achieved. The final determination was by emission spectroscopy.

Countercurrent solvent extraction has this in common with ion-exchange chromatography, that it is a multistage process and allows small differences in distribution ratios to be magnified into acceptable separations.

7.2. Removal of Interfering Ions

This is a straightforward application of an ion-exchange column, but it has many practical uses. As noted in Section 3, anions can be separated from all metallic cations and converted into their respective acids by passing the solution through a cation-exchange resin in its hydrogen form. A somewhat more complicated process is the separation of fluoride ions from interfering cations and anions. This is used in the analysis of ores and phosphate rocks[59] as well as natural and industrial water.[60,61] Where separation from other anions, such as phosphate, is required the fact is used that fluoride ions are absorbed less strongly than other anions and are the first to be released from an anion-exchange resin column by passing hydrochloric acid or sodium hydroxide. By making use of sodium hydroxide, the fluosilicate ion, SiF_6^{2-}, is broken down into silicate and fluoride, and the fluoride is eluted first.

Of course, the practical need for separating fluoride from other ions has been reduced by the development of the lanthanum fluoride membrane electrode.

A somewhat analogous case to the separation of fluoride ions is the separation of lithium ions from other cations. Sulfonated polystyrene resins have a lower affinity for Li^+ than for any other cation, and the difference between Li^{2+} and K^+ is accentuated by adding alcohol. In analyzing water from the Dead Sea, lithium chloride was separated from 1000× its weight of sodium chloride by absorbing the salts in a cation-exchange-resin column and eluting with hydrochloric acid in aqueous methanol. The lithium emerged first.[62]

Ion-exchange resins are sometimes used in batch form. They may be used, for example, to release anions or cations from difficultly soluble salts. Calcium sulfate and even barium sulfate[63] is disproportionated by stirring with a suspension of sulfonated polystyrene resin in the hydrogen form at

80–90° for some hours. Barium and calcium ions go into the resin and may be recovered later if desired. The sulfate ions yield their equivalent in sulfuric acid, which remains in solution and may be titrated as such. This procedure has obvious use in mineral analysis. Another "batch" use of resins is in the determination of traces of boron in rocks.[64] The rock is fused with potassium carbonate and the cake from the fusion is broken up with a slurry of sulfonated polystyrene resin in the hydrogen form, to which mannitol is added. The mannitol keeps the boron in solution as the borate–mannitol complex ion, while most of the cations from the rock and the flux, which would otherwise interfere with the borate determination, enter the resin. After some hours of contact the resin is filtered and washed, while the boron may be determined spectrophotometrically in the filtrate. Ten μg of added boron could be recovered from a 200-mg rock sample. Traces of boron in sedimentary rocks serve as indicators of paleosalinity.

7.3. Systematic Analysis of Silicate Rocks

7.3.1. Ion Exchange and Solvent Extraction

The most comprehensive scheme for the analysis of silicate rocks, based on separations by ion exchange and solvent extraction, is that of Ahrens, Edge, and Brooks.[65] They point out that many minor elements commonly occur in concentrations below the limit of spectrographic detection, and for this reason alone, separation and preconcentration is required. The separation of each individual element in pure form was not necessary, since the final determination was usually made spectrographically. However, some degree of separation was necessary to control matrix effects and interelement interferences.

To dissolve the rocks they used hydrofluoric acid mixed with sulfuric, nitric, hydrochloric, or perchloric acids. In this way they avoided introducing other cations that would compete with the cations sought in the ion-exchange separations. The solid residues were dissolved in $2M$ hydrochloric acid, and the solutions divided into four aliquot portions, which were treated according to the following scheme[65]:

Portion No. 1. Pass through strong-base anion-exchange resin, wash with $2M$ HCl.

Retained on column: (a) Cd, Zn, Bi, Tl, Sn, and Ag; elute with $0.25M$ HNO_3. (b) Platinum metals and gold; elute with $0.25M$ HNO_3 followed by $1M$ NH_3.

Passing: alkali and alkaline earth elements, lanthanides, Al, Fe, Ti,

Mn, Ni, Co, and V. This solution is further analyzed by cation exchange (see Portion No. 4).

Portion No. 2. Treated by solvent extraction: (a) Make 6*M* in HCl, extract with tributyl phosphate; Mo extracted, with some Fe and V. Mo is then back-extracted from tributyl phosphate by shaking with water. (b) Take another portion, make 1.7*M* in HCl and 0.1*M* in HI, pass SO_2 gas to reduce thallium to Tl(I). Extract with ethyl ether: Tl and In are extracted as iodo-complexes.

Portion No. 3. Make 11.3*M* in HCl, pass through strong-base anion-exchange resin; wash with 11.3*M* HCl.
Retained on column: Ti, Zr, Co, Fe, and Ga. These are eluted successively in that order by decreasing concentrations of hydrochloric acid (see Section 4). Ti and Zr are eluted with 8*M*, Co with 4*M*, and Fe and Ga with 1*M* HCl.
Passing: alkali and alkaline earth elements, lanthanides, Al, Mn, Ni, V, Pb, and Ag. This solution is further analyzed by cation exchange (see Portion No. 4).

Portion No. 4. Pass through a column of sulfonated polystyrene cation-exchange resin; wash with 2*M* HCl, increasing the HCl concentration later to speed up the elution. Elements are eluted sequentially in the following order: Al, Ti, Zr, Zn, Sn, and Pb; these elements form negatively charged chloro-complexes and are easily eluted. Li, followed by Na and V [and Fe(III) if this has not already been removed], Mg, Mn, Ni, Co, K, Rb, Cs, Ga, Al, Ca, Sr, and much later Ba.

The original publication[65] gives details of the separation and determination of each element, and explains how the procedure may be modified to suit different rock types.

7.3.2. Countercurrent Solvent Extraction

Boswell and Brooks[58,66,67] have also developed comprehensive separation schemes for metallic elements in rocks that depend entirely on countercurrent solvent extraction. They use a 120-tube Craig extractor. The moving solvent is methyl isobutyl ketone, and the stationary solvent is aqueous hydrochloric acid which starts in the first 20 tubes at 7*M*. HCl is added to the remaining tubes in progressively lower concentrations. The first elements to be extracted from 7*M* HCl are Fe(III), Ga, and Mo, and they move along farthest with the solvent, while such elements as calcium and sodium remain behind in the first few tubes. Roughly, the order of extraction parallels the order of binding by an anion-exchange resin in hydrochloric acid. The big drawbacks to countercurrent solvent extraction as an analytical tool are the cumbersome equipment and the limited number of "theoretical plates," which in this case is the actual number of tubes. The advantages, compared to ion exchange, are that larger quantities of material can be separated, that "tailing" is much less serious, and that the solvents can be purified from trace contaminants more easily than in the case of a resin.

7.3.3. Anion Exchange

Returning to ion exchange, a scheme has been published for silicate rock analysis that uses differential complexation by sulfosalicylic acid.[68] This scheme separates primarily the more abundant elements, Ti, Al, P, Fe, Mg, Mn, Ca, Na, and K, as well as Si. The metals Fe, Al, and Ti form sulfosalicylate complexes and are bound by an anion-exchange resin. They are further separated by anion exchange in hydrochloric acid, where Fe and Ti are separated on a column of carboxylic cation-exchange resin and are eluted in that order.

Anion exchange in hydrochloric acid solution has been used to collect a group of trace elements from rocks and soils.[69] The rock is dissolved in hydrofluoric, nitric, and sulfuric acids, the solution is evaporated, and the metal salts dissolved in 1.5*M* HCl and passed through an anion-exchange resin. The column is washed with 1.5*M* HCl until molybdenum just starts to appear. Then the resin is removed and ashed at a controlled temperature, and the ash is analyzed spectrographically. The elements thus collected are Ag, Bi, Cd, Sb, Sn, Zn, Pb, and Mo.

7.4. Individual Elements in Silicate Rocks

The literature is rich in references to the use of ion exchange as one part of a general separation scheme, where it is used to separate or isolate certain elements. A good source of information is a 10-page table of analytical determinations of major and minor elements in the standard granite and diabase samples, G-1 and W-1, in a review by Fleischer.[70] Many separations by ion exchange are cited, together with even more by solvent extraction. Anion exchange in hydrochloric acid is the most popular ion-exchange method.

For determining trace elements in rocks and meteorites activation analysis is frequently used, and ion-exchange separations often follow the irradiation step. The isolation of potassium, rubidium, and cesium from chondrite meteorites may be cited.[71] The irradiated sample is sintered with sodium peroxide, then dissolved in hydrochloric acid. Inactive rubidium and cesium carriers are added, and K, Rb, and Cs precipitated together with sodium tetraphenylboron. The precipitate is oxidized with nitric and perchloric acids, and after an intermediate scavenging step the three alkali metals are separated by chromatography on ammonium molybdophosphate (AMP).

A general separation of the alkali metals, alkaline earth metals, and rare earths (lanthanides) from chondrites following neutron irradiation has been reported, using a cation-exchange resin.[72] The three groups of cations are eluted by successively increasing concentrations of hydrochloric acid. Scandium[73] is separated from silicate rocks, following hydrofluoric–perchloric acid treatment, by a three-stage ion-exchange procedure, the interesting part of which is that scandium is absorbed on an anion-exchange resin in the sulfate form, then eluted (as its soluble sulfate complex) by an ammonium sulfate–sulfuric acid mixture. The final purification steps are coprecipitation with hydrous ferric oxide, followed by dissolving in $6M$ hydrochloric acid and passing through an anion-exchange resin. Iron and other associated trace elements are retained, while the scandium passes on.

Titanium, niobium, and tantalum are separated as their fluoride complexes by anion exchange,[22,23,74] and if they are present in large amounts they can be separated as chloride complexes with sodium chloride eluents.[75] Zinc,[76] antimony, tin, and mercury[77] have been isolated from meteorites by anion exchange, using hydrochloric acid eluents (zinc is retained very strongly from hydrochloric acid) as well as fluoride (which elutes antimony) and phosphoric acid (which elutes tin). The elements Se, As, Sb, Sn, and Hg have been removed from the matrix after neutron irradiation by distillation as their volatile chlorides.[77] Distillation (or volatilization) is a useful and "clean" method for separating posttransition elements from mixtures, and it has been used frequently in geochemical studies.

Gold and the platinum metals have frequently been separated from meteorites and silicate rocks with the help of ion-exchange techniques. The authoritative work on separations of the platinum metals is that of Beamish.[78] The application to meteorites is illustrated by a recent paper[79] which describes analysis by neutron irradiation followed by the absorption of Pt, Pd, Ir, and Au on an anion-exchange resin which is mixed with cerium dioxide, to ensure conversion of these elements to their highest oxidation states. These four elements are then separated by selective elution with thiourea and hydrochloric acid; Pb, Pd, and Au are eluted as their thiourea complexes, while Ir remains on the column and is later removed by 6*M* HCl. Thiourea is a useful eluent for ion-exchange separations of precious metals and has been used in the determination of these metals by activation analysis in deep-sea manganese nodules.[80] The existence of a chelating resin which is selective for precious metals was noted earlier (see Section 2.1), and now that this resin can be purchased commercially, it may find uses in geochemical analysis.

Many more examples could be cited, but these are representative, not only of the elements that can be separated by ion exchange, but also of the various ways of using ion exchange in conjunction with other separation methods. A compilation of ion-exchange separations of individual elements (many of them of geochemical interest) appears in alternate years in the "Fundamental Reviews" section of *Analytical Chemistry*.[81]

REFERENCES

1. Rosset, R., *Bull. Soc. Chim. France* **1966**, 59.
2. Koster, G., and Schmuckler, G., *Anal. Chim. Acta* **38**, 179 (1967).
3. Pinon, F., Deson, J., and Rosset, R., *Bull. Soc. Chem. France* **1968**, 3454.
4. Gorenc, B., and Kostka, L., *A. Anal. Chem.* **223**, 410 (1966).
5. Prout, W. E., Russell, E. R., and Groh, H. J., *J. Inorg. Nucl. Chem.* **27**, 473 (1965).
6. Albertsson, J. *Acta Chem. Scand.* **20**, 1689 (1966).
7. Bruce, T., and Ashley, R. W. *At. Energy Canada Ltd.*, AECL-2648 (1966).
8. Bauman, A. J., Weetall, H. H., and Weliky, N., *Anal. Chem.* **39**, 932 (1967).
9. Glasö, O. S., *Anal. Chem. Acta* **28**, 543 (1963).
10. Evans, W. H., and Sargent, G. A., *Analyst* **92**, 690 (1967).
11. Greenland, L. P., *Professional Papers USGS* **1967**, No. 575-C, 137.
12. Rengan, K., and Meinke, W. W., *Anal. Chem.* **36**, 157 (1964).
13. Pollard, F. H., Nickless, G., and Spincer, D., *J. Chromatog.* **10**, 215 (1963).
14. Blake, W. E., Oldham, G., and Sumpter, D., *Nature* **203**, 862 (1964).
15. Strelow, F. W. E., Rethemeyer, R., and Bothma, C. J. C., *Anal. Chem.* **37**, 106 (1965).
16. Kraus, K. A., and Nelson, F., *Proc. 1st United Nations Conf. on Peaceful Uses of Atomic Energy* **7**, 113 (1955); *The Structure of Electrolytic Solutions*, W. H. Hamer, ed., John Wiley & Sons, New York, 1959, Chap. 23.
17. Fritz, J. S., and Garralda, B. B., *Anal. Chem.* **34**, 102 (1962).
18. Faris, J. P., *Anal. Chem.* **32**, 521 (1960).
19. Faris, J. P., and Buchanan, R. F., *Anal. Chem.* **36**, 1158 (1964).
20. Strelow, F. W. E., and Bothma, C. J. C., *Anal. Chem.* **39**, 595 (1967).
21. Turner, J. B., Philip, R. H., and Day, R. A., *Anal. Chim. Acta* **26**, 94 (1962).
22. Hague, J. L., and Machlan, L. A., *J. Res. Nat. Bur. Standards* **62**, 53 (1959).
23. Dixon, E. J., and Headridge, J. B., *Analyst* **89**, 185 (1964).
24. Setser, J. L., and Ehmann, W. D., *Geochim. Cosmochim. Acta* **28**, 769 (1964).
25. Fritz, J. S., and Rettig, T. A., *Anal. Chem.* **34**, 1562 (1962).
26. Fritz, J. S., and Abbink, J. E., *Anal. Chem.* **37**, 1274 (1965).

27. Korkisch, J., and Hazan, I., *Anal. Chem.* **37**, 707 (1965).
28. Hazan, I., and Korkisch, J., *Anal. Chim. Acta* **32**, 46 (1965).
29. Korkish, J., *Separation Sci.* **1**, 159 (1966).
30. Korkisch, J., and Ahluwalia, S. S., *Anal. Chim. Acta* **34**, 308 (1966).
31. Korkisch, J., and Hazan, I., *Anal. Chem.* **36**, 2308 (1964).
32. Korkisch, J., *Mikrochim. Acta* **1967**, 401.
33. Hazan, I., Korkisch, J., and Arrhenius, G., *Z. Anal. Chem.* **213**, 182 (1965).
34. Siegel, A., and Degens, E. T., *Science* **151**, 1098 (1966).
35. Shimomura, K., Dickson, L., and Walton, H. F., *Anal. Chim. Acta* **37**, 102 (1967).
36. Tera, F., and Morrison, G. H., *Anal. Chem.* **38**, 959 (1966).
37. Kuroda, R., and Oguma, K., *Anal. Chem.* **39**, 1003 (1967).
38. Tera, F., Ruch, R. R., and Morrison, G. H., *Anal. Chem.* **37**, 358, 1565 (1965).
39. Morrison, G. H., in *Trace Characterization*, W. W. Meinke and B. F. Scribner, eds. U.S. National Bureau of Standards, Monograph 100, 1967.
40. Mangelsdorf, P. C., *Anal. Chem.* **38**, 1540 (1966).
41. Sherma, J., *Separation Sci.* **2**, 177 (1967).
42. Sherma, J., *Anal. Chem.* **39**, 1497 (1967).
43. Lederer, M., *Ann. del. Inst. Superiore de Sanita* **2**, 150 (1966).
44. Lederer, M., and Ossicini, L., *J. Chromatog.* **22**, 200 (1966).
45. Muzzarelli, R. A. A., *Talanta* **13**, 193, 1908 (1966).
46. Campbell, W. J., Spano, E. F., and Green, T. E., *Anal. Chem.* **38**, 987 (1966).
47. Brooks, R. R., *Analyst* **85**, 745 (1960).
48. Biechler, D. G., *Anal. Chem.* **37**, 1054 (1965).
49. Portman, J. E., and Riley, J. P., *Anal. Chim. Acta* **34**, 201 (1966).
50. Feldman, C., and Rains, T. C., *Anal. Chem.* **36**, 405 (1964).
51. Sreekumaran, C., Pillai, K. C., and Folsom, T. R., *Geochim. Cosmochim. Acta* **32**, 1229 (1968).
52. Boni, A. L., *Anal. Chem.* **38**, 89 (1966).
53. Sugimura, Y., and Tsubota, H., *J. Marine Res.* **21**, 74 (1963).
54. Khristova, R., and Krushevska, A., *Anal. Chim. Acta* **36**, 392 (1966).
55. Greenhalgh, R., Riley, J. P., and Tongudai, M., *Anal. Chim. Acta* **36**, 439 (1966).
56. Szabo, B. J., and Joensun, O., *Envir. Sci. Technol.* **1**, 499 (1967).
57. Brooks, R. R., *Talanta* **12**, 505, 511 (1965).
58. Brooks, R. R., *Geochim. Cosmochim. Acta* **29**, 1369 (1965).
59. Glasoe, O. S., *Anal. Chim. Acta* **28**, 543 (1963).
60. Jeffery, P. G., and Williams, D., *Analyst* **86**, 590 (1961).
61. Kelso, F. S., Matthews, J. M., and Kramer, H. P., *Anal. Chem.* **36**, 577 (1964).
62. Ratner, R., and Ludmer, Z., *Israel J. Chem.* **2**, 21 (1964).
63. Osborn, G. H., *Synthetic Ion-Exchangers*, Macmillan, New York, 1956.
64. Fleet, M. E., *Anal. Chem.* **39**, 253 (1967).
65. Ahrens, L. H., Edge, R. A., and Brooks, R. R., *Anal. Chim. Acta* **28**, 551 (1963).
66. Boswell, C. R., and Brooks, R. R., *Talanta* **14**, 655 (1967).
67. Boswell, C. R., and Brooks, R. R., *Mikrochim. Ichnoanal. Acta* **1965**, 814.
68. Maynes, A. D., *Anal. Chim. Acta* **32**, 211 (1965).
69. LeRiche, H. H., *Geochim. Cosmochim. Acta* **32**, 791 (1968).
70. Fleischer, M., *Geochim. Cosmochim. Acta* **29**, 1263 (1965).
71. Smales, A. A., Hughes, T. C., Mapper, D., McInnes, C. A. J., and Webster, R. K., *Geochim. Cosmochim. Acta* **28**, 209 (1964).

72. Shima, M., and Honda, M., *Geochim. Cosmochim. Acta* **31**, 1995 (1967).
73. Shimizu, T., *Anal. Chim. Acta* **37**, 75 (1967).
74. Kallman, S., Oberthin, H., and Liu, R., *Anal. Chem.* **34**, 609 (1962).
75. Smith, F. W., *J. Chromatog.* **22**, 500 (1966).
76. Nishimura, M., and Sandell, E. B., *Anal. Chim. Acta* **26**, 242 (1962).
77. Kiesl, W., *Z. Anal. Chem.* **227**, 13 (1967).
78. Beamish, F. E., *Talanta* **14**, 991, 1133 (1967).
79. Crocket, J. H., Keays, R. R., and Hsieh, S., *Geochim. Cosmochim. Acta* **31**, 1615 (1967).
80. Harris, R. C., Crocket, J. H., and Stainton, M., *Geochim. Cosmochim. Acta* **32**, 1049 (1968).
81. Walton, H. F., *Anal. Chem.* **36**, 51R (1964); **38**, 79R (1966); **40**, 51R (1968).

Chapter 5

COLORIMETRY

Gordon A. Parker

Toledo University
Toledo, Ohio

and

D. F. Boltz

Wayne State University
Detroit, Michigan

1. INTRODUCTION

This chapter consists of a discussion of the fundamental principles, techniques, and methods involved in colorimetric analysis. For many years chemical colorimetry has been a leading analytical method for the determination of small amounts of inorganic constituents, i.e., when the percentage of the desired constituent is in the 10^{-5}–5% by weight concentration range. The determination of elements in geological samples based on the intensity of colored solutions obtained by appropriate chemical treatment of these materials is a well-established quantitative method that is frequently used in geochemical analysis. In addition to the necessity of developing a suitable colored solution, one also has to find a suitable means of relating color intensity to the concentration of a specific element. In the first one-third of the 20th century it was common analytical practice to use the human eye to compare the color intensities of solutions of known concentration with those of similar solutions of unknown concentration. Visual colorimetry is now used only to a very limited extent. Presently, modern photoelectric instrumentation has replaced the visual color comparators, with their inherent limitations involving possible defective visual acuity and human judgement, and provides a more reliable, discriminate, and sensitive means of relating color intensity to concentration. In fact, modern instrumentation has permitted the extension of the spectrophotometric method to the measurement of the absorptivity of solutions in regions of the electromagnetic spectrum not discernible to the human eye. Hence, UV spectrophotometry and IR spectrophotometry (see Chapter 6) are adjunct methods to colorimetry.

2. THEORETICAL CONSIDERATIONS

2.1. Measurement of Light Absorption

Light is but a small portion of the electromagnetic spectrum composed at one extreme of x-ray radiations and at the other extreme of radio waves. Electromagnetic waves are characterized by their wavelength, the distance between corresponding points of a repeating cycle, and by their frequency, the number of complete wave cycles that pass a reference point in a given time. Using λ to represent the wavelength and ν the frequency, the relation

$$\lambda\nu = C = \text{const} = \text{speed of light} \tag{1}$$

holds for all electromagnetic radiant energy. The portion of the electromagnetic spectrum which is discernible to the human eye consists of the wavelength region from approximately 380 to 750 mμ. It varies somewhat among different people and with the same person, depending upon age. The neighboring region consisting of wavelengths shorter than this is the UV region, and the neighboring region consisting of wavelengths longer than this is the IR region. Both wavelength and frequency are commonly expressed in several different units. In addition, the term wave number is used in place of wavelength to represent the location of a point in an electromagnetic spectral region. The wave number, number of wave cycles per unit length, is represented by $\bar{\nu}$ and is defined as

$$\bar{\nu} = \frac{1}{\lambda} \tag{2}$$

where λ is in centimeters.

For a colored solution, the relation of color intensity to concentration of colored component may be obvious to a human observer and one speaks of a colorimetric method of analysis or colorimetry. Analogous methods are possible in the neighboring UV and near IR regions of the electromagnetic spectrum. The relationships are not obvious because the human eye does not perceive radiant energy in these regions. The term colorimetry is often replaced by the more general term spectrophotometry, or a more specific term, light-absorption spectrometry. Principles of spectrophotometry apply equally well to measurements in all three regions of the spectrum, UV, visible, and near IR. The discussion in this chapter will be limited to visible spectrophotometry but the relationships developed and the procedures presented will, provided the proper instrumentation is utilized, apply also to other spectral regions.

To state that a solution is colored means that of all wavelengths of white light incident upon the solution only selected wavelengths are absorbed, depending on the color of the solution; the remaining wavelengths are transmitted. A red solution, for example, appears red because it absorbs the shorter wavelengths of the visible region and transmits the longer wavelengths. Thus, color is attributed to the selective absorption of incident radiant energy of certain wavelengths. The absorption of certain wavelengths of light by colored solutions is due to energy transitions of the electrons within the atoms of the absorbing component. At a given temperature each electron resides in a definite lowest energy state, the ground state, and transitions to higher energy states occur provided the exact amount of energy, corresponding to the difference between the energy levels involved, is available. Each wavelength of radiant energy corresponds to a different amount of energy, that is,

$$E = h\nu = \frac{\hbar c}{\lambda} \tag{3}$$

where E is the amount of energy corresponding to a given frequency of radiant energy ν. The quotient c/λ may be written in place of ν. The term $\hbar$, Planck's constant, relates these quantities. If polychromatic light, radiant energy of several wavelengths, is incident upon a solution, those wavelengths (energies) corresponding exactly to electronic transitions to higher energy states will be absorbed and those which do not correspond to electronic transitions for the particular species in solution will be transmitted. Hence, a red solution does not absorb the longer wavelengths of light because the corresponding energy is unable to cause electronic transitions within the atoms of the absorbing species. A colored solution absorbs radiant energy at those wavelengths at which a solution of complementary color transmits those wavelengths. A red solution absorbs radiant energy in the blue–green region of the spectrum. Table I lists some absorbed and transmitted colors.

Table I

Relationship Between Selective Light Absorption and Color

Observed (transmitted) color (color of solution)	Region of maximum transmittance	Complementary (absorbed) color (color of suitable glass filter)
Red	610–750 mμ	Blue–green
Orange	595–610 mμ	Green–blue
Yellow	570–600 mμ	Blue
Green	500–570 mμ	Purple
Green–blue	480–500 mμ	Red
Blue	440–480 mμ	Yellow
Violet	400–440 mμ	Yellow–green

If monochromatic radiant energy known to be absorbed by a particular species in solution is allowed to irradiate this solution, the extent to which it is absorbed depends both upon the depth of solution through which it passes and the concentration of absorbing species in the solution. These relationships between intensity, light path, and concentration were developed by Bouguer, Lambert, and Beer. They are expressed mathematically as follows[85]:

$$\frac{P}{P_0} = e^{-kbc} = 10^{-abc} \tag{4}$$

where P_0 is the incident, monochromatic, radiant power, P is the transmitted, monochromatic, radiant power, P/P_0 is the transmittance, the fraction of radiant power transmitted, b is the path length, the distance traversed by the light in passing through the sample, c is the concentration of the absorbing species in solution, e is the base of natural logarithms, k is a constant relating the various quantities, and a is a constant equal to $k/2.3$.

Taking the logarithm of both sides of Eq. (4) and inverting to change sign gives

$$A = \log\frac{P_0}{P} = \log\frac{1}{T} = -\log T = abc \tag{5}$$

where A is the absorbance, the negative logarithm of the transmittance. Beer's law is most often expressed in the form $A = abc$. The constant a is the absorptivity. Its numerical value depends upon the choice of concentration units for c. Generally, b, the path length, is in centimeters, and if c is in moles per liter, a becomes the molar absorptivity and is represented by the symbol ε in units of liters per mole per cm. Absorbance is a dimensionless quantity.

In the past various other symbols and names have been used to represent the terms of the Beer's law relation but these are not recommended.[2,62]

2.2. Colorimetric Techniques

Beer's law serves as the basis for the quantitative colorimetric determination of many substances. It relates the absorbance of a solution directly to the concentration of absorbing species. Several different techniques are available for ascertaining concentration in colorimetry.

2.2.1. The Standard Series Technique

This technique requires the preparation of a series of solutions of known concentration in respect to the desired constituent. The absorbance of each solution is measured and a calibration plot of absorbance *vs* concentration is prepared. By measuring the absorbance of the unknown solution, its concentration may be read directly from the calibration curve. If the system under investigation obeys Beer's law exactly over the entire concentration range studied, the calibration graph is linear. If Beer's law is not obeyed and a nonlinear plot is obtained, the data are still valid and the graph serves as a calibration curve for the quantitative determination of the desired constituent.

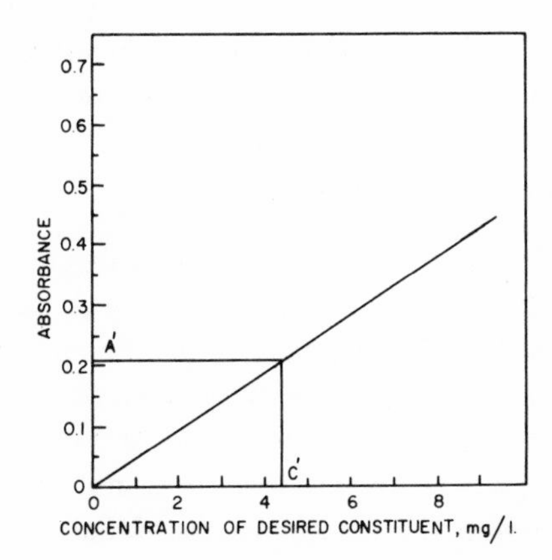

Fig. 1. Beer's law graph: plot of absorbance *vs* concentration. A') absorbance of unknown solution; C') concentration of unknown solution.

Figure 1 shows a typical Beer's law plot expected when using the standard series technique. This method is widely used and reliable results are obtained provided instrumental settings and sample treatment are identical for all known and unknown solutions. A calibration curve is especially useful in repetitive analyses of similar samples. The time necessary to prepare a calibration curve is a disadvantage of this technique, especially if only one or two samples are to be analyzed.

2.2.2. The Absolute Technique

This technique, although most applicable to a single determination, is normally not as reliable. The absorbance of a solution containing a known concentration of the desired constituent in a cell of known path length is measured and the absorptivity is calculated. This value is substituted into the Beer's law equation and solved for the concentration of the unknown solution based upon its observed absorbance. Deviations from Beer's law by the system under study negates the accurate determination of unknown concentrations by this method. Even if the known concentration is nearly equal to the unknown concentration, the possibility of errors based upon a single absorbance measurement may be appreciable.

2.2.3. Indirect Spectrophotometric Method

Some constituents are best determined by an indirect spectrophotometric method. The decrease in absorbance of a colored solution is a measure of the amount of constituent reacting, and may therefore be used to determine the concentration of the desired constituent present.[75]

2.2.4. Simultaneous Determination

Colorimetric methods are not restricted to the determination of a single constituent. The simultaneous determination of two or more constituents in the same sample is feasible. The technique to be adopted will depend on the nature of the particular chemical system. If for a two-component system the absorption spectra of constituents I and II possess an absorbance maximum at wavelengths where the other does not absorb, neither constituent will interfere with the determination of the other—assuming, of course, that no chemical interactions occur. Thus, constituent I can be determined by measurements at the wavelength of its absorbance maximum and constituent II can be determined by measurements at the wavelength of its absorbance maximum. If the absorption spectra of I and II overlap, as

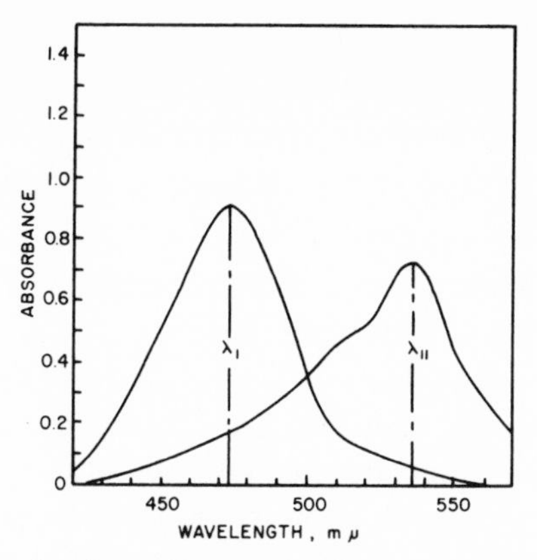

Fig. 2. Absorption spectra of two-component system.

is shown in Fig. 2, the simultaneous determination may still be possible. In cases where the spectra of both constituents obey Beer's law at the wavelengths of interest the absorbance values observed for the mixture will be the sum of the absorbances of each constituent. The optimum wavelength for measurement of each constituent is the wavelength at which the ratio of the absorptivity of one constituent to that of the other is a maximum. Then, the absorptivity of each constituent at all wavelengths selected for measurement are determined. The absorbance of the mixture is measured at these same wavelengths using cells of known path length. From these data the concentration of each constituent may be calculated by solution of a set of simultaneous equations. For constituents I and II and measurement at wavelengths $\lambda(x)$ and $\lambda(y)$ the following equations apply:

at $\lambda(x)$

$$A_{\text{total}, \lambda(x)} = A_{\text{I}, \lambda(x)} + A_{\text{II}, \lambda(x)} = a_{\text{I}, \lambda(x)} b C_{\text{I}} + a_{\text{II}, \lambda(x)} b C_{\text{II}} \tag{6}$$

at $\lambda(y)$

$$A_{\text{total}, \lambda(y)} = A_{\text{I}, \lambda(y)} + A_{\text{II}, \lambda(y)} = a_{\text{I}, \lambda(y)} b C_{\text{I}} + a_{\text{II}, \lambda(y)} b C_{\text{II}} \tag{7}$$

where C_{I} is the concentration of I and C_{II} is the concentration of II.

Equations (6) and (7) are solved simultaneously for C_{I} and C_{II}. Detailed procedures for simultaneous spectrophotometric analysis of polycomponent systems are available.[12] If any or all of the constituents comprising the sample do not obey Beer's law, the above approach must be modified and a more involved treatment is necessary.[132]

2.2.5. *Photometric Titrimetry*

Another interesting aspect of chemical colorimetry is found in photometric titrimetry. The desired constituent is titrated with a suitable reagent and the equivalence point of the titration is detected by measurement of a color change of the solution. Observed data are plotted to obtain titration graphs similar to one of the types shown in Fig. 3. The graphs are linear, except perhaps in the immediate vicinity of the endpoint region. Generally, data are recorded at a few points prior to the endpoint and at a few points beyond the endpoint. The linear portions of the curves based on these points are extrapolated and the point of intersection is taken as the endpoint of the titration. A photometric titration can be performed in a titration flask by periodically removing aliquots for photometric measurements, or a flow system is used whereby the solution in the titration flask is circulated through the absorption cell in the spectrophotometer. The solution is forced to circulate during the titration by means of a pump. Another alternative technique is to perform the titration directly in a vessel located inside the cell compartment of the instrument. Many instruments have cell compartments large enough to permit this method to be used. Photometric titration graphs differ depending upon whether the sample, titrant, or both, absorb at the wavelength chosen to measure the absorbance of the solution. Photometric titration methods for many elements appear in the literature.[56]

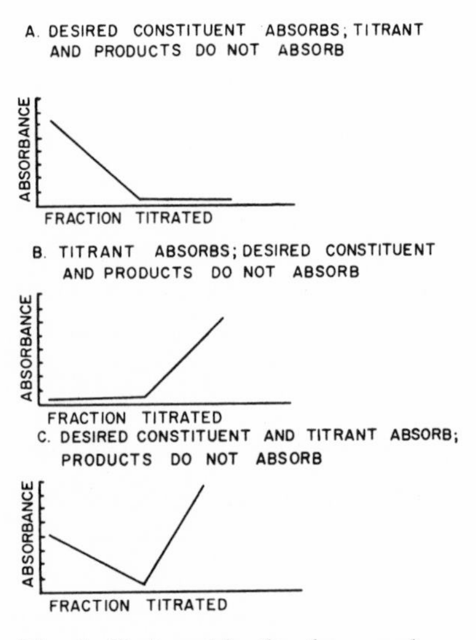

Fig. 3. Photometric titration graphs.

3. INSTRUMENTATION

Color comparators utilizing visual observation to match the color intensity of known and unknown solutions are not as widely used at the present time and have been replaced by photoelectric instruments which are more convenient and reliable. Photoelectric instruments differ widely with respect to their design and capabilities. A device which measures the radiant power or relative radiant power of light beams is called a photometer. If in addition to the photometer a spectrometer or monochromator is employed for isolating a particular spectral region, the instrument is termed a spectrophotometer. Spectrophotometers are described in terms of five components: (a) source of radiant energy; (b) monochromator, with its accompanying entrance and exit slits; (c) absorption cell compartment; (d) detector, which responds to the radiant power incident upon it; and (e) readout device. Spectrophotometers are also classified as single-beam or double-beam instruments. Single-beam instruments have one optical path from the exit slit of the monochromator through the sample cell to the detector. Double-beam instruments have two optical paths, of which one passes through the sample cell and the other through the reference cell. Double-beam instruments may utilize two detectors, one for each light beam, or a single detector, in which case the response from each light beam is alternately allowed to impinge upon the single detector. Many spectrophotometers, depending upon their optical and photometric systems, can be utilized for absorbance measurements in the UV and near IR spectral regions as well as in the visible region. Additional information pertaining to instrumentation is given elsewhere.[73,121,148]

3.1. Filter Photometers

The source of continuous radiant energy in filter photometers and spectrophotometers for the visible region is generally an incandescent tungsten-filament lamp. It emitts radiant energy throughout the visible region, although the intensity of this radiation is not uniform. The spectral emittance of a tungsten lamp is higher near the red region of the visible spectrum, and the relative spectral emittance increases with increasing temperature of the lamp filament. The power source operating the lamp must be stable in as much as fluctuations in source power will cause fluctuations in filament temperature and result in variations in the emittance. Variations in source intensity will cause variations in the detector response and lead to erroneous readings. A battery-operated dc power supply or a stabilized ac power supply is necessary to insure a steady lamp output. With double-beam instruments, lamp fluctuations are compensated for because the light beams passing through the sample and reference solutions are simultaneously subjected to the same variations. With single-beam instruments, in which a time interval elapses between the measurement of the reference solution and sample solution, source fluctuations can be a serious source of error.

Filter photometers employ a filter to isolate a desired portion of the spectrum. The filter usually consists of a piece of colored glass, or dyed gelatin film mounted between glass plates, in order to transmit radiant energy of a limited spectral region. Generally, the width of the wavelength band passed by a filter is about 50 mμ. The location of the transmitted band depends upon the color of the filter. These are available for isolation of almost any desired spectral region. The manufacturer's literature should be consulted for the appropriate filter suitable for a particular wavelength region. A compilation of these data is available.[142] The filter chosen for a colorimetric analysis should transmit radiant energy in the region of maximum absorbance for the colored species being measured. In general, the appropriate filter will have a color complementary to that of the absorbing species (see Table I). Interference filters which isolate a narrower wavelength region are also available and a compilation pertaining to their transmittance characteristics should be consulted for further information.[15]

Glass absorption cells are employed for measurements in the visible spectral region. Silica cells, necessary for measurements in the UV region, may also be used. Round cells are satisfactory if precautions are taken to insure their consistent positioning in the cell holders. Absorption cells with flat, parallel surfaces perpendicular to the optical path are preferred. These cells have a definite path length, a necessary value in Beer's law calculations. Cells should also be optically matched, that is, they should give the same transmittance or absorbance reading when filled with aliquots of the same solution. If unmatched cells are used care must be take to always use the same cell for sample and reference solutions. Cells require careful handling and must be scrupulously clean. Cells of 1-, 2-, and 5-cm path length are common. Larger cells, semimicro cells, microcells, variable-path-length cells, flow-through cells, thermostated cells, high- and low-temperature cells, and numerous other cells for specialized applications are available.

The detector employed in photoelectric instruments consists of a device capable of converting radiant power into an electrical response. Some photometers employ photovoltaic cells, or barrier-layer cells. These cells consist of a layer of conducting and of semiconducting material deposited upon a metallic base. Upon irradiation, a current flows between these layers and a galvanometer attached to the barrier layer cell registers the magnitude of this current. In a direct-reading photometer, the galvanometer scale is calibrated directly in transmittance or absorbance units. In a null-type instrument, the detector output is balanced by an external circuit until no current flows through a galvanometer. The transmittance or absorbance of the sample solution relative to that of the reference solution is read from a scale on the external compensating device. Barrier-layer cells are sensitive to radiant energy over approximately the same region as the human eye and are widely used in filter photometers. The sensitivity of this type of photocell decreases near the IR and UV regions. The electrical current produced by barrier-layer cells is a linear formation of irradiance, provided the external circuit resistance is low. Photovoltaic cells are rugged and relatively inexpensive but because of their low impedance the photocurrent is not easily amplified electronically. Upon continuous, prolonged exposure to radiant energy their response changes. This fatigue effect, the slow response to intermittent irradiation, and the sensitivity to fluctuations in temperature, have seriously limited the use of this type of photocell, except in filter photometers.

Numerous filter photometers are commercially available. The manufacturer's literature should be consulted for the specific details pertaining to a particular instrument. A comparison of some of the more popular photometers is available.[128] The merits of several newer instruments have also been discussed.[92]

3.2. Spectrophotometers

The light source of a spectrophotometer for use in the visible region is also an incandescent tungsten-filament lamp.

A spectrophotometer employs a monochromator consisting of either a prism or grating as a dispersive element. White light composed of all possible wavelengths incident upon the face of a prism is dispersed upon passing through the prism and emerges as a spectrum of rays. The prism material must transmit radiant energy in the region of interest. Glass can be used for the visible region and quartz for the visible, UV, and near IR regions. The dispersion of a prism is not linear and decreases as one approaches longer wavelengths. A grating consists of a series of finely ruled lines upon the face of a transparent or reflecting surface. White light incident upon these rulings is dispersed into its spectrum. The dispersion from a grating is approximately linear with wavelength if the striking angle of the incident radiation is small. Gratings produce second-, third-, and higher order spectra. In some instances a higher order spectrum overlaps another of a lower order. If this occurs the emerging light from the exit slit will no longer be monochromatic. Gratings can be used in regions other than the visible. Replica gratings, reproductions molded from a master grating, are commonly used and are relatively less expensive.

Not only the dispersing element, either prism or grating, but also the entrance and exit slits of the monochromator system determine the ability of the system to isolate a narrow spectral region. A narrower entrance slit will decrease the total intensity of the radiant energy incident upon the dispersing element but, at the same time, increase the spectral purity of the dispersed rays. An even narrower exit slit will further decrease the intensity of the light available to the detector and isolate a narrower wavelength region. Many commercial spectrophotometers have entrance and exit slits of equal width and give a triangular energy distribution about the nominal wavelength. The nominal wavelength is the wavelength value read from the monochromator setting. A spectral band of slightly higher and lower wavelengths is emitted from the exit slit. The spectral slit width, also called the spectral region isolated, encompasses the wavelength region actually emitted from the exit slit. Within this isolated spectral region a narrower spectral region is defined as the effective slit width, which represents the wavelength region within three-fourths of the total radiant energy of the emergent beam. It is desirable that the effective slit width be approximately one-tenth the natural band width, the width of the wavelength region at half the height of the absorbance maximum. A species exhibiting a narrow absorbance maximum in its absorption spectrum will require a narrow effective slit width for optimum resolution, sensitivity, and more likely adherence to Beer's law. Thus, better resolution is obtained by specifying a narrow mechanical slit width for the spectrophotometer used in making the measurements. Many spectrophotometers have provisions for manual or automatic adjustment of the slit width to take into account the requirements for optimum resolution and nonlinear dispersion of prism instruments. Several high-resolution spectrophotometers employ a double-monochromator system with two prisms or gratings or a prism–grating combination to achieve more nearly monochromatic radiant energy.

Spectrophotometers generally employ a photoemissive cell for detection of radiant energy. The detector has a photoemissive cathode which, when exposed to a beam of radiant power, emits electrons. The photoelectrons are attracted to a positive anode present with the cathode in an evacuated glass envelope, thus causing the generation of a photocurrent. The magnitude of the current is proportional to the radiant power of the incident radiant energy. The sensitivity of a photoemissive tube varies with the wavelength of incident radiant energy, and the nature of the photocathode surface. More than one phototube may be necessary in a spectrophotometer, depending upon which spectral region is being utilized. An increased photocurrent may be achieved for a photoemissive tube if the electrons emitted from the cathode are allowed to strike a second cathode within the evacuated envelope. More photoelectrons are emitted from the second cathode than from the first cathode. If these second-stage photoelectrons next impinge upon a third photocathode, etc., an avalanche effect is achieved which results in a gain in amplification within the tube. The acceleration of electrons from one photocathode to the next photocathode (dynode) is achieved by having each successive photocathode at a slightly more positive potential. Such tubes, containing 10–15 dynodes, are known as photomultiplier tubes and they produce a relatively larger output signal than does a simple photoemissive tube. A large amplification is necessary with high-resolution spectrophotometers because the narrow slits required for a high-resolution spectrum attenuate the total radiant energy reaching the detector. Phototubes and photomultiplier tubes have a high impedance and their current output can be increased conveniently by electronic amplifiers. However, higher output signal amplification also results in an increase in electronic noise and this ultimately limits the extent to which meaningful amplification is possible. Photoemissive detectors require a power supply and with the amplifier components contribute to the complexity and cost of a spectrophotometer.

The amplified signal from a photoemissive tube can be directed to an appropriate meter, in which case the instrument is direct reading in either absorbance or transmittance units, depending upon the calibration of the meter scale. The detector signal can also be compensated for by an external signal with a galvanometer null indicator. This compensation can be either optical, perhaps a movable optical wedge in the reference beam, or electrical. In place of a galvanometer null indicator, the imbalance between detector signal and reference signal can be used to drive a recorder pen. If the recorder is coupled to the wavelength drive of the spectrophotometer, the automatic recording of spectra may be achieved. The detector signal can be displayed on the face of a digital meter, adjusted to read directly the absorbance of the sample solution. With proper calibration digital readout of component concentration is possible. The output signal can be recorded on punch cards or magnetic tape and utilized for computer calculation of desired quantities.

Numerous spectrophotometers are available from various manufacturers. The specifications of some of these instruments have been summarized.[73]

3.3. Instrument Selection

Depending upon its intended use, one type of instrument may be more advantageous than another. For routine analysis of only a few constituents, each of which forms a well-characterized species suitable for colorimetric measurement, a filter photometer may be entirely satisfactory. If the colored system under investigation demands the use of nearly monochromatic radiant energy, a spectrophotometer with adequate slit control and sensitive detector response is required. Studies involving new and untried photometric procedures should be made with a high-resolution sensitive spectrophotometer to insure proper delineation of optimum operating conditions. For photometric measurements at a single wavelength, as in the preparation of Beer's law plots, a manual, single-beam instrument is suitable. If one wishes to obtain a complete absorption spectrum, a recording, double-beam instrument is faster and more convenient. For systems showing conformity to Beer's law, meter or digital readout provides the absorbance or transmittance value associated with each measured concentration. When UV or near IR measurements are to be made, an instrument with an exchangeable source and detector system, which allow for measurements in the UV and near IR regions, should be considered. Numerous accessories are available for certain spectrophotometers and the acquisition of some of these accessory units may be advantageous for a particular laboratory. Automatic sampling equipment, for instance, enables one to make measurements on numerous samples automatically. Thermostated cells, provisions for plotting or direct readout in either absorbance or transmittance, and scale expansion over a narrower optical region are possible options. Cell-compartment size for accommodation of a variety of cells may be important. Provisions for flame emission, fluorescence, internal reflection, rapid scanning, dual wavelength, and photometric titration measurements are available with many instruments. One should also consider the initial cost, availability of competent operators, adequate laboratory space, and proper maintenance procedures.

Deviations from Beer's law occur from instrumental causes as well as from chemical changes involving the measured species. The proper choice of measuring instrument minimizes these deviations. Thus, when too wide a slit width is used, an apparent deviation from Beer's law results. A slit width larger than the natural band width of the absorbance maximum being measured will result in a low absorbance value. Narrow slit-width settings are especially necessary for adherence to Beer's law when the absorbance maxima have narrow natural band widths. This finite slit-width effect is the main cause for differences between observed and true absorbance values.

A second instrumental source of error occurs because of reflections of the optical beam from the surfaces of the absorption cell. Internal reflection causes the incident beam to traverse the sample solution more than once, leading to a multiple-reflection-path effect. The beam which finally emerges and proceeds to the detector will produce a higher absorbance reading than that which would have been observed from a single pass through the sample solution. The multiple-reflection-path effect is most pronounced when measuring very dilute solutions having low absorbance values. The use of longer-path-length cells help to minimize this source of error.

Stray polychromatic radiant energy which has been reflected and scattered by the various components within the interior of the instrument may be superimposed on a monochromatic beam. In measuring solutions of high absorbance very little energy reaches the detector and the stray radiant energy can constitute an appreciable fraction of the detector signal. In this case, a lower absorbance value would be observed in the absence of stray radiant energy. Stray radiant energy is minimized by employing a double-monochromator system, a filter, and proper optical design. Additional information pertaining to instrumental causes of error is available.[7,48,87]

Any spectrophotometer, if it is to provide meaningful, reproducible data, must be calibrated and the calibration checked at various intervals. Both wavelength and photometric accuracy are necessary and various calibration procedures have been recommended.[39,49] Comparison and adjustment of the wavelength-scale reading taken when measuring the emission line of a mercury lamp at 546.1 mμ serve to check the wavelength scale. Other emission lines from a hydrogen lamp and the location of absorbance maxima of certain rare-earth oxides are also used. Absorbance comparison and adjustment are made using standard absorbing glasses, perforated screens, or chemical solutions with highly reproducible photometric characteristics. Acidic copper sulfate, acidic cobalt ammonium sulfate, and basic potassium chromate are used for this purpose. The references cited above should be consulted for detailed calibration procedures.

4. SAMPLE PREPARATION

4.1. Range of Optimum Concentration

The exact concentration range over which a constituent can be determined colorimetrically may be limited, depending primarily upon the particular constituent and chromogenic reagent being used. In general, solutions more concentrated than 0.01 mole of desired constituent per liter cannot be conveniently determined. Solutions of this concentration usually have too large an absorbance for accurate measurement.[133] In ascertaining a lower limit of colorimetric detection one must consider the sensitivity of the measuring instrument to small changes in absorbance. Sensitive instruments can measure accurately differences of 0.001 absorbance. The light-absorbing capacity of the desired constituent also determines the sensitivity of a colorimetric method, as does the particle size. A larger cross sectional area of the absorbing species will have a greater probability of interaction with incident radiant energy, and the probability that after interaction an electronic transition occurs within the absorber limits the absorptivity of a species. These limitations set an upper value for the molar absorptivity of the absorbing species. Generally, molar absorptivities do not exceed 10,000 liters per cm per mole, although values larger than this have been reported in favorable cases. Assuming a molar absorptivity of 10,000, a minimum absorbance reading of 0.001, and the use of 1-cm cells, the lower concentration limit will be 10^{-8} *M*.[17,145] This value can be further reduced if cells of longer path length are employed.

Concentration units commonly used in colorimetric measurements are generally moles per liter or weight per unit volume, such as g per liter, g per 100 ml, mg per 100 ml, or ppm. The latter is equivalent to mg per liter or μg per ml. Absorptivity determines the sensitivity of a colorimetric method. The larger the absorptivity the smaller the amount of desired constituent that can be detected. Another measure of detection limit is the sensitivity value. It is defined as the number of μg of desired constituent per cm path length of absorbing solution required to produce an absorbance reading of 0.001.[116] The smaller the sensitivity the smaller the amount of desired constituent that can be detected.

Within the extremes of limiting concentration over which colorimetric measurements are possible, there is a much narrower concentration region which yields the most meaningful data.[11,69,86,117] The limits of this region, the optimum-concentration range, are best found graphically by the method of Ringbom.[3,111] A typical Ringbom plot is shown in Fig. 4. It is a plot of

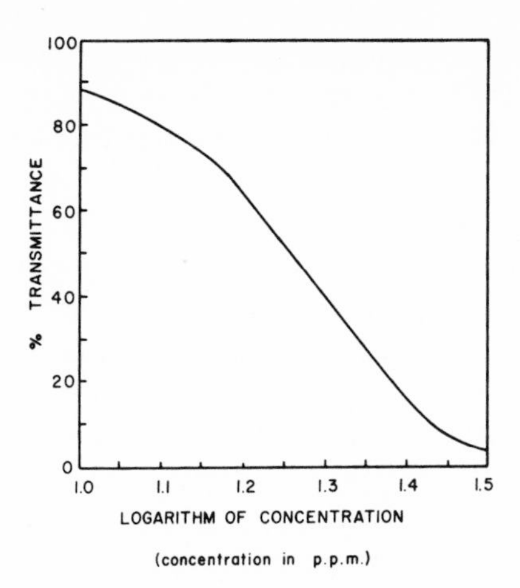

Fig. 4. Ringbom plot.

percent transmittance against the logarithm of the concentration. The linear portion of the plot, having the steepest slope, represents the region of optimum concentration for a spectrophotometric determination of that constituent.

4.2. Preparation of the Colored System

The preliminary treatment of a sample often requires dissolution by fusion or acid treatment. Although details of recommended reagents and techniques are beyond the scope of this chapter, it is essential that every effort be made to avoid contamination, since colorimetric methods are especially sensitive to traces of contaminants. It is advisable to always perform a blank determination so that any trace contaminants in reagents can be detected and compensated for. Likewise, it is convenient to dilute the sample solution to a specific volume. In analyzing for one or more constituents by colorimetric methods, it is very helpful to be able to take smaller or larger aliquots from a sample stock solution.

Depending on the composition of the matrix of the sample, it may be necessary to either isolate the desired constituent prior to development of a colored system or to remove some interfering substance. In analyzing inorganic systems, ion-exchange, extractions, and electrodeposition methods are usually the most applicable. Distillation, precipitation, and various chromatographic methods are occasionally used.

Another approach to eliminate the deleterious effect of a specific ion may be to resort to "masking"—a method of suppressing an undesired reaction by adding a complexing agent which preferentially reacts with the interfering ion. EDTA may be used, for example, to eliminate the interference of certain metal ions in the colorimetric determination of phosphates. In all cases the preparative treatment must be consistent for all samples and standards to ensure reproducible analytical results.

Suspended particles and foreign matter cannot be tolerated in the test solution. Ion-pair formation, polymerization, solvent association, and other undesirable reactions must also be avoided, since they may cause changes in both the location and magnitude of the absorbance maximum.

The final chemical treatment consists of adding a reagent which undergoes a reaction with the desired constituent, producing a color proportional to the concentration of this specific constituent. Color formation should be rapid and the reaction stoichiometric. The resulting color must be stable, reproducible, and very sensitive to variation in concentration of the desired constituent. For these reasons the direct measurement of naturally colored species is seldom used. A single reagent such as a complexant, reductant, or oxidant is often sufficient to develop a colored solution. Some color-forming reactions depend on a combination of reagents or a systematic sequence of reactions. It is advantageous to use high-purity, colorless reagents or those giving a small, reproducible blank. The amount of each reagent used should be measured accurately.

Both the rate of color development and ultimate intensity of the color depend on the temperature and the pH value of the solution. At low temperatures the rate of color formation is too slow, and at high temperatures chemical decomposition of the colored species may occur. Absorbance values may also be temperature dependent because of changes in solution density and refractive index.[36]

The stability of the colored system must be well known. Some colored systems are stable for long periods of time, while other systems will increase gradually in intensity and then remain relatively constant. Some colored systems begin to fade after a limited period of stability, while a few used systems exhibit a gradual but continuous change in absorptivity. The time elapsed between preparation of a colored solution and photometric measurement may be critical. A thorough knowledge of the specificity of the colorimetric method or, more significantly, an understanding of what substances interfere and the nature of their interference is essential.[11]

5. PHOTOMETRIC MEASUREMENTS

5.1. Accuracy and Precision

Assuming, for the moment, that other sources contributing to the error of a photometric determination are negligible, consider the effect of transmittance error from a conventional photoelectric instrument in which the error in transmittance ΔT is constant over the entire transmittance range. This error will cause a corresponding error ΔC in the concentration obtained from the photometric reading. If Beer's law is obeyed for the system under consideration, the relative concentration error $\Delta C/C$ is found by differentiating Beer's law with respect to concentration.[5,58]

$$\frac{\Delta C}{C} = \frac{0.434\,\Delta T}{T \log T} \tag{8}$$

The magnitude of the relative concentration error is determined by assigning a value to the transmittance error suitable for the photometric instrument used and evaluating the right-hand side of Eq. (8). A plot of these data is shown in Fig. 5. The uncertainty associated with a concentration value at the extremes of photometric measurements is large. The position of most accurate transmittance measurement, smallest relative concentration error, is found by setting equal to zero the derivative of Eq. (8) taken with respect to concentration. This point occurs at a percent transmittance of 36.8 (0.434 absorbance). At this point the relative concentration error per unit photometric error is at a minimum

$$\left(\frac{\Delta C/C}{\Delta P}\right)_{\max} = -2.72\%$$

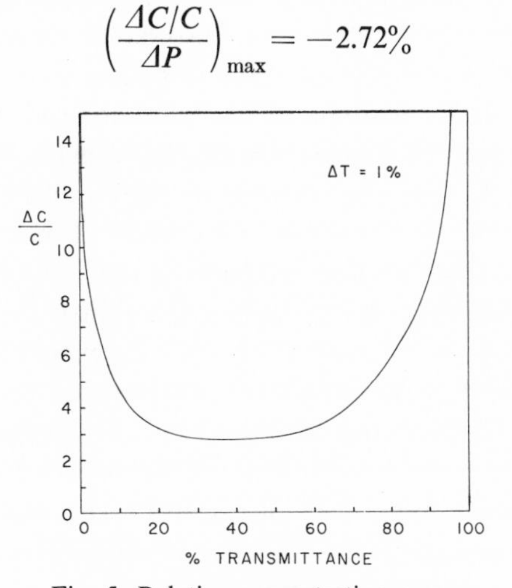

Fig. 5. Relative concentration error.

The relative concentration error does not increase appreciably above its minimum value over the range 20–60% transmittance (0.7–0.2 absorbance). Hence, this region is recommended for photometric studies.[16] Transmittance readings outside this range are possible provided the corresponding increase in relative concentration error is tolerable. Measurements near the limiting extremes of 0 and 100% transmittance should be avoided.

5.1.1. Conventional Method

In conventional photoelectric measurements the lower limit of the transmittance scale is set by adjusting the photometer readout to 0% T with the detector protected from all possible light sources; the upper limit of the transmittance scale is set by adjusting the photometer to read 100% T with only solvent in the reference cell which is placed in the light path. The 0% T setting is referred to as the dark-current setting. Commonly, in place of solvent transmittance at 100% T, a blank or reference solution containing all constituents present in the sample but the desired constituent is used to set the 100% T value. This practice has the advantage of cancelling possible interferences from species other than the desired constituent. These interferences, present in both reference and sample solution, are effectively cancelled by adjustment of the instrument to 100% T (zero absorbance) with the blank solution in the reference cell placed in the light path. By this technique only the absorbance of the desired constituent should be obtained. Thus, after the dark-current and 100% T settings have been made, the sample solution is placed in the sample cell located in the light path of the instrument and the percent transmittance or absorbance measured.

In an effort to improve the precision of photometric measurement, the conventional method of measurement has been modified. Various modifications are possible and the choice of procedure depends upon whether the sample of interest is more concentrated or more dilute than necessary to give ordinary transmittance readings in the range of optimum photometric accuracy. These methods, called precision or differential methods, allow measurements outside the ordinary range of photometric reliability and with a higher precision than is possible by the conventional method.[100,109,136] Precision photometric measurements are accomplished by judiciously fixing one or both ends of the transmittance-scale solutions containing specific concentration of the desired constituent. If the special reference solution is identical to the sample solution in all respects except the concentration of the desired constituent, both the compensation effects of a blank solution and an increase in precision will result.

5.1.2. Transmittance-Ratio Method

The transmittance-ratio method of precision photometry is applicable to concentrated solutions of the desired constituent. With the conventional photometric method and most spectrophotometers the transmittance values observed for concentrated solutions would be too low for reliable measurement. With the transmittance-ratio method the dark current, 0% T, is set as in the conventional method, the detector in total darkness. The 100% T setting is made with a solution slightly less concentrated with respect to the desired constituent than the sample to be analyzed. This method has the effect of expanding the readout scale (see Fig. 6a).

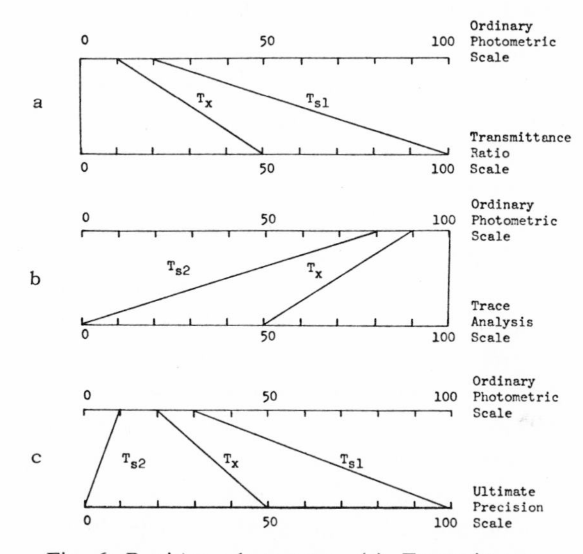

Fig. 6. Precision photometry. (a) Transmittance-ratio method: T_{s1} is determined by a standard slightly less concentrated than the unknown (reading T_x) and is used to fix one limit of the expanded scale. Fivefold expansion. (b) Trace-analysis method: T_{s2} is determined by a standard slightly more concentrated than the unknown (reading T_x) and is used to fix one limit of the expanded scale. Fivefold expansion. (c) Ultimate-precision method: T_{s1} is determined by a standard slightly less concentrated than the unknown (reading T_x) and is used to fix one limit of the expanded scale. T_{s2} is determined by a standard slightly more concentrated than the unknown (reading T_x) and is used to fix the other limit of the expanded scale. Fivefold expansion.

5.1.3. Trace-Analysis Method

The trace-analysis method of precision photometry is applicable to very dilute solutions of the desired constituent. With the conventional photometric method the transmittance values observed in these cases would be too high for reliable measurement. In the trace-analysis method the 100% T limit is set in the ordinary way with a solvent or a reagent-blank solution in the reference cell. The 0% T setting is made with a solution slightly more concentrated in respect to the desired constituent than the sample solution. This method has the effect of further expanding the photometer scale (see Fig. 6b).

5.1.4. Ultimate-Precision Method

The ultimate-precision method of differential photometry is applicable to those solutions where measurements of the highest reliability are desired. With this method the limits of the instrument scale are set using two solutions, one slightly more concentrated than the unknown sample and the other slightly less concentrated. The 0% T limit is set with the slightly more concentrated standard and the 100% T limit is made with the slightly less concentrated standard (see Fig. 6c).

Precision methods have been applied to indirect and multicomponent photometric methods of analysis.[100,136] Precision methods, although applicable at concentration extremes and yielding more accurate results than ordinary photometric methods, are not without disadvantages. The validity of Beer's law cannot be assumed. Calibration graphs with known concentrations of the desired constituent are not always linear. Prior knowledge of the approximate amount of desired constituent is necessary before known standards of the proper concentration can be prepared. The preparation of these standards is time consuming and the spectrophotometric determination may take considerably longer than when the conventional photometric method of measurement is employed. Small differences in the sample cells and in the positioning of the cells in the optical path become increasingly significant in the precision methods.

5.2. Advantages

Colorimetric methods for the quantitative determination of trace constituents in a sample satisfy many of the desirable requirements to be considered in choosing an analytical procedure. The sensitivity is generally good, falling in the ppm range. Spectrophotometric methods are accurate and provide, within the optimum absorbance region and concentration limits described above, a reliable and precise method of analysis. With suitable, sensitive instrumentation ordinary photometric measurements are capable of a relative concentration precision of less than 1%. Even more precise measurements are possible if one uses a differential photometric method. Minor interferences and the interferences from solvent and other reagents are eliminated by the proper choice of a blank solution for setting the transmittance (absorbance) limits of the photometric instrument. If many samples of similar composition are to be analyzed repetitively for the same constituent, a calibration graph will be very useful for subsequent determinations, provided the same procedure and the same instrument are used. If a constituent is not colored itself, colorimetric analysis is still possible provided a suitable reagent can be found which when reacted with the desired constituent will produce a colored product. Simultaneous spectrophotometric determinations of two or more constituents are possible. Major constituents can be analyzed by use of the transmittance-ratio method and minor trace constituents too small to be measured by the conventional photometric method can be determined by the trace-analysis method. Instrumentation for colorimetric analysis is generally available in most chemical laboratories together with specific procedures.

5.3. Limitations

The accuracy and precision of a colorimetric analysis diminishes when measurements are made outside the optimum absorbance and concentration limits. It is possible to dilute a concentrated sample exhibiting too high absorbance (i.e., too low transmittance). Extreme dilution is not recommended for the dilution errors generally become larger than the photometric errors. Very dilute sample solutions can be concentrated. An ion-exchange technique, for example, can be used to concentrate a dilute solution. A large volume of the sample solution is passed through an appropriate ion-exchange column. When a sufficient amount of the desired constituent has been collected on the column, it is eluted with a small volume of appropriate eluting solution. Extraction of a colored species in a relatively small volume of an immiscible organic solvent is another method of concentrating colored systems. Some colored systems are not suitable for colorimetric measurements. They may be unstable, exhibit a large temperature effect on absorbance, or may not adhere to Beer's law. A color-forming reagent may form colored species with several constituents in the sample. The absorbance of other species in the sample may also interfere with that of the desired constituent.

Table II

Typical Spectrophotometric Methods for the Determination of Metals

Metal	Reagent, or method	Reference
Ag	Dithizone	41, 115
	p-Dimethylaminobenzylidenerhodanine	22, 115, 118, 122
Al	Eriochrome cyanine R	57
	Alizarinsulfonic acid (Alizarin Red S)	6, 115
	Ammonium aurintricarboxylate (Aluminon)	115
	Xylenol orange	19, 38, 90, 138
As	Heteropoly blue	13, 91, 135
	12-Molybdoarsenic acid	141
Au	*p*-Diethylaminobenzylidenerhodanine	114
	Bromoauric acid	80
	Rhodamine B	83
	2,2′-Dipyridyl ketoxime	59
Be	Quinalizarin	42
	p-Nitrobenzeneazo-orcinol	29, 140
Bi	Iodide	74, 81, 146
	Dithizone	115
	Ammonium 1-pyrrolidinecarbodithioate	67
	Thiourea	74, 94
Co	Sodium 1-nitroso-2-hydroxynaphthalene-3,6-disulfonate (Nitroso-R-Salt)	105
	1-Nitroso-2-naphthol	79, 93
	Ammonium 1-pyrrolidinecarbodithioate	67
Cr	Chromate	26, 115
	Peroxy-2,2′-bipyridine	103
	1,5-Diphenylcarbohydrazide	106
Cu	2,9-Dimethyl-1,10-phenanthroline (Neocuproine)	45, 126
	Sodium diethylaminecarbodithioate	60
Fe	1,10-Phenanthroline	44, 119
	Thiocyanate	149
Mg	Sodium methylbenzothiazole-(1,3)-4,4′-diazoaminobenzene-(2,2′-disulfonate (Titan Yellow)	50, 76
	O,O′-Dihydroxyazobenzene	37
	1,2-Hydroxy-3,2,4-dimethylcarboxyanilide naphth-1-ylazo-2-hydroxybenzene (Magon)	1, 123
Mn	Permanganate	78, 99, 115, 147
Mo	Thiocyanate	30, 51, 63
	Ammonium-1-pyrrolidinecarbodithioate	67
	4-Methyl-1,2-dimercaptobenzene	53

Table II (*continued*)

Metal	Reagent, or method	Reference
Ni	Dimethylglyoxime	27, 98
Pb	Dithizone	129
	Sodium diethyldithiocarbamate	68
	4-(2-Pyridylazo)-resorcinol	34
Sb	Rhodamine B	107
	Brilliant green	18, 130
	Iodide	40, 81
Sn	4-Methyl-1,2-dimercaptobenzene	28
	Catechol violet	33
Ti	Hydrogen peroxide	144
	Disodium-1,2-dihydroxybenzene-3,5-1 disulfonate (Tiron)	152
U	Hydrogen peroxide	112
	Thiocyanate	32, 96
	1-(2-Pyridylazo)-2-naphthol	4
V	Tungstovanadophosphoric acid	124, 150
	Hydrogen peroxide	137, 151
	Molybdovanadophosphoric acid	64
W	Thiocyanate	31, 47, 108
	4-Methyl-1,2-dimercaptobenzene	23, 89
	Peroxytungstic acid	104
Zn	Dithizone	115, 139

Although suitable colorimetric methods may not be available for the determination of certain constituents, analytical information resulting from the absorption of electromagnetic radiant energy may still be possible. Many chemical systems which exhibit no color have definite and distinct absorbance maxima in the UV or near IR spectral regions. For certain elements spectrometric, flurophotometric, or atomic absorption spectrometric methods can be utilized advantageously.

6. APPLICATIONS

6.1. Determination of Metals

The spectrophotometric method for the determination of metals is one of the most extensively used when trace quantities are to be determined, because the technique is sensitive, specific, rapid, and the instrumentation required is relatively inexpensive. The nature of the matrix of the sample material, the specific metal to be determined, and/or the number of metals in the sample to be analyzed are factors to be considered before selecting a particular colorimetric method.

The excellent treatises of Sandell,[115] Snell and Snell,[127] and Charlot[24] contain numerous colorimetric methods for each metal, and procedures for specific applications to particular materials including minerals, ores, and water. In Table II typical colorimetric methods for 22 common metals are summarized. New reagents, improved methods, and recent applications may be found in reviews on this methodology.[14]

Table III

Typical Spectrophotometric Methods for the Determination of Nonmetals

Nonmetal	Reagent, or method	References
B	Carminic acid	11, 20, 54, 113
	Quinalizarin	11, 82
	1,1′-Dianthrimide	11, 35
	Curcumin	11, 52, 55, 131
F	Zirconium–eriochrome cyanine R	11, 88
	Zirconium–alizarin	11, 72
	Zirconium–SPADNS	9
N(NH_3) or (NH_4^+)	Indirect: Hypobromite	61
	Pyridine–pyrazolone	71
	Phenol–hypochlorite	84, 143
	Nessler	11
(NO_3^-)	Phenoldisulfonic acid	11
	Brucine	97
NO_2^-)	Griess-II	5, 110
	UV: 4-Aminobenzenesulfonic acid	102
P (PO_4^{-3})	Heteropoly blue	11, 13, 77
	Molybdovanadophosphoric acid	70
S (S^{-2})	Methylene blue	43, 65
(SO_4^{-2})	Indirect: Barium chloranilate	10, 120
	Methylene blue	65
Se	3,3′-Diaminobenzidene	25
Si	Heteropoly blue	13, 101, 134
	Molybdosilicic acid	21, 125
Te	Iodotellurite	46, 66
	Thiourea	95

6.2. Determination of Nonmetals

Spectrophotometry is a method of choice in the determination of nonmetals. Several reference works[11,24,127] contain information of specific methods and applications. In Table III typical colorimetric methods for eight common nonmetals are quoted.

REFERENCES

1. Apple, R. F., and White, J. C., *Talanta* **8**, 419 (1961).
2. *ASTM Committee Manual on Recommended Practices in Spectrophotometry*, American Society for Testing and Materials, Philadelphia, Pa., 1966, p. 1.
3. Ayres, G. H., *Anal. Chem.* **21**, 652 (1949).
4. Baltisberger, R. J., *Anal. Chem.* **36**, 2369 (1964).
5. Barnes, H., and Folkard, A. R., *Analyst* **76**, 599 (1951).
6. Barton, C. J., *Anal. Chem.* **20**, 1068 (1948).
7. Bauman, R. P., *Absorption Spectroscopy*, John Wiley & Sons, New York, 1962, p. 149.
8. Bauman, R. P., *ibid.*, p. 378.
9. Bellack, E., and Schoubee, P. J., *Anal. Chem.* **30**, 2032 (1958).
10. Bertolacini, R. J., and Barney, J. E., *Anal. Chem.* **30**, 202 (1958).
11. Boltz, D. F., ed., *Colorimetric Determination of Nonmetals*, Interscience, New York, 1958.
12. Boltz, D. F., *Selected Topics in Modern Instrumental Analysis*, Prentice-Hall, New York, 1952, p. 149.
13. Boltz, D. F., and Mellon, M. G., *Anal. Chem.* **19**, 873 (1947).
14. Boltz, D. F., and Mellon, M. G., *Anal. Chem.* **36**, 256R (1964); **38**, 317R (1966); **40**, 255R (1968); **42**, 152R (1970).
15. Boltz, D. F., and Schenk, G. H., in *Handbook of Analytical Chemistry*, (L. Meites, ed.), McGraw-Hill, New York, 1963, p. 6–9.
16. Boltz, D. F., and Schenk, G. H., *ibid.*, p. 6–16.
17. Braude, E. A., in *Determination of Organic Structures by Physical Methods*, Volume I, (E. A. Braude, and E. C. Nachod, eds.), Academic Press, New York, 1955, p. 131.
18. Burke, R. W., and Menis, O., *Anal. Chem.* **38**, 1719 (1966).
19. Budesinsky, B., *Zh. Anal. Khim.* **18**, 1071 (1963).
20. Callicoat, D., and Wolszon, J. D., *Anal. Chem.* **31**, 1434 (1959).
21. Case, O. P., *Ind. Eng. Chem., Anal. Ed.* **16**, 309 (1944).
22. Cave, G. C. B., and Hume, D. N., *Anal. Chem.* **24**, 1503 (1952).
23. Chan, K. M., and Riley, J. P., *Anal. Chem.* **36**, 220 (1966).
24. Charlot, G., *Colorimetric Determination of Elements. Principles and Methods*, Elsevier, New York, 1964.
25. Cheng, K. L., *Anal. Chem.* **28**, 1738 (1956).
26. Christow, S., *Z. Anal. Chem.* **125**, 278 (1943).
27. Claassen, A., and Bastings, L., *Analyst* **91**, 725 (1966).
28. Clark, R. E. D., *Analyst* **61**, 242 (1936); **62**, 661 (1937).
29. Covington, L. C., and Miles, M. J., *Anal. Chem.* **28**, 1728 (1956).

30. Crouthamel, C. E., and Johnson, C. E., *Anal. Chem.* **26**, 1284 (1954).
31. Crouthamel, C. E., and Johnson, C. E., *Anal. Chem.* **26**, 1288 (1954).
32. Currah, J. E., and Beamish, F. E., *Anal. Chem.* **19**, 609 (1947).
33. Dagnall, R. M., West, T. S., and Young, P., *Analyst* **92**, 27 (1967).
34. Dagnall, R. M., West, T. S., and Young, P., *Talanta* **12**, 583, 589 (1965).
35. Danielsson, L., *Talanta* **3**, 138, 203 (1959).
36. Delahay, P., *Instrumental Analysis*, Macmillan, New York, 1957, p. 208.
37. Diehl, H., Olsen, R., Spielholtz, G. E., and Jensen, R., *Anal. Chem.* **35**, 1144 (1963).
38. Dvorak, J., and Nyvltova, E., *Mikrochim. Acta* **1966**, 1082.
39. Edisbury, J. P., *Practical Hints on Absorption Spectrometry*, Plenum Press, New York, 1967.
40. Elkind, A., Gayer, K. H., and Boltz, D. F., *Anal. Chem.* **25**, 1744 (1953).
41. Fischer, H., Leopoldi, G., and Vol Uslar, H., *Z. Anal. Chem.* **101**, 1 (1935).
42. Fischer, W., and Wernet, J., *Angew. Chem.* **A60**, 729 (1948).
43. Fogo, J. K., and Popowsky, M., *Anal. Chem.* **21**, 732 (1949).
44. Fortune, W. B., and Mellon, M. G., *Ind. Eng. Chem., Anal. Ed.* **10**, 60 (1938).
45. Gahler, A. R., *Anal. Chem.* **26**, 577 (1954).
46. Geiersberger, K., and Durst, A., *Z. Anal. Chem.* **135**, 11 (1952).
47. Gentry, C. H. R., and Sherrington, L. G., *Analyst* **73**, 57 (1948).
48. Gibson, K. S., in *Analytical Absorption Spectroscopy*, (M. G. Mellon, ed.), John Wiley & Sons, New York, 1950, p. 244.
49. Gibson, K. S., *ibid.*, p. 256.
50. Gillam, W. S., *Ind. Eng. Chem., Anal. Ed.*, **13**, 499 (1941).
51. Grimaldi, F. S., and Wells, R. C., *Ind. Eng. Chem., Anal. Ed.* **15**, 315 (1943).
52. Grinstead, R. R., and Snider, S., *Analyst* **92**, 532 (1967).
53. Hamence, J. H., *Analyst* **65**, 152 (1940).
54. Hatcher, J. T., and Wilcox, L. V., *Anal. Chem.* **22**, 567 (1950).
55. Hayes, M. R., and Metcalfe, J., *Analyst* **87**, 956 (1962).
56. Headridge, J. B., *Photometric Titrations*, Pergamon Press, New York, 1961.
57. Hill, U. T., *Anal. Chem.* **28**, 1419 (1956).
58. Hiskey, C. T., *Anal. Chem.* **21**, 1440 (1949).
59. Holland, W. J., and Bozic, J., *Anal. Chem.* **39**, 109 (1967).
60. Howard, J. M., and Spauschus, H. O., *Anal. Chem.* **35**, 1016 (1963).
61. Howell, J. A., and Boltz, D. F., *Anal. Chem.* **36**, 1799 (1964).
62. Hughes, H. K., *Anal. Chem.* **24**, 1349 (1952).
63. Hurd, L. C., and Allen, H. O., *Ind. Eng. Chem., Anal. Ed.* **7**, 396 (1935).
64. Jakubiec, R., and Boltz, D. F., *Anal. Chim. Acta* **43**, 137 (1968).
65. Johnson, C. M., and Nishita, H., *Anal. Chem.* **24**, 736 (1952).
66. Johnson, R. A., and Kwan, F. P., *Anal. Chem.* **23**, 651 (1951).
67. Kalt, M. B., and Boltz, D. F., *Anal. Chem.* **40**, 1086 (1968).
68. Keil, R., *Z. Anal. Chem.* **229**, 117 (1967).
69. Kirkbright, G. F., *Talanta* **13**, 1 (1966).
70. Kitson, R. E., and Mellon, M. G., *Ind. Eng. Chem., Anal. Ed.* **16**, 379 (1944).
71. Kruse, J. M., and Mellon, M. G., *Anal. Chem.* **25**, 1188 (1953).
72. Lamar, W. L., *Ind. Eng. Chem., Anal. Ed.* **17**, 148 (1945).
73. Lewin, S. Z., *J. Chem. Educ.* **37**, A401, A507, A637 (1960).
74. Lisicki, N. M., and Boltz, D. F., *Anal. Chem.* **27**, 1722 (1955).
75. Lothe, J. J., *Anal. Chem.* **27**, 1546 (1955).

76. Ludwig, E. E., and Johnson, C. R., *Ind. Eng. Chem., Anal. Ed.* **14**, 895 (1942).
77. Lueck, C. H., and Boltz, D. F., *Anal. Chem.* **28**, 1168 (1956).
78. Luke, C. L., *Anal. Chim. Acta* **34**, 302 (1966).
79. Lundquist, R., Markle, G. E., and Boltz, D. F., *Anal. Chem.* **27**, 1731 (1955).
80. McBryde, W. A., and Yoe, J. H., *Anal. Chem.* **20**, 1094 (1948).
81. McChesney, E. W., *Ind. Eng. Chem., Anal. Ed.* **18**, 146 (1946).
82. MacDougal, D., and Biggs, D. A., *Anal. Chem.* **24**, 566 (1952).
83. MacNulty, B. J., and Woollard, L. D., *Anal. Chim. Acta* **13**, 154 (1955).
84. Mann, L. T., Jr., *Anal. Chem.* **35**, 2179 (1963).
85. Meehan, E. J., in *Treatise on Analytical Chemistry*, (I. M. Kolthoff, and P. J. Elving, eds.), Interscience, New York 1964, Part I, Vol. 5, p. 2755.
86. Meehan, E. J., *ibid.*, p. 2758.
87. Meehan, E. J., *ibid.*, p. 2765.
88. Megregian, S., *Anal. Chem.* **26**, 1161 (1954).
89. Miller, C. C., *Analyst* **69**, 109 (1944).
90. Miyajima, T., *Bunseki Kagaku* **13**, 1042 (1964).
91. Morris, H. J., and Calvery, H. O., *Ind. Eng. Chem., Anal. Ed.* **9**, 447 (1937).
92. Muller, R. H., *Anal. Chem.* **40** (6), 109A (1968).
93. Nichol, W. E., *Can. J. Chem.* **31**, 145 (1953).
94. Nielsch, W., and Boltz, G., *Z. Anal. Chem.* **142**, 321 (1954).
95. Nielsch, W., and Giefer, L., *Z. Anal. Chem.* **145**, 347 (1955); **155**, 401 (1957).
96. Nietzel, O. A., and DeSesa, M. A., *Anal. Chem.* **29**, 756 (1957).
97. Noll, C. A., *Ind. Eng. Chem., Anal. Ed.* **17**, 426 (1945).
98. Norwitz, G., and Gordon, H., *Anal. Chem.* **37**, 417 (1965).
99. Nydahl, F., *Anal. Chim. Acta* **3**, 144 (1949).
100. O'Laughlin, J. W., and Banks, C. V., in *The Encyclopedia of Spectroscopy*, (G. L. Clark, ed.), Reinhold, New York, 1960, pp. 19–33.
101. Pakalns, P., and Flynn, W. W., *Anal. Chim. Acta* **38**, 403 (1967).
102. Pappenhagen, J. M., and Mellon, M. G., *Anal. Chem.* **25**, 341 (1953).
103. Parker, G. A., and Boltz, D. F., *Anal. Chem.* **30**, 420 (1968).
104. Parker, G. A., and Boltz, D. F., *Anal. Letters* **1**, 679 (1968).
105. Pascual, J. N., Shipman, W. H., and Simon, W., *Anal. Chem.* **25**, 1830 (1953).
106. Pilkington, E. S., and Smith, P. R., *Anal. Chim. Acta* **39**, 321 (1967).
107. Ramette, R. W., and Sandell, E. B., *Anal. Chim. Acta* **13**, 455 (1955).
108. Reef, B., and Doge, H. G., *Talanta* **14**, 967 (1967).
109. Reilley, C. N., and Crawford, C. M., *Anal. Chem.* **27**, 716 (1955).
110. Rider, B. F., and Mellon, M. G., *Ind. Eng. Chem., Anal. Ed.*, **18**, 96 (1946).
111. Ringbom, A., *Z. Anal. Chem.* **115**, 332 (1939).
112. Rodden, C. J., ed., *Analytical Chemistry of Manhattan Project*, Chap. 1 by C. J. Rodden and J. C. Warf, McGraw-Hill, New York, 1950.
113. Ross, W. J., and White, J. C., *Talanta* **3**, 311 (1960).
114. Sandell, E. B., *Anal. Chem.* **20**, 253 (1948).
115. Sandell, E. B., *Colorimetric Determination of Traces of Metals*, Interscience, New York, 1959.
116. Sandell, E. B., *ibid.*, p. 80.
117. Sandell, E. B., *ibid.*, p. 105.
118. Sandell, E. B., and Neumayer, J. J., *Anal. Chem.* **23**, 1863 (1951).
119. Saywell, L. G., and Cunningham, B. B., *Ind. Eng. Chem., Anal. Ed.* **9**, 67 (1937).

120. Schafer, H. N. S., *Anal. Chem.* **39**, 1719 (1967).
121. Schilt, A. A., and Jaselskis, B., in *Treatise on Analytical Chemistry*, (I. M. Kolthoff, and P. J. Elving, eds.), Interscience, New York, 1964, Part I, Vol. 5, Chap. 58.
122. Schoonover, I. C., *J. Res. Natl. Bur. Standards* **15**, 377 (1935).
123. Shafran, I. G., and Solover, E. A., *Tr. Vses. Nauch.-Issled. Inst. Khim. Reaktivov Osobo Chist Khim Veshihestv*. **28**, 96 (1966).
124. Sherwood, R. M., and Chapman, F. W., *Anal. Chem.* **27**, 88 (1955).
125. Shink, D. R., *Anal. Chem.* **37**, 764 (1965).
126. Smith, G. F., and McCurdy, W. H., *Anal. Chem.* **24**, 371 (1952).
127. Snell, F. D., and Snell, C. T., *Colorimetric Methods of Analysis*, 3rd Ed., Van Nostrand, New York, 1959, Vols. II and IIA.
128. Snell, F. D., and Snell, C. T., in *Encyclopedia of Industrial Chemical Analysis*, (F. D. Snell, and C. L. Hilton, eds.), Interscience, New York, 1966, Vol. I, pp. 358–369.
129. Snyder, L. J., *Anal. Chem.* **19**, 684 (1947).
130. Soldatova, L. A., Kilina, Z. G., and Kataev, G. A., *Zh. Anal. Khim.* **19**, 1267 (1964).
131. Spicer, G. S., and Strickland, J. D. H., *Anal. Chim. Acta* **15**, 231 (1958).
132. Stearns, E. I., in *Analytical Absorption Spectroscopy*, (M. G. Mellon, ed.), John Wiley & Sons, New York, 1950, Chap. 7.
133. Strobel, H. A., *Chemical Instrumentation*, Addison-Wesley, Reading, Mass., 1960, p. 153.
134. Sturton, J. M., *Anal. Chim. Acta* **32**, 394 (1965).
135. Sutzaberger, J. A., *Ind. Eng. Chem., Anal. Ed.* **15**, 408 (1943).
136. Svehla, G., *Talanta* **13**, 641 (1966).
137. Telep, G., and Boltz, D. F., *Anal. Chem.* **23**, 901 (1951).
138. Tikhonov, V. N., *Zh. Anal. Khim.* **20**, 941 (1965).
139. Vallee, B. L., *Anal. Chem.* **26**, 914 (1954).
140. Vinci, F. A., *Anal. Chem.* **25**, 1580 (1953).
141. Wadelin, C., and Mellon, M. G., *Analyst* **77**, 708 (1952).
142. Weast, R. C., ed., *Handbook of Chemistry and Physics*, 48th Ed. The Chemical Rubber Co., Cleveland, Ohio, 1967, pp. E-165-E-181.
143. Weatherburn, M. W., *Anal. Chem.* **39**, 971 (1967).
144. Weissler, A., *Ind. Eng. Chem., Anal. Ed.* **17**, 695 (1945).
145. West, T. S., *Analyst* **91**, 69 (1966).
146. Wiegand, C. S. W., Lann, G. H., and Kalich, F. V., *Ind. Eng. Chem., Anal. Ed.* **13**, 912 (1941).
147. Willard, H. H., and Greathouse, L. R., *J. Am. Chem. Soc.* **39**, 2366 (1917).
148. Willard, H. H., Merritt, L. L., and Dean, J. A., *Instrumental Methods of Analysis*, Van Nostrand, Princeton, N. J., 1965, Chap. 3.
149. Woods, J. T., and Mellon, M. G., *Ind. Eng. Chem., Anal. Ed.* **13**, 551 (1941).
150. Wright, E. R., and Mellon, M. G., *Ind. Eng. Chem., Anal. Ed.* **9**, 251 (1937).
151. Wright, E. R., and Mellon, M. G., *Ind. Eng. Chem., Anal. Ed.* **9**, 375 (1937).
152. Yoe, J. H., and Armstrong, A. R., *Anal. Chem.* **19**, 100 (1947).

Chapter 6

INFRARED SPECTROPHOTOMETRY

W. M. Tuddenham and J. D. Stephens

Kennecott Copper Corporation
Salt Lake City, Utah

1. INTRODUCTION

The wavelengths at which radiation is absorbed or emitted by minerals in the IR region can be related to the interatomic vibrations in the molecules or crystals. Infrared measurements, therefore, have definite theoretical significance. The widest application to date, however, has been empirical. The very fact that vibrations within molecules of a sample are related to the frequencies absorbed results in no two minerals giving exactly the same pattern when transmission of radiation is plotted against wavelength. Not only does each member of a mixture of minerals give a distinctly characteristic pattern, but the features of each mineral are seen in the absorption curve of the mixture. Furthermore, the depth of an individual absorption band can be related to the concentration of the material responsible for it. The method is thus suited for quantitative work. By combining quantitative and qualitative capabilities, IR spectrometry provides a powerful tool for crystal chemistry studies.

1.1. Historical

Developments within the past 20 years have made possible the addition of IR analysis to the list of instrumental techniques which may be used routinely in the analysis of minerals. A review of the work of Coblentz[1] published in 1906 emphasizes that the primary barrier to earlier recognition of IR as an important geochemical tool was the result of inadequate instrumentation rather than lack of demonstrated capability of the technique as a tool in studying minerals. Lack of a really acceptable sample preparation technique further delayed wide application of the method. Coblentz utilized thin sections in most of his work and was consequently faced with difficulty in sample preparation and limited in the resolution available in the areas of high absorption.

A review of all mineral-related IR studies from the time of Coblentz to 1949 indicates an average of less than one paper per year. A marked rise in the number of such papers published commenced in 1950 and has continued through the present. A good analysis of these publication trends is presented in a critical bibliography by Lyon[2] that covers published work through 1962.

One of the most significant papers describing application of IR techniques to the solution of geochemical and mineralogical problems was published in 1950 by Hunt and coworkers.[3] This paper was followed in 1952 by a catalog of IR spectra published by Miller and Wilkins[4] and a significant paper on the IR absorption spectra of silicate minerals by Launer.[5]

1.2. Theory

A thorough coverage of the theoretical principles of IR absorption or emission is beyond the scope of this study. For this reason, reference to the literature is recommended to those interested in a theoretical background. Williams[6] has presented the fundamentals of infrared in a readily understood manner. A more sophisticated treatment is found in Herzberg's classic work *Infrared and Raman Spectra*[7] and Nakamoto[8] has given particular attention to the principles as they apply to inorganic compounds specifically.

1.3. Instrumentation

Modern IR instrumentation has advanced substantially since the time Coblentz made point-by-point measurements utilizing a rock-salt prism for energy dispersion and a radiometer as a detector. The increased convenience and reliability of commercially available instruments have been the critical factors in developing IR as a major analytical tool. Herscher[9] has given good coverage of the subject of spectrometer development. He also compares available systems and lists their individual advantages and disadvantages. The reader is further referred to instrument manufacturers for information on the most recent developments in commercially available instruments.

Until recently the majority of instruments used rock-salt prisms which had a wavelength range of about 2–15 μ (5000–650 cm^{-1}). This range was extended in some instances by means of prisms of potassium bromide to 25 μ, cesium bromide to 38 μ, or thallous bromide–iodide (KRS-5) to 40 μ. The energy dispersion was sometimes improved in the short-wavelength regions by the use of lithium fluoride to 6 μ, or with calcium fluoride to 9 μ.

With the availability of high-quality gratings, grating and prism-grating instruments came into use, and at present these allow studies out to over 700 μ (14 cm^{-1}). The grating instrument has the inherent advantage over the prism instrument of giving better resolution. This, however, is not considered to be too critical in the case of mineral studies because of the rather different nature of the absorption bands of minerals as compared to those of organic compounds.

A new approach, multiple-scan interferometry, uses a combination of the Michaelson interferometer with a computer. This instrumentation promises further advances in the use of IR for mineralogical examinations.[10] This is especially true in situations where location and sampling are insoluble by other means. The interferometer technique allows studies at wavelengths normally considered impractical for conventional spectrometer.

For the geochemist or mineralogist, equipment cost must be taken into consideration as well as versatility. Prism instruments are disappearing from the scene, although a number of relatively inexpensive prism instruments covering the 2.5–15 μ range are still available. Instrument costs cover a wide range and are dependent upon the degree of sophistication. Limited-spectral-range instruments sell for as little as $ 3000. Highly sophisticated grating and prism-grating instruments are available at prices up to $ 30,000, while a complete multiple-scan interferometer system having its own analog computer could cost over $ 40,000. The equipment utilized to prepare the IR curves in this chapter cost about $ 6400.

2. TECHNIQUES OF SAMPLE PRESENTATION

2.1. Absorption

The primary technique available for mineral studies involves absorption of IR radiation by the sample. Important techniques developed for the preparation and presentation of samples for IR analysis by absorption measurement include the following.

2.1.1. Thin Sections

The sample is ground and polished to a thin section following conventional preparative techniques.[11] Because of the low IR transmission of most minerals and rocks, sample preparation and control of thickness with this technique is difficult. This method was used in early studies but is little used at present because of the difficulties involved.

2.1.2. Oil Mulls

The finely ground sample is dispersed in mineral oil or other viscous liquid with a high degree of transparency in the areas of interest. The resulting slurry is sandwiched between two rock-salt or potassium bromide windows. The technique is applied easily but suffers because of difficulty in quantization. Interference from absorption bands present in the dispersing liquids adds to the problems. Another inherent difficulty is related to nonmatching of refractive indices of the oil and the samples. This causes radiation scattering particularly in the 2–4 μ region.

2.1.3. Deposited Powder

In this technique, described by Hunt *et al.*,[3] the ground sample is dispersed in water and allowed to settle to separate coarse particles. The supernatant slurry containing the fine material is removed and the water is evaporated. The deposited powder is dispersed in a volatile, organic liquid which is allowed to evaporate on an IR transparent plate. The IR radiation is passed through the plate and the thin deposit. Quantization is difficult and the technique primarily is restricted to qualitative measurements. Mixtures of minerals may also undergo segregation during the separation of the finely ground material.

2.1.4. Alkali Halide

Of all methods, the alkali halide disc technique has contributed the most to the increased application of IR analyses to problems in geochemistry. This technique was described independently by Stimson and O'Donnel,[12] and by Schiedt and Reinwein.[13] The technique involves mixing a small amount of finely ground sample with high-purity potassium bromide or other alkali halide, and then pressing the mixture in an evacuated die. Since the potassium bromide has no absorption spectrum in the region of interest, the resulting clear or translucent disc yields an absorption curve characteristic of the sample. When using this technique one should be alert to possible anion exchange between reactive minerals and the alkali halide.[14] This effect is very evident with chalcanthite ($CuSo_4 \cdot 5H_2O$). The authors have observed it in practice with very few other natural minerals. Details of the recommended potassium bromide method are as follows.[15]

Place in a mullite mortar 20 mg of sample prepared to pass through a 200-mesh screen. Add approximately 2 ml of ethyl alcohol and grind with an automatic grinder for 15 min. Weigh out 2.5 mg of dry material after scraping down the entire grinding area and add it to 1 g of high-purity potassium bromide (IR quality). Blend the sample and the potassium bromide thoroughly; this is conveniently accomplished by using a dentist's amalgamator. The use of a plastic vial or ball in this operation may contaminate the sample[16] and give extraneous absorptions. Prepare the sample disc by weighing the appropriate amount of the blended mixture into a vacuum die. The weight of material should be adjusted to yield a disc 1–1.5 mm thick. Evacuate the die for 5 min and, while evacuated, apply a pressure of 70 tons/in.^{2} for 1 min. The sample thickness can be adjusted to yield the best balance between details in areas of strongest absorption and weaker absorption. Discs can be prepared without evacuation prior to pressing, but these will not remain permanently clear.

2.2. Attenuated Total Reflection (ATR)

Attenuated total reflection, which was introduced in 1959 by Fahrenfort,[17] is a new approach to sampling which may in time prove to be of considerable importance in the IR examination of minerals. A sample held in close contact with a prism surface acts to absorb a portion of the radiation striking the prism surface. The reflected spectrum is characteristic of the sample. This technique has found considerable use in examining materials that normally are opaque to IR. Problems observed in utilizing this approach are the difficulty of obtaining intimate contact of the sample with the prism surface and an inability to get quantitative results with presently available equipment. The use of a thin, silver chloride film between the sample and prism has reportedly overcome some of the contact problems.[18] These problems will probably diminish with time and the literature should be watched closely for developments in this field. Harrick[19] has described the technique in his book *Internal Reflection Spectroscopy*.

2.3. Reflection

In contrast with ATR, reflection deals with specular or diffuse reflection from polished or powdered samples. Specular reflection has been used to some extent in studying polished surfaces,[20–22] but generally is considered to have limited usefulness in mineral examination. Low[23] made a qualitative study of the IR reflection of small and rough specimens using a multiple-scan interference spectrometer of the type discussed in Section 1.3. The phenomenon studied in this case was not diffuse reflection but rather specular reflection from very small areas. The success of the method, therefore, is dependent upon the unique features of the interferometer, wherein multiple-scan techniques can be used to strengthen the recorded spectrum.

The second technique which involves diffuse reflectance would require the utilization of an integrating sphere or its equivalent.[24] Lyon and Burns[25] reported some success in obtaining spectra from powders and rough surfaces using this approach. Equipment to perform this type of study has also been suggested by manufacturers of interferometric apparatus. The technique appears to be restricted to qualitative spectra.

2.4. Emission Spectroscopy

Emission as well as absorption of radiation by a molecule occurs when there is a change in its dipole moment during a vibration period. To the earthbound geochemist or mineralogist, emission IR spectroscopy is of minor importance. The technique has been used extensively by astronomers and its utilization in space exploration is of importance. A paper by Lyon and Burns[25] covers emittance from rough and powdered samples.

3. QUANTITATIVE APPLICATIONS IN GEOCHEMISTRY

Approaches to quantitative mineralogy have been made in the past utilizing optical and chemical techniques, but these techniques are slow and yield incomplete results. Infrared spectrometry has added a new dimension to quantitative mineralogy. The results obtainable by this technique are competitive with more laborious methods and generally the results are better than any simple method devised thus far. Infrared analysis is considered to be complementary to x-ray diffraction, differential thermal analysis, and optical examination in the solution of mineralogical problems.[26] Some specific advantages of the IR technique may be summarized as follows.

(a) Relatively inexpensive instrumentation is utilized.
(b) The technique is adaptable to the analyses of small samples.
(c) Both chemical and crystallographic data may be obtained.
(d) Infrared techniques have a high sensitivity for many economically important minerals.
(e) Both qualitative and quantitative data may be obtained.

Some disadvantages are as follows.

(a) Particle size and optical effects can influence spectra.
(b) Some important minerals are opaque in readily accessible regions of the IR spectrum.
(c) Some important minerals are difficult to determine at low concentration levels.
(d) Special techniques must be used with reactive minerals.

Development of alkali halide disc sample preparation was a major step in quantization of mineralogical analyses by IR. With halide discs, effective sample thickness and concentration can be controlled so that correlation of mineral concentration and strength of absorption bands is possible.

Accurate estimation of the contribution of a component to the total absorbance at a diagnostic wavelength is critical in quantitative work. The "baseline" technique is a useful approach for analyzing mineral mixtures. A line is drawn tangentially to the curve and as close as possible to points of minimum absorbance immediately adjacent to the diagnostic wavelength.

The absorbance measured on this line at the diagnostic wavelength is then subtracted from the total absorbance of the band being measured. This method corrects for radiation scattering and attenuation from causes other than absorption by the species being measured. These "baseline" absorption values have been found to be additive in most instances. When substantially all of the contribution to an absorption band is from a single component, calibration curves can be prepared directly using baseline absorbancy values. When more than one component substantially contributes to the wavelength being measured, the value may be related to absorption values for bands characteristic of the contributing species to develop simultaneous equations from which concentrations of the components can be determined. The strengths and weaknesses of the method have been discussed in some detail by Kendall *et al.*[27]

Lyon and coworkers[28] demonstrated the capability of the technique by analyzing a standard granite sample. They showed that by this method the mineralogy of α sample could be determined quantitatively, regardless of the alteration the sample had undergone. It was not possible to determine rock texture and thereby deduce the full petrographic history of a sample. The present composition of the rock could, however, be resolved, and the method was shown to be sensitive to minerals formed by alteration of the initial mineral constituents. The IR method is useful in determining the carbonate content of phosphate samples[15] and can be utilized in the determination of trace quantities of anions, such as sulfate in ground-water systems.[29,30]

The most critical problem encountered in quantitative analyses of minerals by IR techniques is the particle size of the sample to be analyzed. This is illustrated by studies with quartz and clay minerals.[15] Experimental results with quartz relating absorption at 12.5, 12.8, and 14.4 μ to particle size are shown in Fig. 1. Working with a quartz plate, Stein[31] found that the transmission minimum from the ordinary ray occurs at 12.4 μ. The minimum for the extraordinary ray occurs at 12.8 μ. In Fig. 1 it is shown that the effect of the ordinary ray becomes more marked as particle size decreases.

Using a theoretical approach, Phillippi[31a] developed the proposition that purely optical processes are responsible for a significant fraction of the inconsistencies in the IR absorption spectra of powdered solid materials

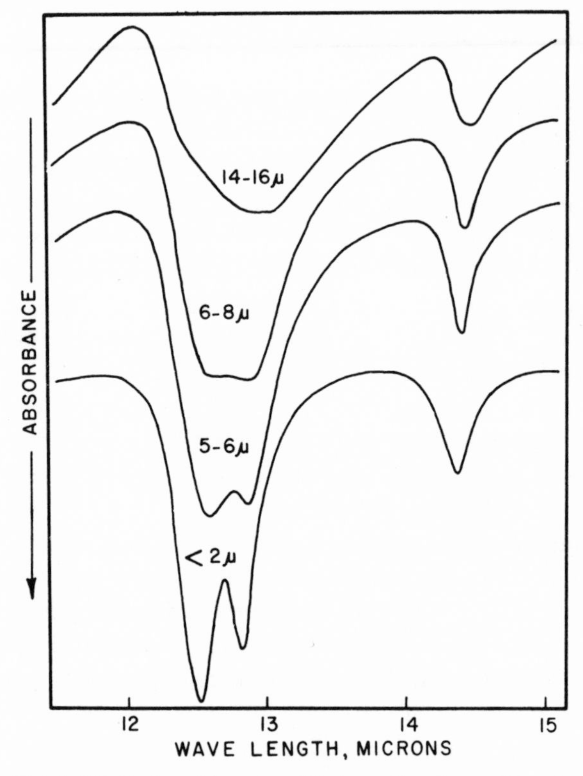

Fig. 1. Effect of particle size on IR absorption curve of quartz. Reprinted by permission from *Anal. Chem.* **32** (12), 163 (1960).

suspended in potassium bromide or Nujol. He showed that for certain classes of materials scattering reflection losses at the surfaces of the particles dominate the form of their powder spectra. Where such factors become important, quantitative estimation and identification by group frequency, as described in Section 4 would be compromised. Fortunately, these inconsistencies have done little to reduce the usefulness of the IR method in mineralogical studies.

Grinding experiments with clay minerals have shown that the structures of some minerals can be disrupted enough to drastically alter the IR curves. This structural breakdown can be avoided by grinding under ethyl alcohol or some other suitable liquid. This is demonstrated for kaolinite by the IR and x-ray diffraction curves in Fig. 2.[15] In order to do reproducible work with minerals it is necessary to standardize grinding procedures. Automatic grinders allow the control of grinding pressure and time, and are superior to hand grinding in most cases.

The sensitivity of minerals to IR analysis depends upon the strength of their individual absorption bands and freedom from interference by

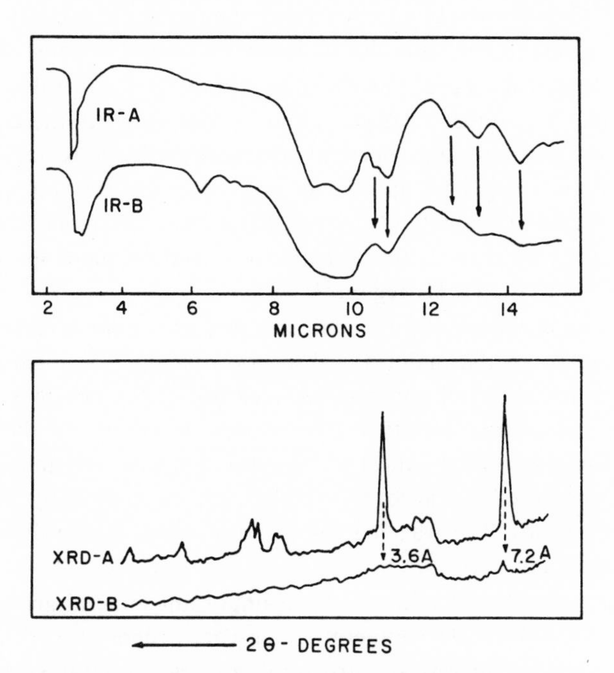

Fig. 2. Comparison of IR and X-ray diffraction curves of wet (A) and dry (B) ground kaolinite. Reprinted by permission from *Anal. Chem.* **32** (12), 163 (1960).

associated species. Sulfate and carbonate minerals can often be determined to less than 1% in their natural environment. Clay minerals such as montmorillonite are more difficult to determine and would be limited to a lower limit of 15% because of compositional variations and interference by other minerals. The sulfides and many oxides cannot be determined in the region from 2 to 15 μ. At longer wavelengths they show relatively low sensitivity because of the broad, shallow absorption bands obtained. Precision of the method is related to the above limitations as well as to the instrumental problems involved.

4. MINERAL IDENTIFICATION WITH INFRARED

Infrared absorption spectra have been characterized as "fingerprints" of molecules and can be utilized in much the same manner for identification. The necessity of compiling personal sets of standard spectra has probably contributed to slow adoption of IR analysis for mineral studies. Perhaps the most comprehensive set of mineral spectra published to date is that of Moenke.[32] Compilation of spectral data for minerals in an easily used form, such as is available for x-ray diffraction work, would be most useful.

The unique nature of the IR spectrum of any mineral species is of prime value. The direct relation of the IR spectrum to the chemistry of minerals is also important. Spectral structure correlations have been considered in some detail for inorganic compounds by Lawson[33] and Nakamoto.[8] With the availability of high-quality grating instruments and Fourier spectrometry, researchers have been able to extend inorganic spectra correlation studies into the far IR region.[34]

The compositions of principal complex anions or structural groups generally relate to the major IR absorption bands. The ionic radius and coordination number of the positive member of the complex anion and the mass and charge density of the nearest neighbors to the anions or structural groups directly affect the spectra. For any one anion or group, the wavelengths of the absorption bands are controlled by the mass of associated cations. The wavelengths of absorptions due to different structural groups may overlap.

Spectra of selected minerals from important mineral groups are presented in Figs. 3–15. The mineral groups are ordered according to their principal complex anions or structural groups. Spectral correlations in the 2.5–25-μ region are discussed below as they relate to these selected minerals.

4.1. Minerals Containing H_2O, OH^-, and Hydrogen Bond

These minerals generally have a first significant absorption band in the wavelength region between 2.7 and 3.5 μ. When bonded H_2O groups are present, a second set of significant absorption bands occurs between 5.8 and 6.5 μ.

Figure 3 illustrates examples of some of these minerals. The spectrum at the top of the figure was obtained with brucite, $Mg(OH)_2$. The first major absorption band in this spectrum occurs at 2.75 μ and is related to the presence of the OH^- groups in the mineral. Other major absorption bands occur in this spectrum at 17.6 and 22.7 μ. The cause of these bands has not been determined, but absorptions in this wavelength region are generally considered to be related to crystal-lattice vibrational frequencies. Minor absorption bands at 2.9 and 6.2 μ are related to water absorbed in the potassium bromide mounting medium.

The lower two curves in Fig. 3 show the isostructural minerals goethite, FeO(OH), and diaspore, AlO(OH). Infrared spectra for these minerals and many of the clay minerals were investigated by Adler *et al.*[35] The sharp

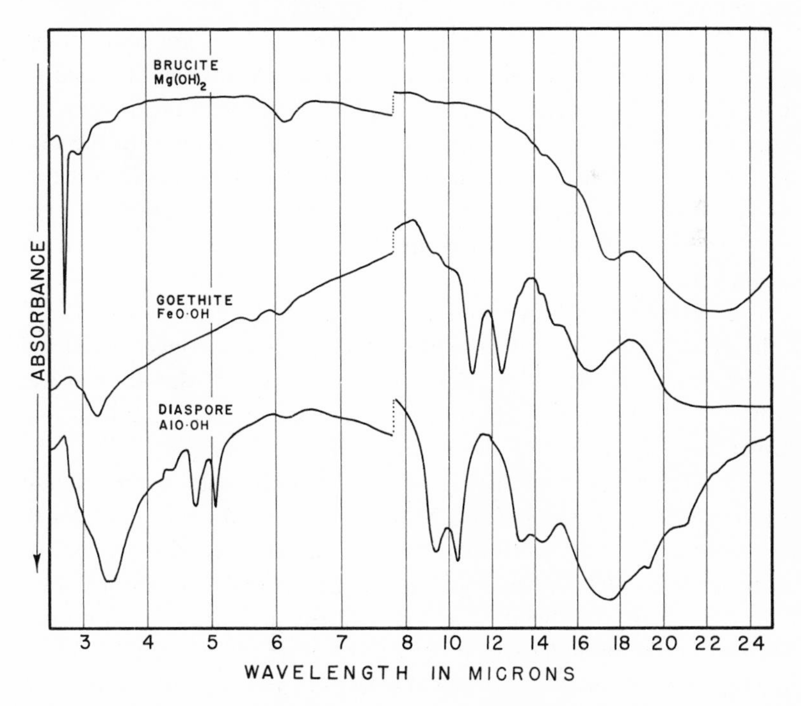

Fig. 3. Infrared spectra of typical minerals containing H_2O, OH^-, and hydrogen bonding.

bands at 2.85 and 2.95 μ are probably related to the O–H stretching vibrations while the broad band at 3.4 μ is related to the hydrogen bonding. Simple metal-oxide hydrates are uncommon as mineral species. Because of this rarity they are not shown in Fig. 3. Some excellent examples of the spectral effects caused by the presence of bonded H_2O in silicate mineral lattices, however, may be seen in Fig. 11. The IR spectra obtained with the two zeolite minerals, chabazite and scolecite, are shown in this figure. Both of these minerals show strong absorption bands near 3 and 6 μ which are related to the presence of bonded H_2O. The spectra obtained with these minerals will be discussed in detail in the section relating to the tektosilicate minerals.

4.2. Carbonate Minerals

All carbonate minerals have one or more absorption bands at or near 7 μ. A study of the IR spectra of aragonite and calcite was made by Adler and Kerr.[37] These same authors studied the spectra of other carbonate minerals in subsequent publications.[38,39] An earlier study of the IR spectra of carbonate minerals was made by Huang and Kerr.[40]

In Fig. 4 a series of IR spectra are shown which are obtained with three different carbonate minerals. The first spectrum was obtained with calcite, $CaCO_3$. Besides calcium, the simple carbonates of magnesium, iron, manganese, cobalt, zinc, and cadmium also form minerals with the calcite structure. In many cases there are solid solution systems between the minerals of these species and, as will be discussed in the section on crystal chemistry, IR spectrometry can be most useful in the study of such solid solution series.

The second curve shown in Fig. 4 was obtained with cerussite, $PbCO_3$. All the absorption bands of cerussite are displaced to longer wavelengths than the equivalent bands of calcite. A major cause of this shift in wavelengths of the absorption bands is the increased mass of the lead ion in cerussite as compared with the calcium ion in calcite.

The bottom spectrum in Fig. 4 was obtained with hydromagnesite, $Mg_4(OH)_2(CO_3)_3 \cdot 3H_2O$. Hydromagnesite is structurally more complex than cerussite or calcite and this is reflected in the complexity of its IR spectrum. The presence of OH^- or H_2O groups in the mineral is confirmed by the presence of absorption bands near 3 μ, and the presence of H_2O groups is confirmed by the shoulder which occurs at 6 μ on the major CO_3^{2-} absorption band. The presence of a strong absorption band at about 3.4 μ suggests that there is hydrogen bonding in the mineral.

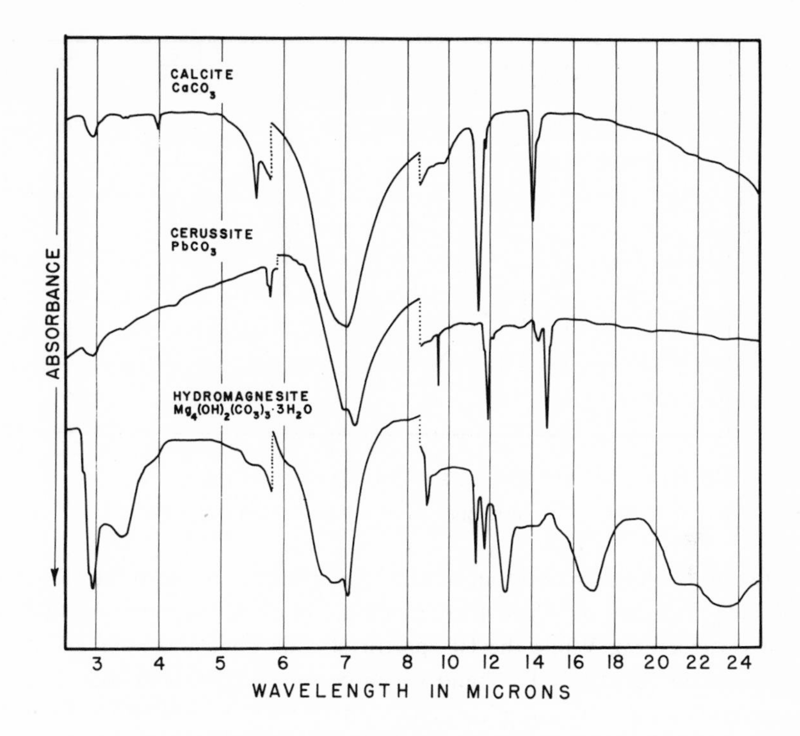

Fig. 4. Infrared spectra of selected carbonate minerals.

As is the case with all carbonates, the strongest band of hydromagnesite is near 7 μ. This band is quite complex, which implies that the CO_3^{2-} vibrations in hydromagnesite are modified by attached atoms or atom groups.

4.3. Nitrate Minerals

All nitrate minerals have one or more absorption bands at or near 7.2 μ. Menzies[41] studied the absorption spectrum of the nitrate ion and pointed out the similarity and relations between the spectra of the carbonate and nitrate ions. The crystal structures of the nitrate minerals are closely related to those of the carbonates. This relation is apparent when the IR spectra of the simple nitrates are compared with those of the simple carbonates.

The two most common nitrate minerals are soda niter, $NaNO_3$, and niter, KNO_3. The IR spectra of these two minerals are shown in Fig. 5. Soda niter is isostructural with calcite, and there are obvious similarities

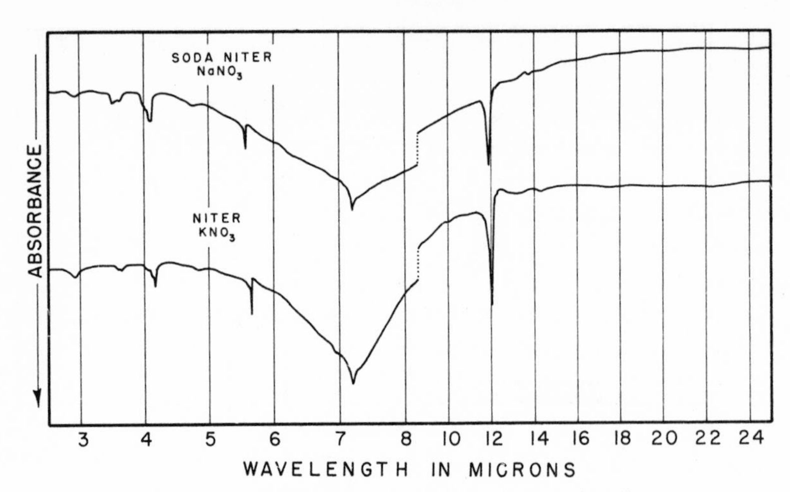

Fig. 5. Infrared spectra of selected nitrate minerals.

between its spectrum and that of calcite. Similarities between the IR spectra of soda niter and niter are also apparent, although these two minerals are not isostructural. While the IR spectra of soda niter and niter are similar they are not identical. The band which occurs at 11.90 μ in soda niter has been shifted to 12.05 μ in niter. This shift seems to be related to the greater mass of the potassium ion when compared with the sodium ion in soda niter.

4.4. Borate Minerals

The first major IR absorption band of borate minerals generally occurs in the wavelength region between 7 and 8 μ. The BO_3^{3-} anion is similar to the CO_3^{2-} and NO_3^- anions, but there are few other areas of similarity between the borate minerals and the carbonate or nitrate minerals. Christ[42] studied the borate minerals and proposed a structural classification for them relating to the composition and shape of the complex anions and polyions. A further study of the borates and borosilicates was made by Tennyson[43] who classified all known species according to their known or assumed structures, as based on the BO_3^{3-} triangle, the BO_4^{5-} tetrahedron, or polyions formed by linking these groups in various ways.

In Fig. 6 tracings of the IR spectra obtained with a number of borate minerals are shown. The top spectrum was obtained with ludwigite, $(Mg, Fe^{2+})_2Fe^{3+}BO_5$. The IR spectrum of ludwigite is atypical of the borates in its simplicity, and it is suspected that the structure of ludwigite is markedly different from that of the other borate minerals.

The remaining three spectra in the illustration are typical of the borate minerals in the complexity of their IR spectra. The second spectrum in Fig. 6 was obtained with boracite, $Mg_3B_7O_{13}Cl$. Neither ludwigite nor boracite have strong absorption bands near 3 μ, indicating that they are anhydrous. The weak absorptions at 3 and 6 μ in the boracite spectrum are due to water absorbed in the KBr mounting medium.

The third spectrum in Fig. 6 was obtained with szaibelyite, $Mg(BO_2)(OH)$. In contrast with ludwigite and boracite, the szaibelyite spectrum does have a strong O–H stretching absorption band at about 2.70 μ. There is no sign of a band near 6.2 μ, confirming that this mineral

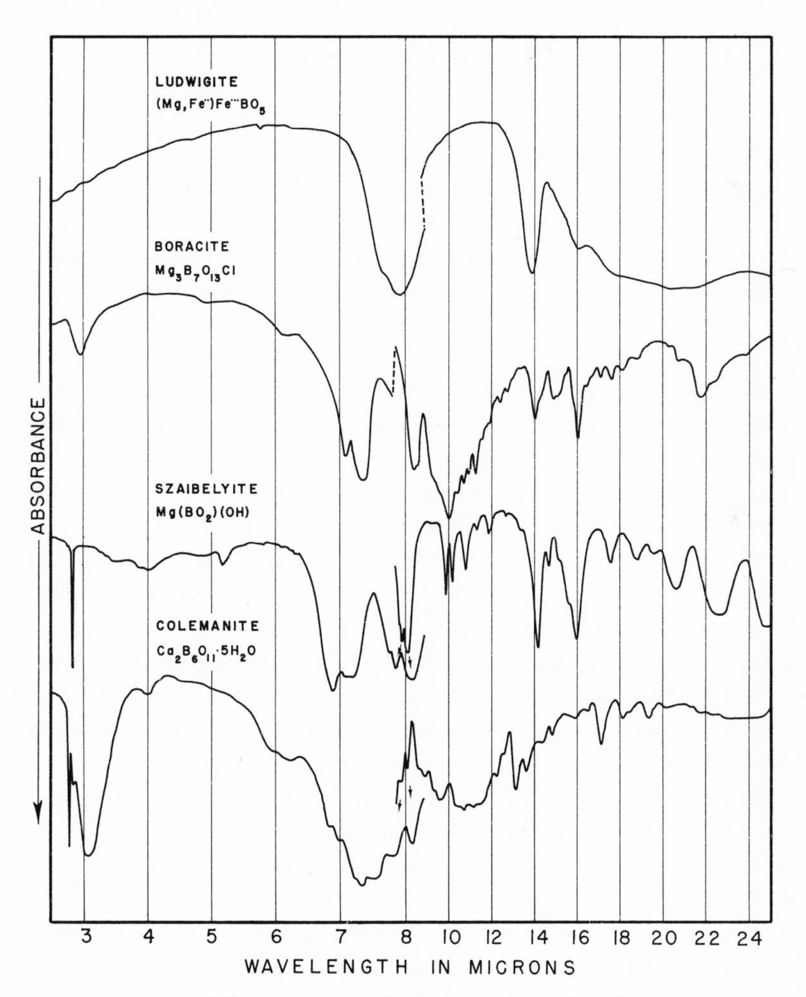

Fig. 6. Infrared spectra of selected borate minerals.

does contain OH^- groups rather than H_2O groups. An unusual aspect of the szaibelyite spectrum is a major absorption band occurring at 6.90 μ. This is a shorter wavelength than is normal for the borate group minerals and overlaps absorptions of the carbonate and nitrate minerals.

The final spectrum in Fig. 6 was obtained with colemanite, $Ca_2B_6O_{11} \cdot 5H_2O$. This spectrum shows complex O–H stretching bands near 2.7 and 2.8 μ, and also a broad band at about 3.1 μ. The colemanite spectrum also has two bands near 6 μ. These bands, coupled with the bands between 2.7 and 3.1 μ, indicate that bonded H_2O groups are present in the colemanite crystal structure.

4.5. The Sulfate Group Minerals

The IR spectra of the sulfate minerals are characterized by one or more major absorption bands in the wavelength region between 8.5 and 9 μ. These absorption bands are related to vibrations of the SO_4^{2-} complex anion.

The IR spectra of many of the sulfate minerals were studied by Omori and Kerr[44] and additional sulfate species were investigated by Adler and Kerr.[45] These latter investigators related the IR spectral configuration to the crystal chemistry of the sulfate species, and the manner in which this chemistry modifies the vibrational character of the SO_4^{2-} group.

In Fig. 7 the IR spectra of a number of typical sulfate minerals are shown. The top two spectra in this figure were obtained with anhydrite, $CaSO_4$, and anglesite, $PbSO_4$, respectively. These two minerals are crystallographically dissimilar but their IR spectra show some similar characteristics which are typical of simple anhydrous sulfates.

The third spectrum was obtained with jarosite, $KFe_3(SO_4)_2(OH)_6$. A very strong O–H stretching band occurs near 3 μ. This band, coupled with the absence of major bands near 6 μ, confirms the presence of OH^- groups in the mineral. The position and to a lesser extent the configuration of absorption bands in the wavelength region from 8 to 25 μ may vary with different jarosite samples. This results from substitutions of other ions for potassium and iron in the mineral's crystal structure. All of the jarosite group minerals, however, yield IR spectra that are obviously related.

The fourth spectrum in Fig. 7 is of hanksite, $Na_{22}K(SO_4)_9(CO_3)_2Cl$. The mineral is rather uncommon but its spectrum illustrates how chemical information can be obtained from IR studies.

The absence of O–H stretching or bending bands near 3 and 6 μ confirms the fact that hanksite is anhydrous. The bands which occur at 6.85, 7, and 11.1 μ are caused by the presence of the CO_3^{2-} anion in the hanksite

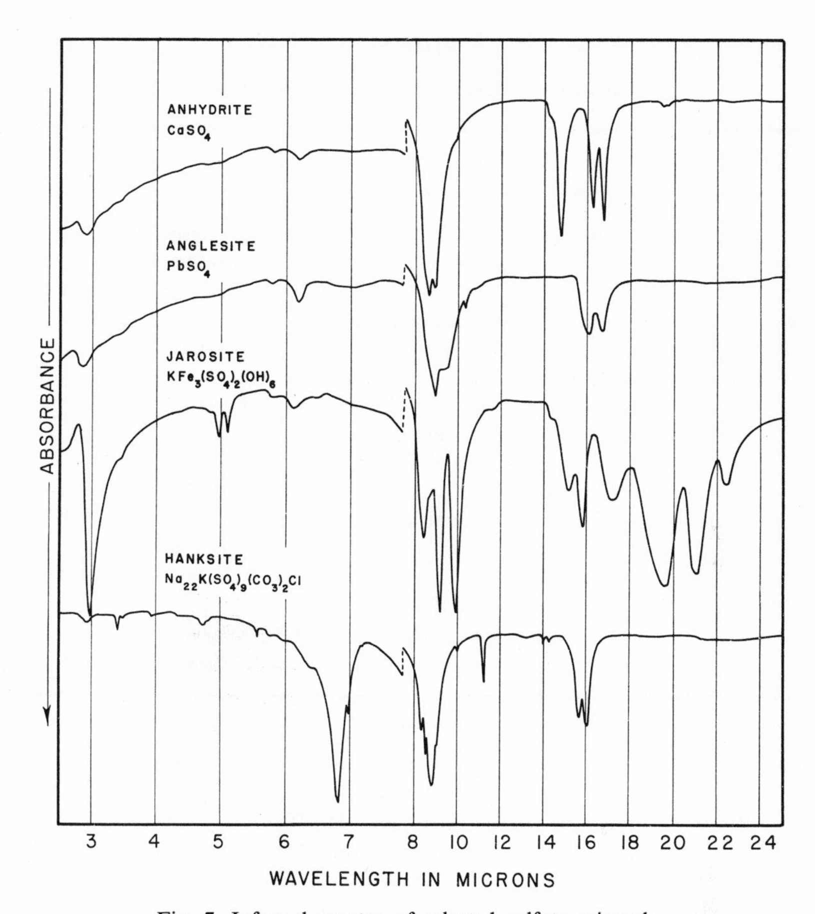

Fig. 7. Infrared spectra of selected sulfate minerals.

structure. The bands occurring between 8.4 and 9 μ and those occurring near 16 μ are caused by the presence of the SO_4^{2-} anion in the hanksite structure.

4.6. The Silicate Minerals

With the exception of very high-pressure phases, all silicates contain silicon tetrahedrally coordinated with oxygen. The tetrahedral groups may occur as isolated units or they may share their corners to form pairs, rings, or infinitely extending chains, sheets, or three-dimensional frameworks. These different units form the basis of several structural classifications of silicate minerals. One of the better classifications is that of Berry and Mason,[46] as reproduced in Table I.

Table I

Structural Classification of the Silicates[a]

Classification	Structural arrangement	Si:O	Examples
Nesosilicates	Independent tetrahedra	1:4	Forsterite, Mg_2SiO_4
Sorosilicates	Two tetrahedra sharing one oxygen	2:7	Hemimorphite, $Zn_4Si_2O_7(OH)_2 \cdot H_2O$
Cyclosilicates	Closed rings of tetrahedra each sharing two oxygens	1:3	Beryl, $Be_3Al_2Si_6O_{18}$
Inosilicates	Continuous single chains of tetrahedra each sharing two oxygens	1:3	Enstatite, $MgSiO_3$
	Continuous double chains of tetrahedra sharing alternately two and three oxygens	4:11	Anthophyllite, $Mg_7(Si_4O_{11})_2(OH)_2$
Phyllosilicates	Continuous sheets of tetrahedra each sharing three oxygens	2:5	Talc, $Mg_3Si_4O_{10}(OH)_2$ Phlogopite, $KMg_3(AlSi_3O_{10})(OH)_2$
Tektosilicates	Continuous framework of tetrahedra each sharing all four oxygens	1:2	Quatz. SiO_2 Nepheline, $NaAlSiO_4$

[a] From Berry, L. G., and Mason, Brian, *Mineralogy: Concepts, Descriptions, Determinations*, W. H. Freeman, San Francisco, 1959.

Aluminum may also fit in tetrahedral coordination with oxygen, and partial substitution of aluminum for silicon often occurs in the silicates. Since aluminum ions have a 3+ valence charge as compared with a 4+ valence charge for silicon, this substitution effectively increases the negative charge of the silicate structural group. These charge unbalances must be compensated for elsewhere in the crystal lattices of the minerals. Substitutions of this type give rise to the aluminosilicate minerals.

The first major IR absorption bands related to vibrations of the tetrahedral SiO_4^{4-} groups or linked silicate groups occur between 8.5 and 12 μ.

4.6.1. Nesosilicate and Sorosilicate Minerals

The nesosilicates are minerals in which the complex anion is the isolated SiO_4^{4-} group. Important minerals in this group include the olivines and garnets. Aluminum generally does not substitute for silicon in the nesosilicates. Most nesosilicates have divalent or trivalent elements in the cation positions. Nesosilicates containing major amounts of alkali elements or hydroxyl groups are rare.

The sorosilicates are minerals in which the structural unit is the $Si_2O_7^{6-}$ group. This structural group is formed by linking two SiO_4^{4-} tetrahedra together at adjacent corners, with both tetrahedra sharing one oxygen between them.

In Fig. 8 the IR spectra obtained with a number of nesosilicate and sorosilicate minerals are shown. The first three spectra in this figure are nesosilicates and the last spectrum is a sorosilicate. The top spectrum is that of zircon, $ZrSiO_4$. This is one of the simplest of the silicate minerals since the single zirconium cation balances the valence charge of the single SiO_4^{4-} tetrahedral anion.

The second spectrum in Fig. 8 was obtained with forsterite, Mg_2SiO_4. This is also a relatively simple silicate since the two magnesium cations balance the negative charge of the single SiO_4^{4-} anion. Forsterite is a member of a much larger group of minerals, the olivine group. The configuration

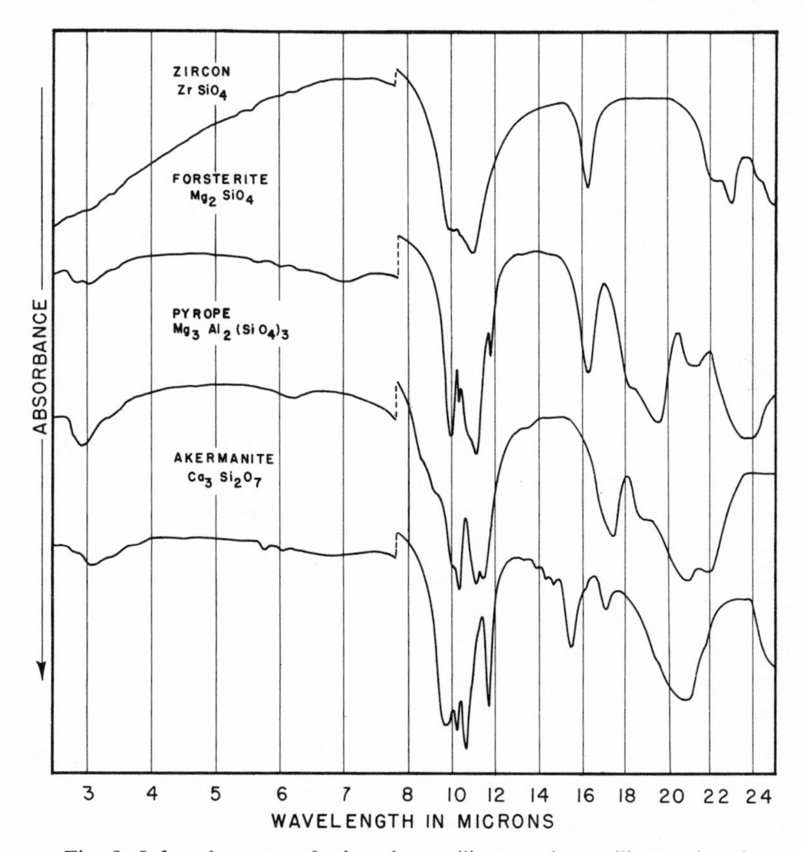

Fig. 8. Infrared spectra of selected nesosilicate and sorosilicate minerals.

of the forsterite spectrum in the 10–12-μ region is typical for olivine-group minerals. Specific absorption bands of the different olivine species will occur at slightly different wavelengths. The slight absorption band at 7 μ in the forsterite spectrum indicates the presence of a trace of carbonate impurity in the mineral sample.

The third spectrum in Fig. 8 was obtained with pyrope, one of the garnet group minerals. In this mineral the negative charge of three isolated SiO_4^{4-} groups is balanced by three magnesium and two aluminum cations. The configuration of the pyrope spectrum between 8 and 12 μ is typical for garnet group minerals. It is difficult to differentiate members of the spessartite–almandite–pyrope family by these bands only. Systematic wavelength shifts of bands in the 16–25-μ region makes differentiation of them possible. More detailed studies of solid solution relations in the garnet group minerals have been made by Hunt *et al.*[3] and by Lyon.[47] The weak absorption bands at 8.5 and 9.2 μ in the pyrope spectrum probably represent a trace of quartz impurity in the mineral sample.

The final spectrum in Fig. 8 was obtained with akermanite, $Ca_3Si_2O_7$. Minerals which contain only the $Si_2O_7^{6-}$ group are relatively rare, but species which contain both SiO_4^{4-} groups and $Si_2O_7^{6-}$ groups are more common. Some examples of these minerals are the epidote group and vesuvianite. The spectra of minerals containing both nesosilicate and sorosilicate groups do not necessarily resemble those of either group and each species must be interpreted individually.

4.6.2. Inosilicate and Cyclosilicate Minerals

In Fig. 9 the tracings of the IR spectra obtained with four inosilicate and cyclosilicate minerals are shown. The top spectrum was obtained with wollastonite, $CaSiO_3$, and the second spectrum is that of diopside, $CaMg(SiO_3)_2$. Both wollastonite and diopside have single-chain structures. Diopside belongs to the pyroxene group minerals while wollastonite belongs to the pyroxenoid group minerals. The diopside spectrum is typical for the monoclinic pyroxenes. The spectra of augite, hedenbergite, fassaite, pigeonite, and clinoenstatite are quite similar to it and can be easily recognized as members of the same mineral family. The wollastonite spectrum is typical for the pyroxenoid group minerals, and rhodonite, bustamite, and fowlerite can be recognized as members of this same group by similarities in their spectra. Pectolite, $HNaCa(SiO_3)_3$, is a hydrated mineral considered to be structurally related to wollastonite. Distinct differences appear in the spectra of the two minerals, but there is enough similarity to support the theory of structural relationship.

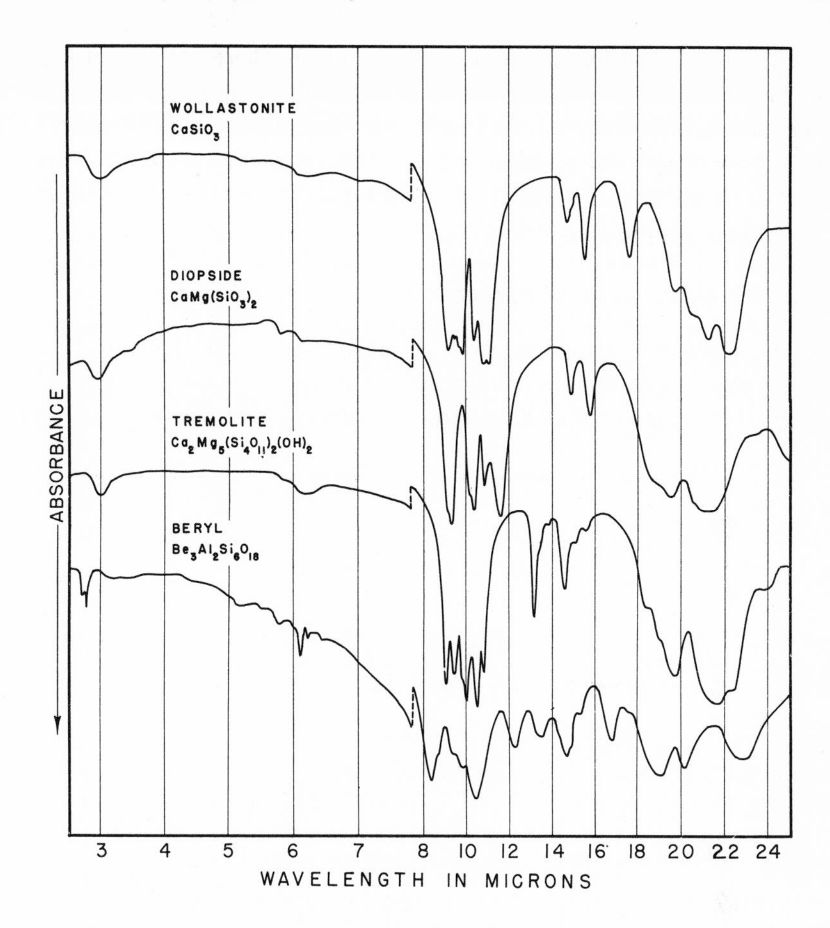

Fig. 9. Infrared spectra of selected inosilicate and cyclosilicate minerals.

The third spectrum in Fig. 9 was obtained with tremolite, $Ca_2Mg_5(Si_4O_{11})_2(OH)_2$. Structural OH^- is known to be present in tremolite and it is interesting that the presence of this OH^- group is not indicated by an absorption band near 3 μ. The fact that this absorption band is not evident probably relates to the orientation of the OH^- group in the mineral's crystal structure. A phenomenon of this type was noted by Bassett[48] in the IR spectrum of phlogopite, the magnesium mica.

The final spectrum in Fig. 9 was obtained with beryl, $Be_3Al_2Si_6O_{18}$. The IR spectra obtained with beryl have been studied separately by Wickersheim and Buchanan[49] and by Plyusina.[50] Although the presence of bonded water is not indicated in the beryl formula, the IR spectrum clearly shows that such bonded water is present in the mineral. There are several O–H stretching bands near 3.7 μ and several H_2O bending bands near 6 μ. Water groups along with other gas molecules and large alkali ions are known to occupy tube or channellike holes in the $Si_6O_{18}^{12-}$ ring structure.[51] The nature of the IR bands indicates that the water is bonded in the crystal lattice and is not simply absorbed in the holes.

4.6.3. Phyllosilicate Minerals

The phyllosilicates are minerals in which the silicate structure forms an indefinitely extending sheet which has an average composition of $Si_4O_{10}^{4-}$. In Fig. 10 tracings of the IR spectra are shown which are obtained with

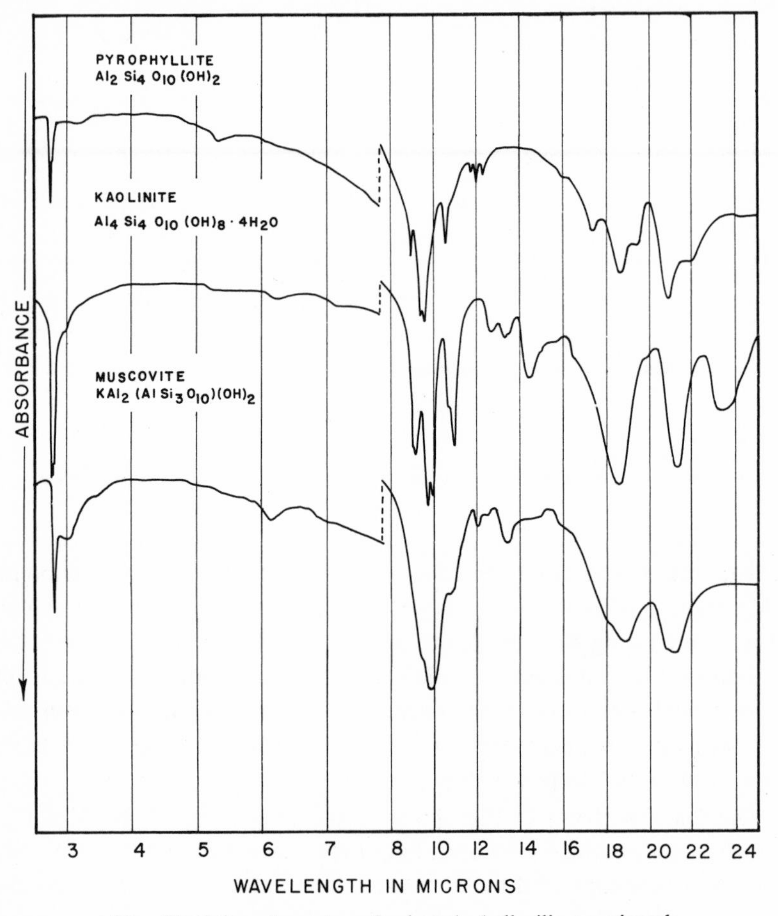

Fig. 10. Infrared spectra of selected phyllosilicate minerals.

three phyllosilicate minerals. The first two spectra were obtained with pyrophyllite, $Al_2Si_4O_{10}(OH)_2$, and kaolinite, $Al_4Si_4O_{10}(OH)_8$, respectively. Both of these minerals have the basic $Si_4O_{10}^{4-}$ sheet structure. The final spectrum was obtained with muscovite. In muscovite one-fourth of the silicon in the $Si_4O_{10}^{4-}$ sheet has been replaced by tetrahedrally coordinated aluminum. The most obvious effect of the substitution of aluminum for silicon in the sheet structure is a decrease in the intensity of most of the absorption bands.

While pyrophyllite, kaolinite, and muscovite vary considerably in their crystal structures, comparison of the spectra in Fig. 10 clearly shows a basic similarity in their IR patterns. Although the spectra differ in detail, the similarities between them are such that one would intuitively feel that the minerals are related. The IR spectra of all of the kaolin group minerals are examined in more detail in Fig. 18 and discussed in Section 5.

4.6.4. Tektosilicate Minerals

The tektosilicates are very common and important mineral species. Included in this group are the silica minerals, the feldspars, the feldspathoids, and the zeolites. It is impossible to obtain framework silicate structures with compositions other than SiO_2 without changing the silicon-to-oxygen ratio. The most common modification of the tektosilicate structure is obtained by substituting aluminum for silicon in some of the SiO_4^{4-} tetrahedra. This gives the framework structure a negative charge which must be balanced by cations in other positions in the crystal lattice.

The IR spectra obtained with many of the tektosilicate minerals were studied by Milkey.[52] Although stilbite was misidentified as natrolite and natrolite was identified incorrectly as mesolite in Milkey's work, this is fundamentally a good source of information about IR analyses of the tektosilicates. Additional information about IR analysis in studying solid solution relationships in the plagioclase feldspar minerals is found in the work of Thompson and Wadsworth.[53] The IR spectra of many of the silica group minerals have been published by Lyon.[47]

In Fig. 11 the IR spectra obtained with a number of the tektosilicate minerals are shown. The top spectrum was obtained with alpha quartz, SiO_2. Most members of the silica group minerals, including alpha quartz, alpha cristobalite, alpha tridymite, and coesite, have spectral characteristics somewhat similar to each other. An exception to this general similarity is seen in stishovite, which has an IR pattern totally unlike those of other members of the silica group. This dissimilarity of the stishovite spectrum from those of the other members of the group is to be expected, since silicon

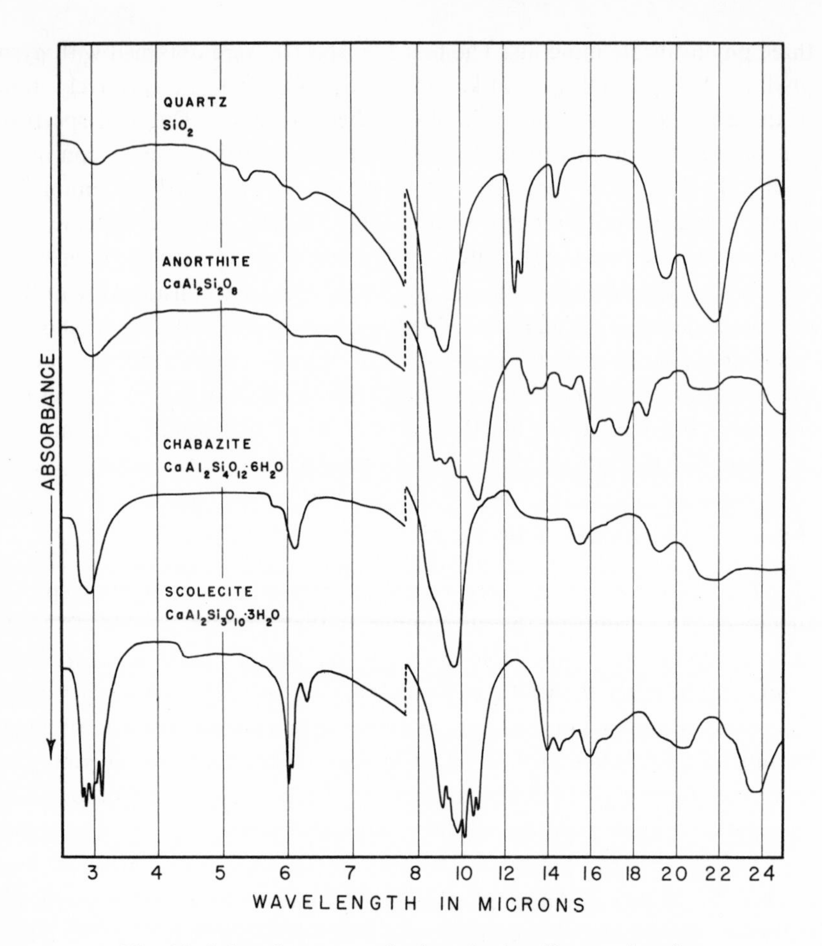

Fig. 11. Infrared spectra of selected tektosilicate minerals.

coordinates six oxygen atoms in stishovite as compared with the coordination of four oxygen atoms in all other silica minerals.

The second spectrum in Fig. 11 was obtained with anorthite, $CaAl_2Si_2O_8$, the calcium end member of the plagioclase feldspar series. This spectrum is typical of the spectra obtained by IR analysis of calcic plagioclases through the composition range from 85 to 100% anorthite, but because of crystal-structure modifications it is quite different from spectra obtained with the sodic plagioclases. According to Deer, Howie, and Zussman[54] there are six structural modifications in the plagioclase feldspar series. Three of these can be clearly distinguished by IR analyses, as was shown by Thompson and Wadsworth.[53]

The last two spectra in Fig. 11 were obtained with the two zeolite minerals, chabazite, $CaAl_2Si_4O_{12} \cdot 6H_2O$, and scolecite, $CaAl_2Si_3O_{10} \cdot 3H_2O$. Structural relationships of the zeolite minerals are complex and will not be discussed in detail. Basically, however, the zeolite minerals are aluminosilicate framework structures in which various percentages of the silicon atoms have been replaced by aluminum. All zeolites have long tube- or channellike holes in their structures, and the negative charge caused by substitution of aluminum in the framework is balanced by alkaline ions or alkali-earth ions, part of which occupy the holes. The cations occupying the channellike holes may be exchanged by other cations from aqueous solution, so long as the charge balance is maintained. Water or other gaseous molecules may also occupy the holes in the zeolite structure.

The presence of water in chabazite and scolecite is shown clearly by the absorption bands in their IR spectra near 3 and 6 μ. The great difference in detail of the water bands of these two minerals, however, leads one to suspect that there is a difference in the manner in which the water molecules are bonded into their structures. This suspicion has been confirmed by dehydration studies and differential thermal analysis of the two minerals, as reported by Koizumi.[55]

4.7. Phosphates, Vanadates, and Arsenates

Minerals containing phosphorus, vanadium, and arsenic often show structural similarities which are related to similarities of their complex anions. The general formula for all these complex anions is XO_4^{3-}, where X is pentavalent P, V, or As. Linking of the XO_4^{3-} groups to form more complex structural groups is well known in inorganic chemicals, but is relatively rare in naturally occurring minerals.

In Fig. 12 the IR spectra obtained with a number of phosphate minerals are shown. The first major absorption band related to the PO_4^{3-} anion in minerals generally occurs between 9 and 10 μ. The first two spectra in this figure are those which were obtained with apatite, $Ca_5(PO_4)_3F$, and pyromorphite, $Pb_5(PO_4)_3Cl$. Although they differ considerably in composition, these two minerals are isostructural and this relationship can be seen clearly from similarities of their IR spectra.

Apatite varies in composition, and these variations affect the IR spectra. Variations in the IR spectra of apatite samples and how these relate to composition have been investigated separately by Coles[56] and by Fischer and Ring.[57] For example, hydroxyl groups or chloride ions may substitute for

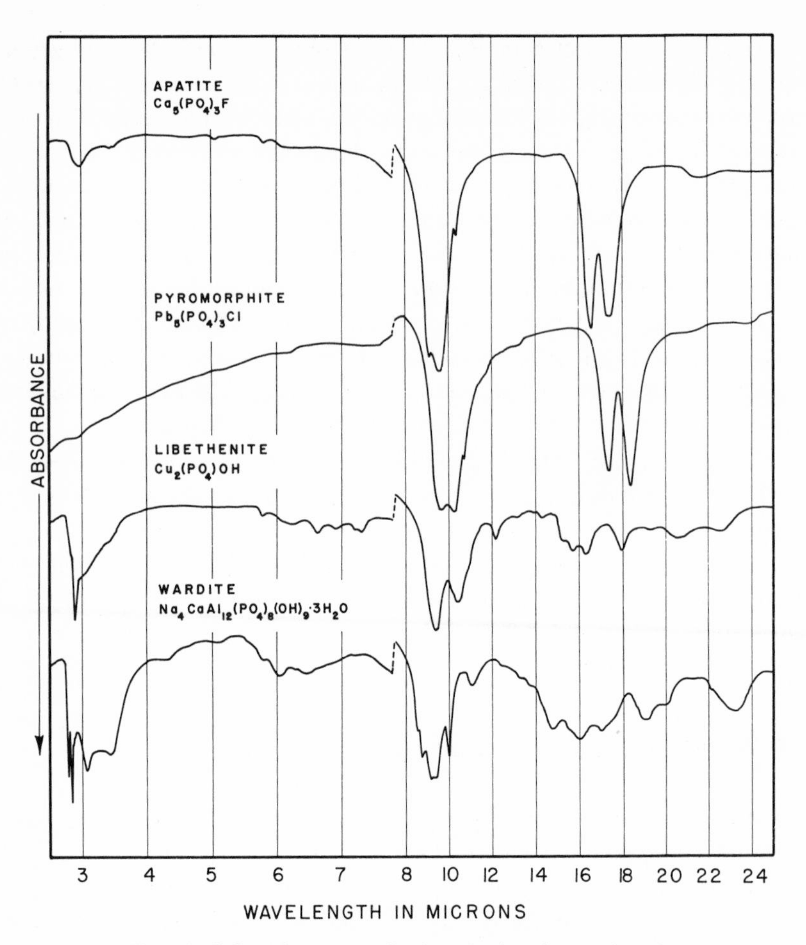

Fig. 12. Infrared spectra of selected phosphate minerals.

fluoride in the apatite structure. As would be expected, when hydroxyl groups are substituted for the fluoride, a sharp O–H stretching band is noted at 2.80 μ. When chloride ions are substituted into the structure the bands near 17 μ are modified. In addition to these substitutions, carbonate ions frequently substitute for part of the phosphate in the apatite structure. When this occurs a sharp double absorption band is always present near 7 μ.

The composition of pyromorphite also varies widely. Up to 8% calcium may substitute for lead in this mineral, and both vanadate and arsenate ions may substitute for the phosphate. The substitution of phosphate and vanadate in this case is limited, but there is a complete solid solution series between pyromorphite and mimetite, $Pb_5(AsO_4)_3Cl$. The IR spectrum for mimetite may be seen in Fig. 13. Intermediates in the pyromorphite–mimetite system have not been investigated, but their IR spectra are expected to show intermediate characteristics between those of the end-member minerals.

The third spectrum in Fig. 12 was obtained with libethenite, $Cu_2(PO_4)OH$. This mineral is isostructural with olivenite, $Cu_2(AsO_4)OH$, whose spectrum may be seen in Fig. 13. The presence of OH^- groups in libethenite is demonstrated clearly by the presence of sharp bands at about 2.9 μ. While not indicated in the formula, the presence of bonded H_2O is suggested by the presence of a broad band near 3 μ and by the presence of

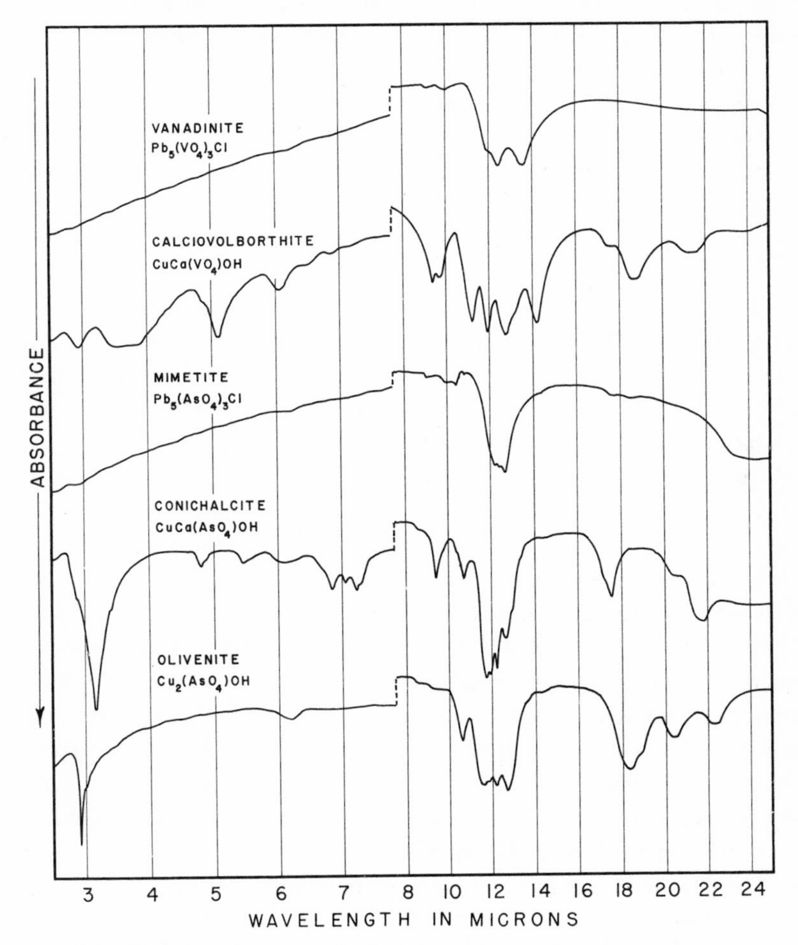

Fig. 13. Infrared spectra of selected vanadate and arsenate minerals.

a band near 5.85 μ and another at about 6.2 μ. In addition, a minor amount of bonded CO_3^{2-} in the mineral is suggested by the series of small but well-defined absorption bands between 6.6 and 7.3 μ.

The final spectrum in Fig. 12 was obtained with the mineral wardite, $Na_4CaAl_{12}(PO_4)_8(OH)_9 \cdot 3H_2O$. The two sharp bands near 2.75 and 2.8 μ give clear evidence of O–H stretching vibrations. In addition, the presence of bands at 3.1 and 3.5 μ indicate the presence of hydrogen bonding. Coupled with these bands, a series of bands centered about 6 μ indicate that bonded H_2O groups are present.

In Fig. 13 the IR spectra which were obtained with a series of vanadate and arsenate minerals are shown. The first major absorption band caused by the presence of the VO_4^{3-} anion in minerals occurs between 9 and 12 μ, and the first major band related to the AsO_4^{3-} anion generally occurs in this same wavelength region. The first spectrum in Fig. 13 was obtained with vanadinite, $Pb_5(VO_4)_3Cl$. Pyromorphite, vanadinite, and mimetite are isostructural with each other and with apatite. While they are isostructural, comparison of the IR spectra of the three minerals shows that the spectral configuration of each is distinctly different. From this it is evident that the complex anion in a mineral plays a stronger role than crystal structure in controlling its spectral configuration.

The second spectrum in Fig. 13 was obtained with calciovolborthite, $CuCa(VO_4)OH$. The calciovolborthite spectrum shows a number of unusual aspects. Although the presence of an OH^- group is indicated in the mineral formula, there is little indication of this OH^- group in the form of an absorption band near 3 μ. The relatively broad band which occurs near 2.9 μ in this spectrum resembles a band which is caused by absorbed water rather than one which is caused by a functional OH^- group. The presence of a band near 6.1 μ also suggests that there may be H_2O in the sample or in the mounting medium. Another unusual aspect of this spectrum is the series of shallow but well-defined bands which occur between 3.4 and 5.5 μ. This region is generally reserved for organic species and it is uncommon for minerals to have absorption bands in this spectral region. The major absorption region between 9 and 15 μ for calciovolborthite is much more complex than the relatively simple spectrum of vanadinite in this region. This difference in the major absorption bands of calciovolborthite and vanadinite may be related to a linkage of the VO_4^{3-} groups to form a more complex structural unit in calciovolborthite.

The third spectrum in Fig. 13 was obtained with mimetite, $Pb_5(AsO_4)_3Cl$. This spectrum and its relation to vanadinite, pyromorphite, and apatite have already been discussed.

The fourth spectrum in Fig. 13 was obtained with conichalcite, $CuCa(AsO_4)OH$. Although they are compositionally similar except for their principal complex anions, calciovolborthite and conichalcite are not structurally related. As was the case with calciovolborthite, however, there are some unusual aspects to the conichalcite spectrum. At least two O–H stretching bands are noted in the conichalcite spectrum. The first of these is at 2.9 μ and the second at 3.15 μ. The long wavelength of the band at 3.15 μ suggests that this band may be due to hydrogen bonding rather than to a normal OH^- group. There are no sharp, well-defined H_2O bands near 6 μ, but a broad band similar to those caused by absorbed water is present. A series of small but well-defined bands near 7 μ indicate that there is some bonded CO_3^{2-} in this conichalcite sample.

The final spectrum in Fig. 13 was obtained with olivenite, $Cu_2(AsO_4)OH$. This mineral is structurally related to libethenite, whose spectrum was shown in Fig. 12. While the minerals are structurally related, there is little similarity between their IR spectra. This indicates once again that the complex anion or structural group is the major factor which controls the configuration of the IR spectrum.

4.8. Molybdates and Tungstates

The molybdate and tungstate minerals contain complex anionic groups with the molybdenum or tungsten cations at the center of a distorted tetrahedron of oxygen ions. Minerals containing molybdate or tungstate groups are often structurally related, and extensive solid solutions exist between members of certain molybdate and tungstate minerals. The first major absorption band caused by the MoO_4^{2-} or the WO_4^{2-} anion in minerals generally occurs between 11 and 13 μ.

In Fig. 14 the IR spectra are shown, which are obtained with a number of molybdate and tungstate minerals. The first spectrum in this figure was obtained with powellite, $CaMoO_4$. Powellite is isostructural with scheelite, $CaWO_4$, whose IR spectrum is also presented in the figure. A complete solid solution system exists between powellite and scheelite. The powellite end member of the series is relatively uncommon. Most material which is called powellite in field identifications is actually a variety of scheelite which contains a relatively minor amount of molybdenum. The material used to obtain the powellite spectrum in Fig. 14, however, was essentially a pure calcium molybdate which formed as the oxidation product of molybdenite, MoS_2.

The similarity of the IR spectra of powellite and scheelite is striking.

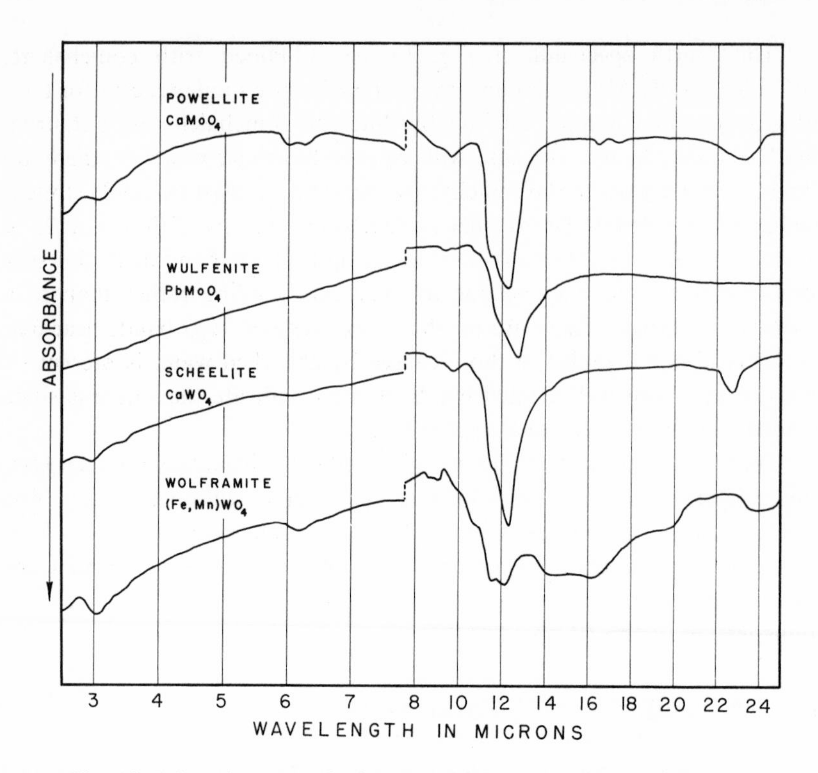

Fig. 14. Infrared spectra of selected molybdate and tungstate minerals.

Examination of other isostructural species showed that changes in the complex anion generally caused major changes in configuration of the IR spectrum. The remarkably close similarity of the IR spectra of powellite and scheelite suggests that the MoO_4^{2-} and WO_4^{2-} complex anions must also have very similar bond and vibrational characteristics.

The second spectrum in Fig. 14 was obtained with wulfenite, $PbMoO_4$. Wulfenite is structurally related to powellite and scheelite, but is not isostructural with them since it lacks a mirror-plane crystal symmetry element that is present in them. The structural similarity of wulfenite to powellite and scheelite is suggested by similarities of their IR spectra. Extensive solid solution relations exist between wulfenite and each of the other minerals. Analytical data presented in Dana's *System of Mineralogy*[58] shows that wulfenite may contain at least 6.88% calcium, which would give it a composition intermediate between wulfenite and powellite. Other data from this same source shows that at least half of the molybdenum in wulfenite may be replaced by tungsten, giving a composition intermediate between wulfenite and its lead tungstate analog, stolzite. These intermediate mineral species probably have IR spectra with characteristics intermediate between their respective end members. The displacement of the major absorption band of wulfenite as compared with the major bands of powellite and scheelite is almost certainly related to the increased mass of lead in wulfenite as compared with calcium in powellite and scheelite.

The final spectrum in Fig. 14 was obtained with wolframite, (Fe, Mn)WO_4. Wolframite is crystallographically dissimilar from the previously discussed molybdate and tungstate minerals, and this dissimilarity is reflected by differences in its IR spectrum. Wolframite samples show a wide range in composition, relating to the ratio of their iron and manganese contents. In separate studies Grubb[59] and Chang[60] showed that minor amounts of calcium may substitute into wolframite, and that consequently there is some degree of solid solution between wolframite and scheelite. Solid solution of this type would presumably produce IR spectra intermediate between those of scheelite and wolframite.

4.9. Oxide Minerals

The oxide minerals do not contain complex anions or other complex structural groups. They therefore cannot be conveniently included in a systematic classification that is based on the relations between complex chemical or structural groups and the IR absorption bands that are characteristic of them. Metal ions which form oxides vary widely in charge and in ionic radius. The different cations therefore, coordinate different numbers of oxygen ions in their oxide structures. As a result the structures and IR spectra of oxide minerals also vary widely. The first major absorption band for oxide minerals may occur anywhere in the wavelength region from 8.5 to 20.0 μ.

A study of the IR spectra of simple metal oxides was made by McDevitt and Baun.[61] They showed that while the IR spectra of metal oxides were generally relatively simple and lacking in sharp absorption bands, they were characteristic and showed systematic relations with structurally related compounds.

In Fig. 15 the IR absorption spectra are shown which were obtained with three oxide minerals. There is a wide range in spectral complexity of the patterns obtained with these three minerals. The first spectrum in Fig. 15 was obtained with rutile, TiO_2. Other minerals having the rutile structure have quite different IR spectra.

The second spectrum was obtained with corundum, Al_2O_3. This pattern

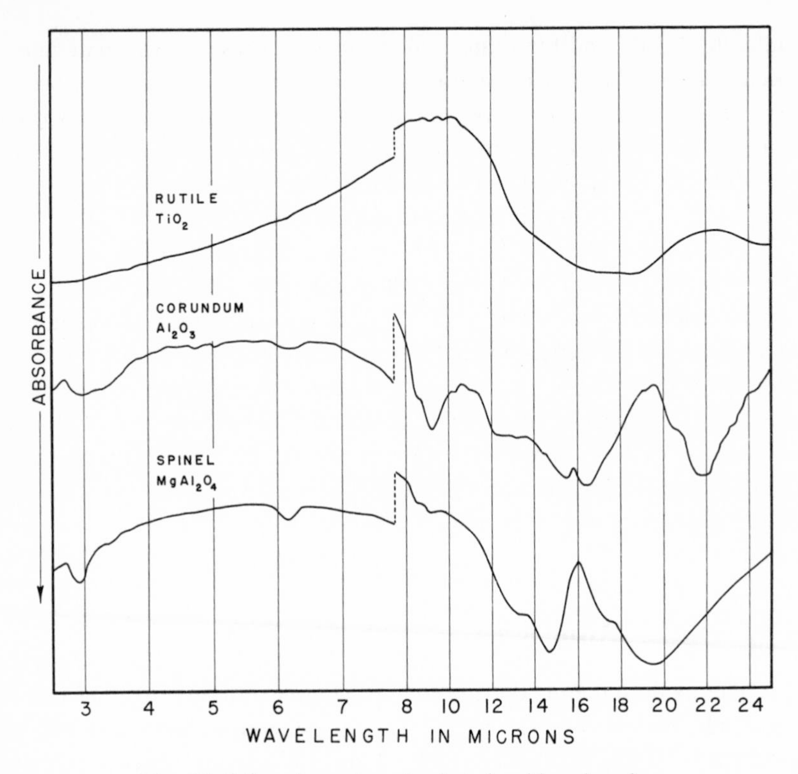

Fig. 15. Infrared spectra of selected oxide minerals.

is typical of those obtained with both natural and synthetic corundum samples, but individual spectra vary slightly. The bands that occur between 8 and 10 μ vary in their intensities. The spectrum shown illustrates approximately the maximum intensity for these bands, but in some cases they are so poorly developed that they are difficult to detect. The bands between 20 and 22 μ show variations in intensity and wavelength. These variations may relate to the degree of ordering of the mineral's crystal lattice. Other polymorphous forms of Al_2O_3 have different patterns that are distinctive and characteristic of their species.

The third spectrum in Fig. 15 was obtained with spinel, $MgAl_2O_4$. Extensive solid solution systems occur between members of spinel group minerals. Divalent iron, zinc, and manganese may substitute for magnesium; and trivalent iron, trivalent manganese, chromium, vanadium, and titanium often substitute for aluminum in spinel. This mineral system has not been studied in detail by the authors, but the IR spectra of the series may be expected to vary systematically with changes in composition.

5. CRYSTAL CHEMISTRY STUDIES

5.1. Solid Solution Series

Infrared absorption spectroscopy offers a particularly useful technique for the study of solid solution series. Studies have been made with chlorites, feldspars, and olivines, among others. Tuddenham and Lyon[62] found that by relating the IR absorption spectra to chemical analyses of 21 chlorites, both the degree of substitution of aluminum for silicon and the total iron content of the mineral could be estimated. The structural type (7 or 14 Å) could also be deduced readily from the IR curve. Stubican and Roy[63] used IR and x-ray diffraction in a study of synthetic chlorites, but were unable to note the effects of the Al^{3+} for Si^{4+} substitutions. This suggests that the synthetic minerals were in disequilibrium as compared to the natural minerals. Hayashi and Oinuma[64] used IR to study the Si–O and OH^- absorption bands of chlorites.

Thompson and Wadsworth[53] demonstrated that there is a systematic variation in the IR spectra of plagioclase feldspars throughout the albite–

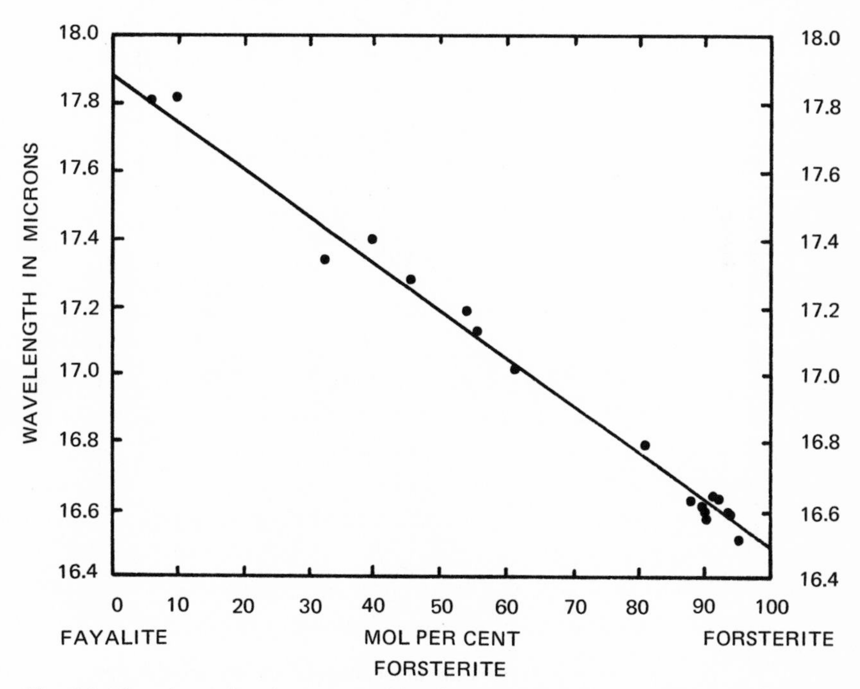

Fig. 16. Plot of wavelength *vs* composition for one absorption band from the forsterite–fayalite solid solution series.[65]

anorthite series. They found definite correlation between the band positions, number of bands and composition. At least two modifications of the plagioclase crystal structure were indicated by the IR data.

Duke and Stephens[65] investigated the relations between the IR spectra and compositions of members of the olivine-group minerals. They demonstrated a continuous displacement of the IR absorption maxima of members of the forsterite, Mg_2SiO_4, to fayalite, Fe_2SiO_4, solid solution series toward longer wavelengths as the iron contents of the samples increased. This relation is shown in Fig. 16. Duke[66] confirmed earlier conclusions that there is a continuous solid solution system between forsterite and tephroite, Mn_2SiO_4.

A solid solution series extends from adamite, $Zn_2(OH)AsO_4$, through

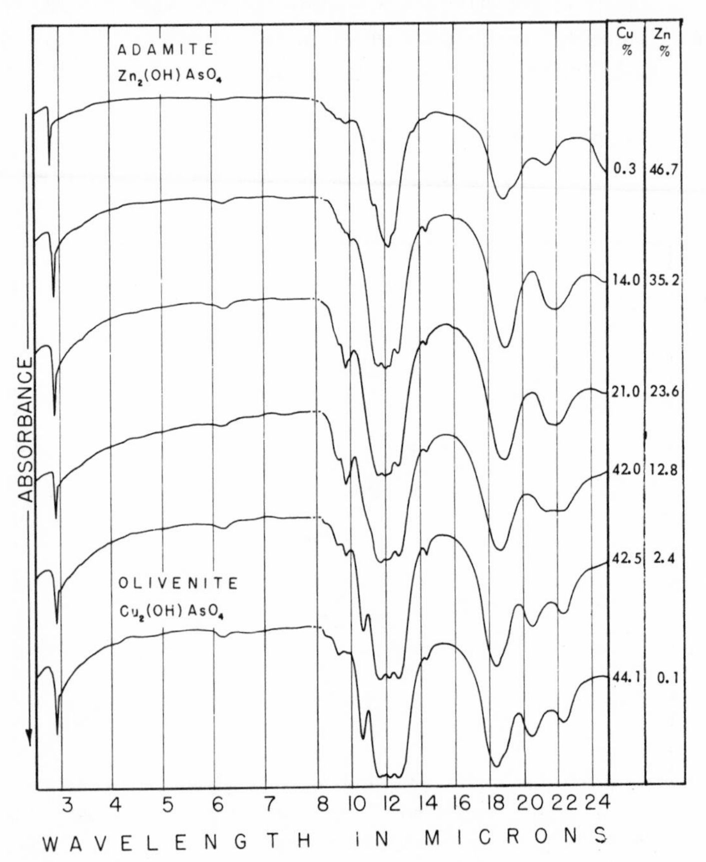

Fig. 17. Infrared absorption spectra of members of the adamite–olivenite series. Note systematic changes of spectral configuration in contrast with the wavelength shifts exhibited by the forsterite-fayalite system (see Fig. 16)

cuproadamate, Zn-olivenite, to olivenite, $Cu_2(OH)AsO_4$. Composition of the samples chosen varied continuously with respect to their copper and zinc contents.

Infrared spectra obtained with members of this series are shown in Fig. 17. Unlike x-ray diffraction spectra, the IR spectra vary continuously throughout the entire composition range. In contrast with the forsterite–fayalite series, the systematic variations are not in wavelength but rather in configuration of the spectra. The changes in spectral configuration are probably associated with progressive distortion of the crystal lattice which is caused by the isomorphous substitutions of copper and zinc. These distortions were also indicated by the discontinuous variations of x-ray diffraction spectra.

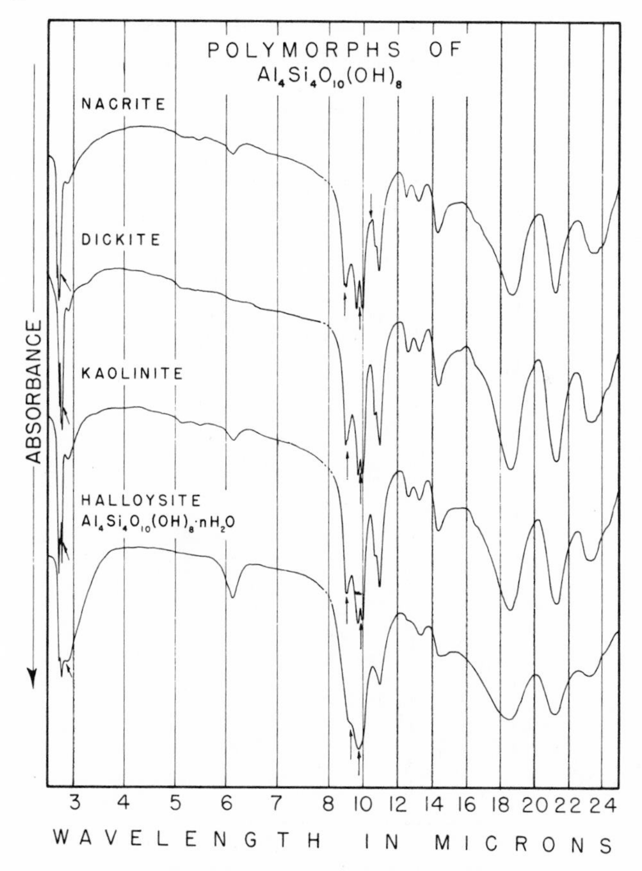

Fig. 18. Infrared spectra of kaolin group minerals. Arrows point to absorptions which show significant variations in the IR spectra. (Nacrite: U. S. National Museum, Cat. No. 3918.)

5.2. Polymorphous Series

The kaolin group of minerals are polymorphous forms of $Al_4Si_4O_{10}(OH)_8$ and consist of nacrite, dickite, kaolinite, and halloysite. Halloysite also contains excess H_2O. Arrangement of the spectra in logical sequence of decreasing spectral complexity produces the series seen in Fig. 18. It seems more than a coincidence that this is also the order into which these minerals fall when arranged in the sequence of increasing complexity of crystal structure and decreasing number of stacked and ordered kaolin sheets.[67]

Changes in the crystal structure of the kaolin-group minerals are so slight that differentiating them by x-ray diffraction analysis is difficult. Because of the compositional and structural similarities of the kaolin group

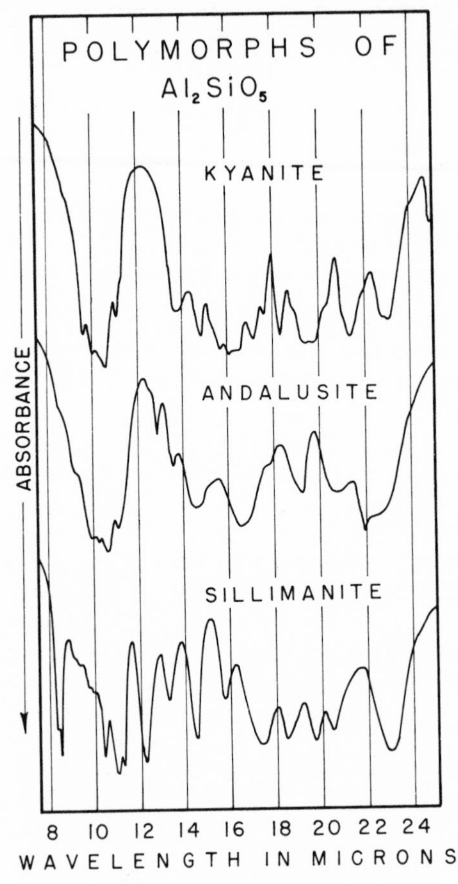

Fig. 19. Infrared spectra of aluminum silicate trimorphs.

minerals, it would be expected that their IR spectra would also be similar. The spectra in Fig. 18 show that this is indeed the case, but small, reproducible differences can be seen as indicated by arrows in the figure. The differences, particularly in the O–H stretching region near 2.75 μ, could be related to changes in nearest neighbors as the stacked kaolin sheets shift.

Another well-known series of this type is the trimorphous forms of Al_2SiO_5—kyanite, andalusite, and sillimanite. Tracings of the IR spectra obtained with these minerals are shown in Fig. 19. Similarities between the spectra of kyanite and andalusite are apparent. The relation between these minerals and sillimanite, however, is not so clear. The IR spectra become more complex proceeding through the series from kyanite to andalusite to sillimanite. The complexity of the spectra is probably related to structural changes in the crystal lattices. This conclusion is supported by studies[54] showing structural modifications of the three minerals to occur in the same order as that established empirically from IR spectra.

5.3. Leaching Studies

In a study of loughlinite, $Na_2O \cdot 3MgO \cdot 6SiO_2 \cdot 8H_2O$, Fayhey, Ross, and Axelrod[68] showed that it was transformed to sepiolite, $4MgO \cdot 6SiO_2 \cdot 8H_2O$, by leaching sodium from its structure with distilled water. This decomposition was followed using IR analysis. Figure 20 shows that the IR spectra of the 60-day leach product resembles that of a natural sepiolite, but it lacks the sharp definition of a natural mineral. This lack of detail is probably due to disordering caused by removal of sodium from the loughlinite lattice.

5.4. Firing Studies

The IR study of fired serpentine minerals illustrated in Fig. 21 points out the applicability of the technique to this type of study. In the chrysotile series no reactions took place up to 500°C. From 500 to 600°C the mineral decomposed with loss of structural hydroxyl, as indicated by the disappearance of the OH^- absorption bands at 2.7 μ. With dehydration the chrysotile crystal lattice broke down leaving a material nearly amorphous to x-ray diffraction studies. This structureless state persisted through the 700°C temperature interval. From 700 to 800°C recrystallization occurred with the formation of a mixture of forsterite and enstatite. This mixture remained stable through 1000°C with only a little additional enstatite forming at higher temperatures.

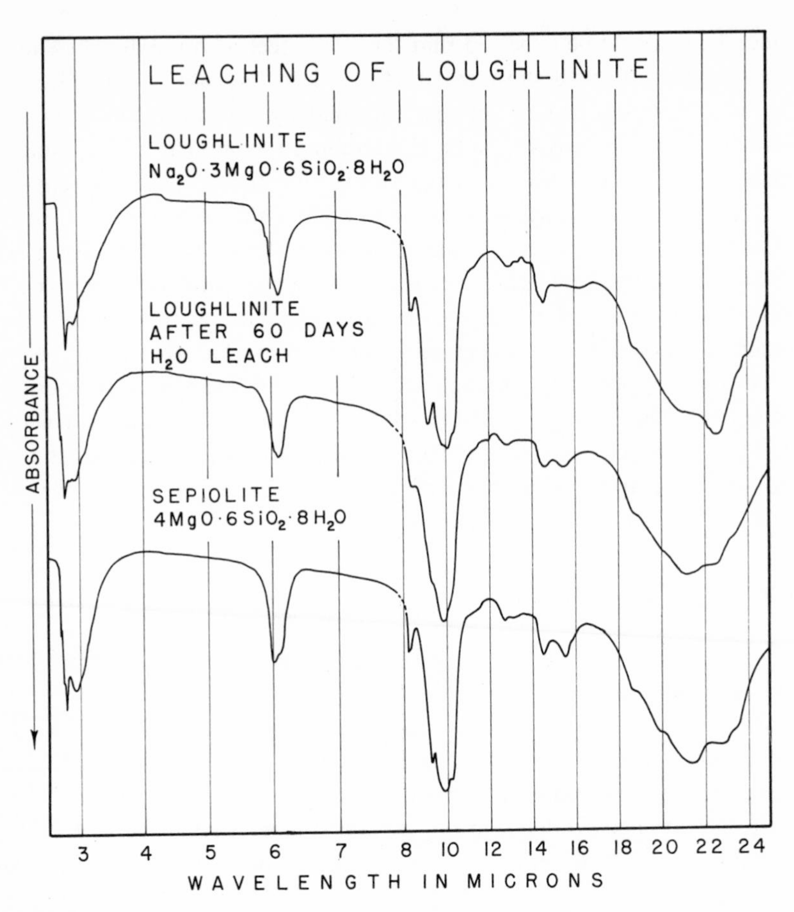

Fig. 20. Infrared spectra illustrating effects of water leaching of loughlinite.

Antigorite showed a tendency to become disordered at a relatively low temperature, as indicated by the broadening and loss of detail of its spectrum after firing at 400–500°C. The structureless interval noted in the case of chrysotile did not occur in the firing of antigorite. Recrystallization to forsterite apparently occurred simultaneously with the destruction of the antigorite crystal lattice. Some bonded hydroxyl remained in the 600°C spectrum as evidenced by absorption at 2.7 μ. Forsterite absorptions were apparent in the same spectrum. Enstatite, which apparently formed simultaneously with forsterite in the thermal recrystallization of chrysotile, did not form in the antigorite series until 900°C. As with the chrysotile series, the final reaction products were enstatite and forsterite. Similar techniques have been utilized in the study of quartz, kaolin, and microcline.[69]

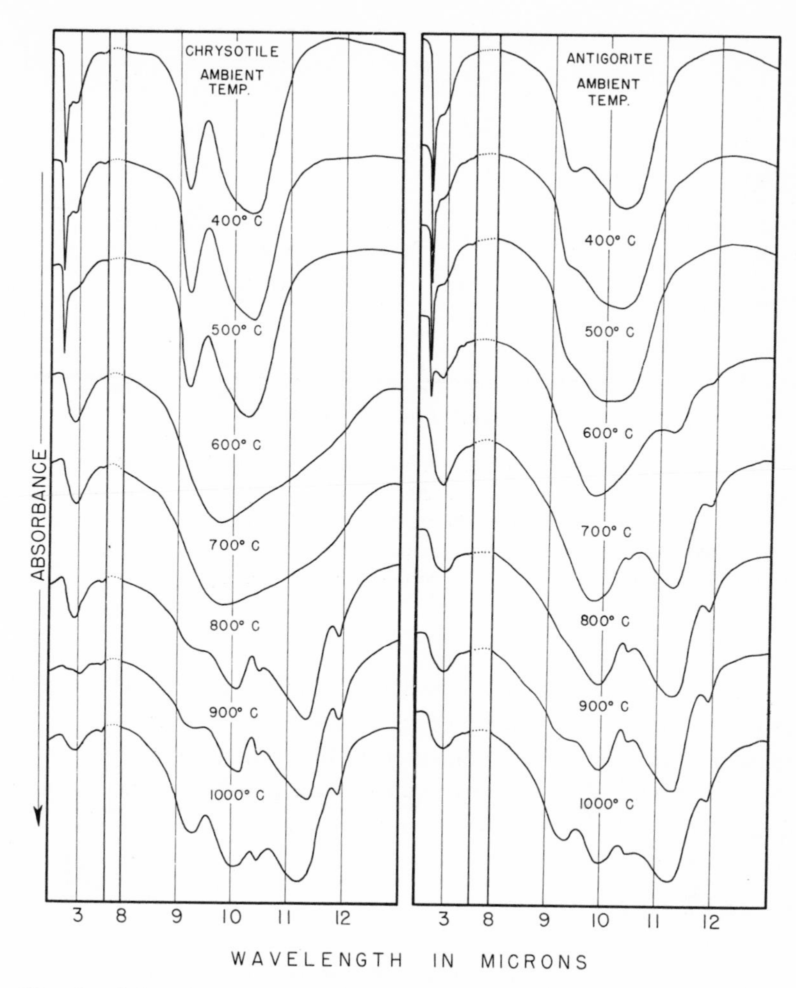

Fig. 21. Infrared spectra showing the pattern of thermal recrystallization followed by the serpentine minerals, chrysotile and antigorite.

5.5. Alteration Studies

Finding new economic mineral deposits requires many approaches. One approach is based on the study of mineral alteration in areas surrounding ore bodies. Bleached and altered areas are important clues to the presence of sulfide mineralization. Geologists often use these halos of altered rock to determine drilling sites; but to make best use of the alteration halo, they must map in detail the degree of alteration. These mineral changes are often difficult to follow with conventional methods, so special techniques

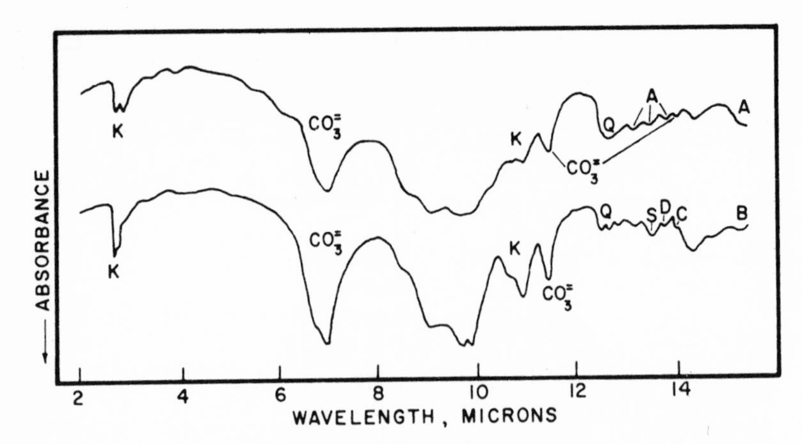

Fig. 22. Alteration studies with IR. Absorption curve A shows presence of 20% kaolinite (K), 25% quartz (Q), and 20% calcite (CO_3^{2-}). Curve B, for a 3-mg sample from a thin vein crossing rock A, shows 34% kaolinite (K) and 40% quartz (Q); siderite (S), dolomite (D), and calcite (C) make up a total of about 25% CO_3^{2-} in the sample. [Reprinted by permission from *Eng. Min. Journ.* **161** (7), 93 (1960)].

must be used. The IR method is especially valuable in this respect. Figure 22 shows the IR curve of an altered rock in which the minerals were so fine grained that accurate optical estimation of the relative concentrations was impossible.[70] With IR it was easy to identify the important minerals and to measure their amounts.

REFERENCES

1. Coblentz, W. W., *Investigation of Infrared Spectra*, Part III: 1906; IV: 1906; V: 1908, VI: 1908; VII: 1905; Carnegie Institution of Washington, republished by the Coblentz Society and the Perkin-Elmer Corp., 1962.
2. Lyon, R. J. P., *Minerals in the Infrared*, Stanford Research Institute, Menlo Park, Calif., 1962.
3. Hunt, J. M., Wisherd, M. P., and Bonham, L. C., *Anal. Chem.* **22**, 1478 (1950).
4. Miller, F., and Wilkins, C. H., *Anal. Chem.* **24**, 1253 (1952).
5. Launer, P. J., *Am. Min.* **37**, 764 (1952).
6. Williams, V. Z., *Rev. Sci. Instr.* **19**, 135 (1948).
7. Herzberg, G., *Molecular Spectra and Molecular Structure, II. Infrared and Raman Spectra of Polyatomic Molecules*, Van Nostrand, New York, 1945.
8. Nakamoto, K., *Infrared Spectra of Inorganic and Coordination Compounds*, John Wiley & Sons, New York, 1963.
9. Herscher, L. W., Instrumentation, in *Applied Infrared Spectroscopy*, (D. N. Kendall, ed.), Reinhold, New York, 1966, pp. 88–135.

10. Low, M. J. D., *Int. Science and Techn.*, February (1967).
11. Rogers, A. F., and Kerr, P. F., *Optical Mineralogy*, McGraw-Hill, New York, 1942, pp. 3–8.
12. Stimson, M. M., and O'Donnel, M. J., *J. Am. Chem. Soc.* **74**, 1805 (1952).
13. Schiedt, U., and Reinwein, H., *Z. Naturforsch.* 7b, 270 (1952).
14. Meloche, V. W., and Kalbus, G. E., *J. Inorg. Nucl. Chem.* **6**, 104 (1958).
15. Tuddenham, W. M., and Lyon, R. J. P., *Anal. Chem.* **32**, 1630 (1960).
16. Lyon, R. J. P., *Am. Min.* **48**, 1170 (1963).
17. Fahrenfort, J., *Spectrochim. Acta* **17**, 698 (1961).
18. Hirschfeld, T., *Appl. Spectry.* **21**, 335 (1967).
19. Harrick, N. J., *Internal Reflection Spectroscopy*, Interscience, New York, 1967.
20. Lyon, R. J. P., and Burns, E. A., *Econ. Geol.* **58**, 274 (1963).
21. Hovis, W. A., *Applied Optics* **5**, 245 (1966).
22. Simon, J., and McMahon, H. O., *J. Chem. Phys.* **21**, 23 (1953).
23. Low, M. J. D., *Applied Optics* **6**, 1503 (1967).
24. Kropotkin, M. A., and Kozyrev, B. P., *Optics and Spect.* **17**, 136 (1964).
25. Lyon, R. J. P., and Burns, E. A., Infrared Emittance and Reflectance Spectra of Rough and Powdered Rock Surfaces, Presented at Meeting of the Working Group on Extraterrestrial Resources, Cape Kennedy, Florida, November, 1964.
26. Lyon, R. J. P., and Tuddenham, W. M., *Min. Eng.* **11**, 1233 (1959).
27. Kendall, D. N., Survey of Practical Information, pp. 59–60; R. D. Moss and W. J. Potts, Jr., Infrared on the Chemist's Bench, pp. 216–217; J. L. Koening, Application of Infrared Spectroscopy to Polymers, pp. 262–263; in *Applied Infrared Spectroscopy*, (D. N. Kendall, ed.), Reinhold, New York, 1966.
28. Lyon, R. J. P., Thompson, C. S., and Tuddenham, W. M., *Econ. Geol.* **54**, 1047 (1954).
29. Tai, H., and Underwood, A. L., *Anal. Chem.* **29**, 1430 (1957).
30. Citron, I., and Underwood, A. L., *Anal. Chim. Acta* **22**, 338 (1960).
31. Stein, W., *Annal. Physick.* **36**, 462 (1939).
31a. Phillippi, C. M., AFML-TR-67-437, Air Force Materials Laboratory, Wright Patterson Air Force Base, Ohio, April 1968.
32. Moenke, H., *Mineral Spectren*, Deutsche Akademie der Wissenschaften Zu Berlin, Akademie Verlag, Berlin, 1962.
33. Lawson, K. E., *Infrared Absorption of Inorganic Substances*, Reinhold, New York, 1961.
34. Ferraro, J. R., *Anal. Chem.* **40**, 24A (1968).
35. Adler, H. H., *et al.*, *Am. Petr. Inst.*, Project **49**, 146 (1950).
36. Busing, W. R., and Levy, H. A., *Acta Cryst.* **11**, 798 (1958).
37. Adler, H. H., and Kerr, P. F., *Bull. Geol. Soc. Am.* **72** (1961); *Am. Min.* **48**, 700 (1962).
38. Adler, H. H., and Kerr, P. F., *Am. Min.* **48**, 124 (1963).
39. Adler, H. H., and Kerr, P. F., *Am. Min.* **48**, 839 (1963).
40. Huang, C. K., and Kerr, P. F., *Am. Min.* **45**, 311 (1960).
41. Menzies, A. C., *Proc. Roy. Soc.* (*London*) **A134**, 265 (1931).
42. Christ, C. L., *Am. Min.* **45**, 334 (1960).
43. Tennyson, C., *Fortschr. Mineral.* **41**, 64 (1963).
44. Omari, K., and Kerr, P. F., *Bull. Geol. Soc. Am.* **74**, 709 (1963).
45. Adler, H. H., and Kerr, P. F., *Am. Min.* **50**, 132 (1965).

46. Berry, L. G., and Mason, B., *Mineralogy*; *Concepts*, *Descriptions*, *Determinations*, W. H. Freeman, San Francisco, 1959, p. 467.
47. Lyon, R. J. P., Evaluation of Infrared Spectrophotometry for Compositional Analysis of Lunar and Planetary Soils, Stanford Research Institute Project No. PSU-3943, 1962.
48. Bassett, W. A., *Bull. Geol. Soc. Am.* **71**, 449 (1959).
49. Wickersheim, K. A., and Buchanan, R. A., *Am. Min.* **44**, 440 (1959).
50. Plyusina, I. I., *Geokhimiya* **1**, 31 (1964).
51. Deer, W. A., Howie, R. A., and Zussman, J., *Rock Forming Minerals*, *Vol. 1. Ortho and Ring Silicates*, Longmans, London, 1963.
52. Milkey, R. G., *Am. Min.* **45**, 990 (1960).
53. Thompson, C. S., and Wadsworth, M. E., *Am. Min.* **42**, 334 (1957).
54. Deer, W. A., Howie, R. A., and Zussman, J., *Rock Forming Minerals*. *Vol.* 4. *Framework Silicates*, Longmans, London, 1963, p. 96.
55. Koizumi, M., *Miner. Jour.* **1**, 36 (1963).
56. Coles, J. L., *A Study of Some Synthetic Apatites*, Ph.D. dissertation, Dept. of Mineralogy, University of Utah, 1963.
57. Fischer, R. B., and Ring, C. E., *Anal. Chem.* **29**, 431 (1957).
58. Dana, J. D., and Dana, E. S., *The System of Mineralogy*, *II*, 7th Ed., as rewritten and enlarged by C. Palache, H. Berman, and C. Frondel, John Wiley & Sons, New York, 1951.
59. Grubb, P. L. C., *Am. Min.* **52**, 418 (1967).
60. Chang, L. L. Y., *Am. Min.* **52**, 427 (1967).
61. McDevitt, N. L., and Baun, W. L., A Study of the Absorption Spectra of the Simple Metal Oxides in the Infrared Region 700–240 cm^{-1}, Technical Documentary Report No. RTD-TDR-63-4172, A. F. Materials Laboratory Research and Technology Division, Wright-Patterson Air Force Base, Ohio, Project #7360, Task No. 736005, 1964.
62. Tuddenham, W. M., and Lyon, R. J. P., *Anal. Chem.* **31**, 377 (1959).
63. Stubican, V., and Roy, R., *J. Am. Ceram. Soc.* **44**, 625 (1961).
64. Hayashi, H., and Oinuma, K., *Am. Min.* **52**, 1210 (1967).
65. Duke, D. A., and Stephens, J. D., *Am. Min.* **49**, 1388 (1964).
66. Duke, D. A., *Infrared Investigation of the Crystal Chemistry of Olivine and Humite Minerals*, Ph.D. dissertation, Department of Mining and Geological Engineering, University of Utah, 1962.
67. Deer, W. A., Howie, R. A., and Zussman, J., *Rock Forming Minerals*, *Vol. 3. Sheet Silicates*, Longmans, London, 1963.
68. Fahey, J. J., Ross, M., and Axelrod, J. M., Loughlinite, *Am. Min.* **45**, 270 (1960).
69. Stephens, J. D., and Tuddenham, W. M., *Am. Ceram. Soc. Bul.* **46**, 725 (1967).
70. Tuddenham, W. M., and Zimmerly, S. R., *Eng. and Min. J.* **161** (7), 92 (1960).

Chapter 7

OPTICAL EMISSION SPECTROSCOPY

Armin P. Langheinrich and D. Blair Roberts*

Kennecott Copper Corporation
Salt Lake City, Utah

1. INTRODUCTION

Optical emission spectroscopy includes the fields of flame-, arc-, and spark-induced emission phenomena in the UV, visible, and near IR regions of the electromagnetic spectrum. As a technique, it furnishes the analytical investigator with qualitative and quantitative information on the elemental composition of matter through simultaneous multielement determinations. Detection limits in the low-ppm range are quite common.

Emission spectroscopy, often referred to as spectrochemistry, was the first direct instrumental analytical technique to be widely used in geochemical investigations. In spite of the development of new methods and techniques and the ready availability of corresponding instrumentation, emission spectroscopy remains today an indispensable tool for the geochemist.

1.1. History

Approximately 1800 years passed between Seneca's observation of the colors of the visible spectrum through a primitive prism[1] and the development of a practical spectroscope.[2] In 1860 Bunsen and Kirchhoff[3] initiated spectral analyses. By 1907, 13 elements had been discovered by spectroscopic means.[2] Though the use of this instrument became widely accepted during

* Presently at Westinghouse Georesearch Laboratory Boulder, Colorado.

this period, its application to quantitative work remained limited until Gerlach,[4] in 1925, reported the application of internal standards. The result was a dramatic improvement in the precision and accuracy of the technique. From 1929 to 1936 Goldschmidt[5–10] demonstrated the great potential of spectrochemistry in geochemical work. He and his coworkers made element-abundance studies, the results of which are still quite acceptable today.

Historical introductions are presented in most textbooks and treatises on emission spectroscopy. An interesting chapter on the subject was written by Weise.[1] Recent presentations have included the history of flame emission by Fassel,[11] of the spark discharge by Walters,[12] and of the dc arc by Strock.[13] Since L. W. Strock himself was a member of Goldschmidt's team, his paper adds considerably to the knowledge of that era.

1.2. Principles

Detailed discussions of basic theory and the origin of atomic spectra are beyond the scope of this chapter and can be found in textbooks on modern physics and spectroscopy, such as Herzberg,[14] Richtmyer, Kennard, and Lauritsen[15] and McNally.[16] It is appropriate, however, to include the following summary of the underlying principles.

Optical emission spectra may be observed when transitions of the outer energy-level electrons occur within an atom. This phenomenon is illustrated diagrammatically in Fig. 1, where ground-state conditions are compared with excitation and emission steps. An electron in a given atom is defined by a set of quantum numbers and is associated with a definite amount of energy. When electrical, optical, or thermal energy is introduced into the atom, electrons are transferred from low-energy to high-energy quantum states, i.e., the "ground-state" atom is now in an "excited state." After extremely short time intervals the energy-rich electrons undergo transitions into states of lower energy. The resulting energy difference is emitted in the form of electromagnetic radiation. From the law of conservation of energy the energy of the emitted photon $\hbar\nu$ must equal ΔE, the difference in energy content of the electron prior to and after transition, where $\hbar$ is Planck's constant (6.6×10^{-27} erg-sec) and ν is the frequency in vibrations per sec.

According to the Bohr theory electrons are stable in definite quantum states. Therefore, only specific energy differences and frequencies are possible. This explains the occurence of a definite line spectrum for each atom. A considerable number of transitions are possible, and a large number of spectral lines can be found for most elements. In the case of iron, for example, nearly 5000 lines are listed in the M.I.T. wavelength tables.[17] Since

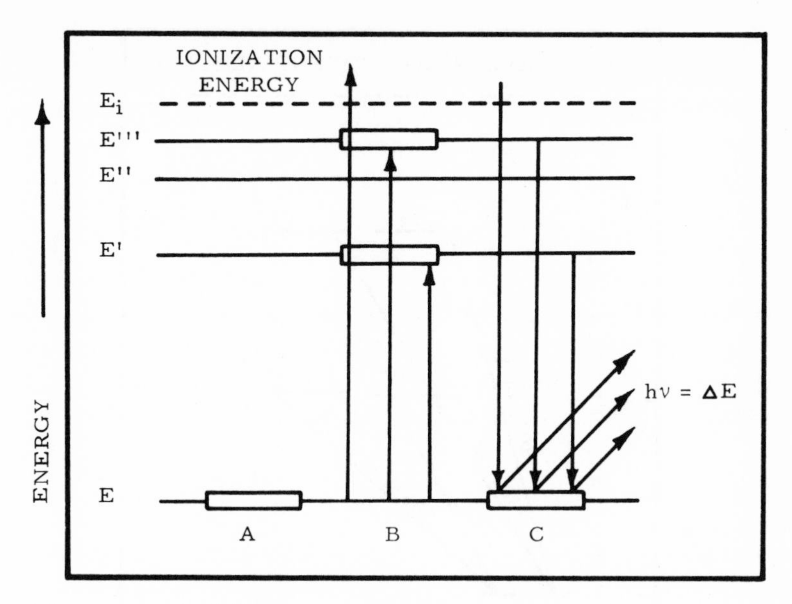

Fig. 1. Energy-level diagram showing electronic transitions. (A) Ground-state conditions. (B) Excitation. (C) Emission.

the probabilities of these transitions vary, certain lines are found to dominate a spectrum. In most cases these dominating lines are used to establish the presence of a particular element.

Excitation takes place as soon as the "excitation potential" of a specific transition has been reached. The excitation potential is a numerical expression of the minimum energy required to cause excitation, and is usually reported in volts or electron volts. When the exciting energy is increased further, ionization may occur. In such a case an electron is removed completely from the influence of the parent nucleus of atomic number Z. The resulting ion may also be excited, but its spectrum differs significantly from the atom spectrum of element Z, and resembles the atom spectrum of element $Z - 1$. Similarly, the ion spectrum of a doubly ionized element Z resembles the atom spectrum of element $Z - 2$. and so on. In addition to atoms and ions, certain molecules can be excited under spectrochemical conditions. The resulting molecular band spectra often interfere with line spectra and are of limited analytical value. Figure 2 represents a partial energy-level diagram of the Grotrian type.[18] It shows transitions for the four most sensitive lines of the neutral sodium atom. Excitation and ionization potentials (in volts) are taken from Ahrens.[19]

In spectrochemical analyses samples are vaporized and excited by means

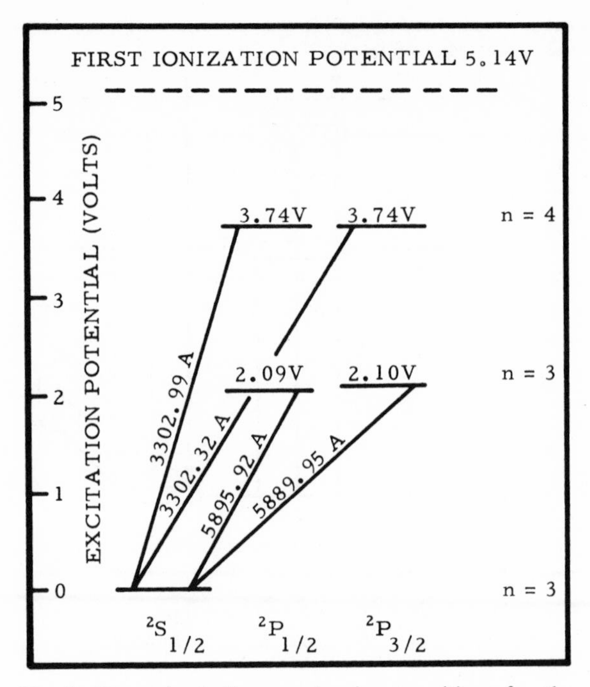

Fig. 2. Energy-level diagram showing transitions for the four most sensitive lines of the neutral sodium atom.

of flames, arcs, sparks, or lasers. The resulting atomic, ionic, and molecular radiation is focused on a narrow slit, dispersed by gratings or prisms, and observed visually or recorded photographically or electronically. The projected spectrum or photographic record consists of a series of lines where each line represents an image of the slit and corresponds to a specific wavelength of radiation. The determination of line position, and hence of wavelength, allows qualitative identification. By measuring the intensity of the emitted characteristic radiation, a quantitative result may be obtained.

1.3. Relation to Other Techniques

Spectrochemical analyses may be used independently of other analytical techniques. However, it must be realized that no analytical instrument is available that will meet all requirements. In many cases information obtained from two or more complementary methods must be correlated to arrive at a complete answer. This is, however, not often possible within the desired limits of speed, accuracy, precision, and practicality. Spectrochemical procedures can contribute significantly to the solution of other instrumental and chemical problems. The spectrograph may furnish details on impurity levels in reagents and process intermediates, in "insoluble" residues and in precipitates. In many instances the spectrographic results provide significant "survey" information prior to the selection of final geochemical or analytical procedures.

Laboratories with emission spectrographic equipment often have other analytical facilities available. The spectrochemist should utilize these. He can benefit greatly from information obtained by other methods. For example, IR spectra yield information on anions, on specific minerals, and on mineral types. X-ray diffractometer scans provide information on major concentrations of chemical compounds and minerals. X-ray emission determinations furnish elemental compositions in a nondestructive way. Atomic absorption, flame photometric, and colorimetric techniques assist where external standards are being generated or verified. Such information will enable the spectrochemist to make proper selection of sample size, internal standards, external standards, dilution materials, operating conditions, etc.

Although the multimethod approach is certainly not necessary in every case, its potential value to both spectrochemist and geochemist must not be underrated. It may lead to significant shortcuts and it is of inestimable value where the history of a sample is uncertain.

2. EQUIPMENT AND FACILITIES

In the following section specific instruments are not discussed, since excellent advertising material from a number of firms adequately covers this area. Lists of suppliers of complete analytical systems as well as of major and auxiliary components have been published.[20,21] Manufacturers have generally been found helpful in assisting in the solution of specific instrumental and even analytical problems. Their help, however, is no full substitute for experience and sound knowledge on the part of the spectroscopist.

2.1. Components and Functions

Basically, an emission spectrographic system consists of three components: (a) the excitation source, (b) the dispersing unit, and (c) detection equipment. The analytical sample, properly positioned between the electrodes of an arc–spark stand, is excited. It emits radiation characteristic of the elements present in the sample. The spectrograph separates and isolates these radiant energies prior to detection and recording.

2.1.1. Excitation Sources

In addition to flame, several other excitation techniques have been utilized. These include the dc arc, the ac arc, the high-voltage ac spark, and various modified arc–spark sources, such as the interrupted or ignited arc. Recently, laser vaporization with auxiliary arc excitation has been added to this list. The dc arc is the popular excitation source for most types of analyses encountered in geochemical and related work. With the dc arc good sensitivity may be achieved, and numerous semiquantitative and quantitative determinations have been performed successfully. Ahrens and Taylor,[19] for example, based their well-known textbook on spectrochemical analyses solely on the use of this source. The ac arc, most widely used for some time, has been largely replaced by the high-voltage ac spark or by various combinations of ac-spark sources. Good precision is obtainable with the spark source, but the corresponding sensitivity is generally poorer than that obtainable with the dc arc. The spark source is widely used for the quantitative analysis of metal and metal-alloy systems. The laser may be used for the primary vaporization of small sample volumes prior to arc excitation of the vaporized sample. Its usefulness in geochemistry, especially in mineralogical studies of small inclusions or small mineral grains, has already been recognized.

2.1.2. Dispersing Units

Spectral dispersion is accomplished by prisms or diffraction gratings. The use of prisms is declining, mainly as a result of improved technology in the production of high-quality gratings. Whenever prisms are used in geochemical work they should be of the quartz type since both the visible and the UV regions of the spectrum are of interest. Prism instruments have the advantage of large dispersions in the UV region. They produce no overlapping spectral orders. Grating instruments, on the other hand, provide nearly constant dispersion at all wavelengths. Higher orders are produced and are utilized when large dispersion is a necessity. Prism mounts (e.g., Littrow and Cornu) and grating mounts (e.g., Eagle, Ebert, Wadsworth, Paschen–Runge) have been discussed adequately in the literature.[2,22] All these types are currently in use.

2.1.3. Detection Equipment

The detection of dispersed line spectra can be accomplished visually, by photographic means, or electronically with photomultiplier tubes. Photo-

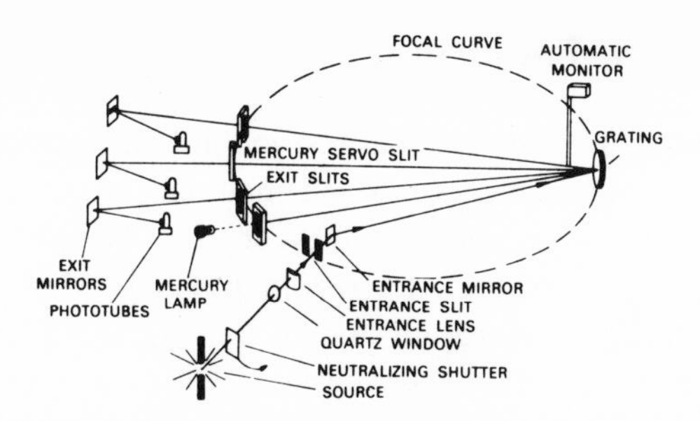

Fig. 3. Optics diagram of a direct reader. Courtesy of Baird-Atomic, Inc.

graphic methods, when applied to quantitative work, necessitate some kind of plate or film calibration to relate line intensities of the photographic image to light intensities originating from the excited sample. The application of direct readers, i.e., instrumentation with electronic detection devices, is expanding rapidly in rock and mineral work. Commercial units are available that can detect about 60 elements simultaneously plus up to eight in the vacuum region below 2000 Å. A diagram of a direct-reader arrangement is presented in Fig. 3. The dispersed radiation must pass through properly positioned exit slits prior to reaching the photomultiplier tubes. These convert light into electrical energy. The resulting signals are integrated during the period of arcing and finally amplified and recorded. This approach is especially valuable where much routine work has to be done. It has by no means replaced the photographic technique. Variations often encountered in geochemical work can be met with maximum flexibility by the latter. In this case permanent records of all elements present are produced. The photometric step is done on microdensitometers available commercially in various sizes and complexities.

2.1.4. Auxiliary Equipment and Materials

Commercially available spectrochemical materials allow the operator to concentrate on his analytical work load. Except in the case of unusual applications, it is no longer necessary to spend valuable time in preparing and shaping electrodes and other time-consuming preparatory steps. Items to be considered in equipping a laboratory include the following: electrodes, electrode shapers, milling and grinding units, sample presses, gas supplies and atmosphere controllers, photographic film, plates and chemicals, and calculating accessories.

2.2. Laboratory Facilities

Cost and space are the principal restrictions when purchasing spectrographic equipment. Instrumentation and materials are available in a wide range of complexity, price, and usefulness for specific applications. Simple visual comparators are available for less than $ 2000. Highly sophisticated direct-reading spectrometers for air and vacuum applications, coupled with computer-controlled operation and determination, may cost up to $ 200,000. Among read-out options one can choose strip-chart recorders, digital voltmeters, punched paper tape, electric typewriters, and completely computer-controlled spectrometer systems or interfaces to operate with most digital computers. When photographic detection is used film or plate processing facilities are needed. The laboratory should have controlled temperature and humidity, and it should be kept scrupulously clean.

2.3. Field Facilities

The general instrumentation is not different from laboratory units, although space restrictions are much more severe. Field instruments, by necessity, have to be small to allow mounting in mobile laboratories. Daylight developing equipment must be used if darkroom facilities are not available. With respect to sample preparation, i.e., crushing, grinding, and homogenizing, the operator has to depend on his own ability and apparatus. All equipment must be able to withstand effects of vibration and changes in temperature and humidity.

One manufacturer features a small-size spectrograph that can be mounted in an area of less than 3 × 6 ft. Its reciprocal linear dispersion in the second order is listed as 6.4 Å/mm. Although equipment of this type is somewhat limited in its usefulness, it can provide rapid on-the-spot information of great value. An example of a field instrument is shown in Fig. 4.

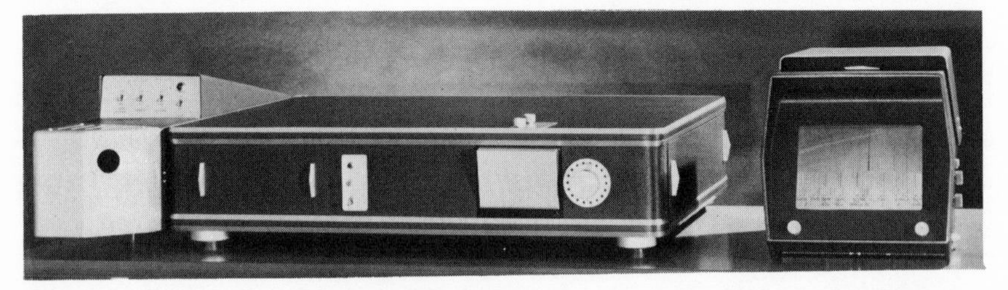

Fig. 4. Table-top ARL spectrograph with reader. Courtesy of Applied Research Laboratories.

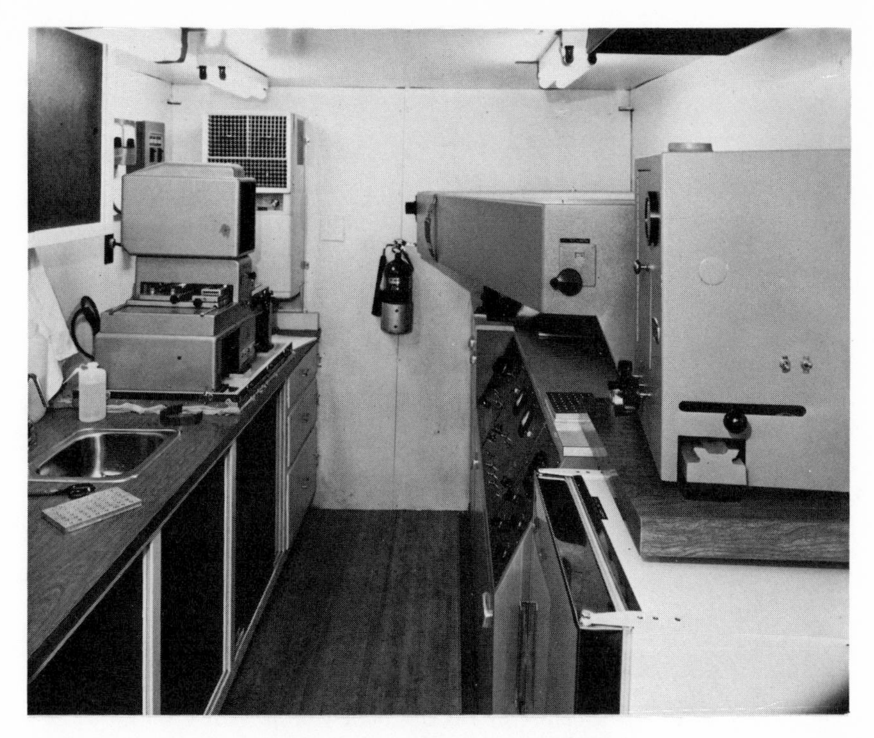

Fig. 5. Interior of mobile laboratory. Right, 1.5-m JACO Spectrograph. Left, ARL microdensitometer. Courtesy of the Jarrell-Ash Co.

Mobile laboratories are gaining in popularity. They provide desirable close liaison with geologic mapping or prospecting. Considerable experience in field applications of emission spectrographs has been gained by the U. S. Geological Survey. Their laboratories, mounted on 2.5-ton trucks, are equipped with 1.5-m spectrographs. Each unit can handle up to 100 samples per day if 30 elements are to be determined.[23] The interior of one of these laboratories is shown in Fig. 5.

2.4. Special-Purpose Equipment

The experienced analyst will undoubtedly develop his own variations of procedure and equipment. He may look beyond a minimum instrument package. Special attachments are available which allow arcings under controlled atmospheric conditions.[24,25] Automated sample presentation systems have been developed.[26] Lasers, as shown in Fig. 6, are used in direct excitation and in primary excitation to feed the normal arc. Automatic develop-

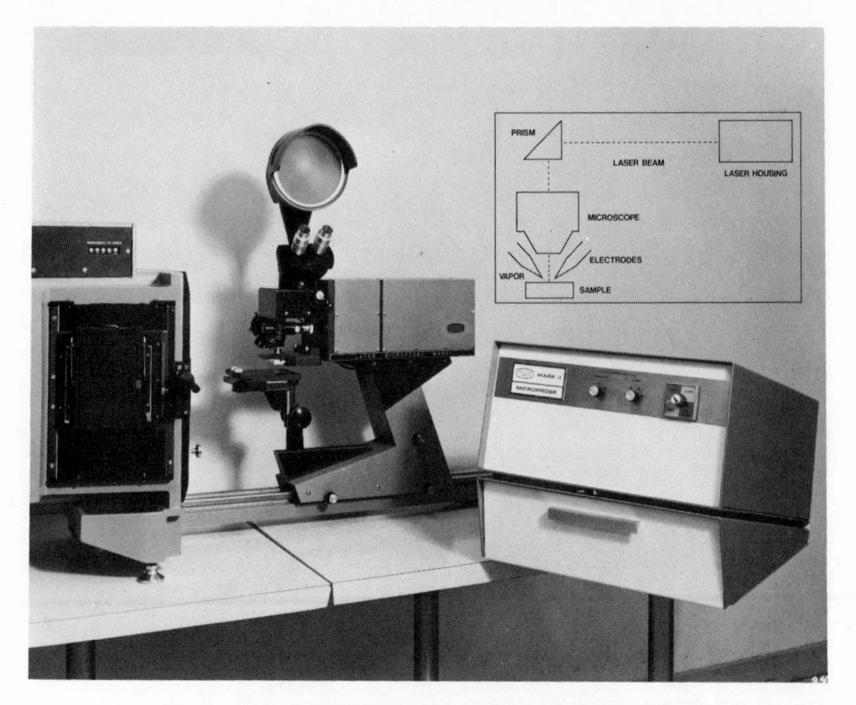

Fig. 6. Laser microprobe by JACO. Courtesy of the Jarrell-Ash Co.

ing equipment facilitates the photographic steps. Vacuum attachments allow the analytical evaluation of the most sensitive lines of antimony, arsenic, carbon, mercury, phosphorus, selenium, sulfur, and zinc. Combination spectrometer–spectrographs are available, which permit changeover from the direct-reading to the photographic method, and vice versa, in less than 1 min. Finally, computers, including small desk computers, can swiftly perform time-consuming calibration and working-curve calculations.

3. TECHNIQUES

Spectrochemical techniques vary widely from laboratory to laboratory, even from operator to operator. In spite of such variations certain basic steps are common to all methods. The following discussion is divided into qualitative, semiquantitative, and quantitative analyses, although clearly defined divisions do not exist. Whether an analytical result is called quantitative or semiquantitative depends on the particular requirements of the project and on the type of laboratory available. By convention, however, the term "semiquantitative" usually refers to results that lie within $\frac{1}{3}$–3× the amount actually present.

Since the dc arc is the common excitation source in geochemical work, the text refers primarily to this source. Ordinarily the sample is used as the anode or positive electrode. The positive ions from the discharge migrate towards the cathode or negative electrode. In certain techniques, however, the sample may be made the negative electrode.[27] A narrow layer of highly concentrated ions and atoms (through the recombination of ions and electrons) will then form near the cathode. When this region is focused on the slit increased sensitivities may be obtained, especially for elements that ionize readily in the discharge. This procedure, however, is quite sensitive to operational variations.

A block diagram explaining the various stages of a spectrochemical determination is given in Fig. 7. The steps involved include sample preparation, excitation, and data accumulation and treatment. Important aspects of the diagram are discussed more fully in the text.

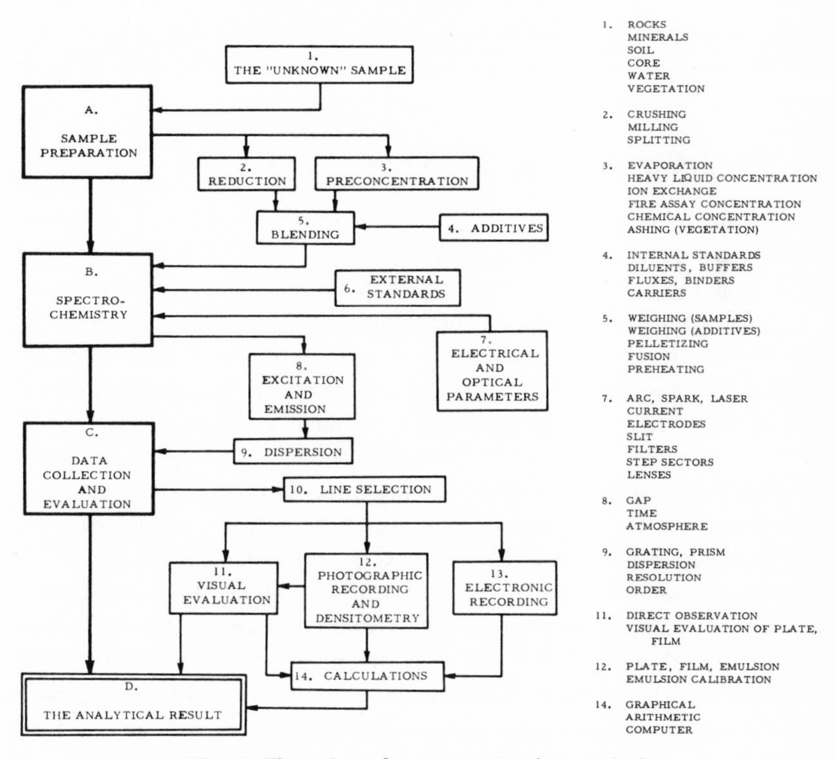

1. ROCKS
MINERALS
SOIL
CORE
WATER
VEGETATION

2. CRUSHING
MILLING
SPLITTING

3. EVAPORATION
HEAVY LIQUID CONCENTRATION
ION EXCHANGE
FIRE ASSAY CONCENTRATION
CHEMICAL CONCENTRATION
ASHING (VEGETATION)

4. INTERNAL STANDARDS
DILUENTS, BUFFERS
FLUXES, BINDERS
CARRIERS

5. WEIGHING (SAMPLES)
WEIGHING (ADDITIVES)
PELLETIZING
FUSION
PREHEATING

7. ARC, SPARK, LASER
CURRENT
ELECTRODES
SLIT
FILTERS
STEP SECTORS
LENSES

8. GAP
TIME
ATMOSPHERE

9. GRATING, PRISM
DISPERSION
RESOLUTION
ORDER

11. DIRECT OBSERVATION
VISUAL EVALUATION OF PLATE, FILM

12. PLATE, FILM, EMULSION
EMULSION CALIBRATION

14. GRAPHICAL
ARITHMETIC
COMPUTER

Fig. 7. Flow sheet for spectrochemical analysis.

3.1. Qualitative Analyses

An experienced spectrochemist should be able to detect and report the presence of about 50 elements in less than 1 hr. This estimate is based on a good instrument used independently of other methods. Standard reference spectra are commercially available for comparison with unknown spectra. In the simplest case identification is accomplished by visual inspection of the emission spectrum projected onto a diffusion screen. In the more general case the spectrum is retained on film or plate and is studied with the unaided eye, with the help of hand magnifiers, or with projection equipment. For purposes of direct comparison elements or compounds of elements of interest are normally recorded on the same plate or film as is used for the unknown(s). With purely electronic detection devices the data output is dependent upon the number and positions of the direct-reading channels.

In actual practice a strictly qualitative analysis is seldom asked for. Identification of elements without distinguishing between trace and major constituents is normally of little value. Consequently, most analyses, even those called qualitative, include a means of estimating concentrations by simple classification, such as trace, minor, or major. The value of this approach increases when the categories are defined numerically, or when the above groupings are subdivided.

3.2. Semiquantitative Analyses

The upgrading and refining of qualitative work eventually leads to the so-called semiquantitative analysis. In many geochemical applications this approach is more important than either qualitative or quantitative work.

Visual comparisons in semiquantitative analyses are commonly replaced by densitometric measurements of spectral lines and, quite often, by background measurements in the vicinity of these lines. Line-to-background intensity ratios may improve the quality of final results through compensation of uncontrolled variables. Widely practiced visual techniques have also been reported. Among these are the three-step and six-step methods reported by the U. S. Geological Survey.[28,29] In these techniques standards are prepared of all elements to be determined, in three or six different concentrations, for several orders of magnitude. In the three-step method, for example, concentrations of approximately 1, 3, and 7 ppm, 10, 30, and 70 ppm, 100, 300, and 700 ppm, etc., are selected. The unknown spectra are compared visually with the standard spectra on a projection microdensitometer. Concentration estimates are obtained by bracketing. This procedure has given results that compare well with other semiquantitative techniques. Emulsion calibrations are ordinarily omitted, but comparison standards are recorded on each plate or film.

Semiquantitative work normally requires the addition of diluents to the samples. Through this step the matrix and, hence, excitation conditions are approximated for all samples and standards under investigation. The effect of buffer materials on the dc arc has been studied by Decker and Eve.[30] Additives in general have been discussed briefly by Leys and Dehm.[31] In powdered rock samples, powdered graphite in comixture with an alkali-metal salt is often used. The addition of high-purity quartz is also fairly common. Quartz not only serves as a diluent, but it is also important in promoting volatilization of the sample. Quartz additions are generally permissible since many commonly analyzed samples contain major amounts of silicon, so that the spectrochemical determination of this element is of little interest.

Since the semiquantitative method is based on the comparison of samples with standards of known concentration, any bias may be reduced if the standards and samples are similar in composition. When samples ranging from acidic or intermediate igneous rocks to limestones or dolomites are analyzed with the same set of standards and working curves, the resulting errors may be quite large. Matrix changes in the samples must, therefore, be borne in mind when spectrochemical data are to be evaluated.

In normal procedures these errors rarely exceed a factor of two; but in volatile-element techniques they may change the result by an order of magnitude. This effect becomes especially apparent in the determination of tungsten and tin. These elements can be reported in volatile-element analyses of silicate rocks, but not in volatile-element analyses of limonite–pyrite–magnetite samples, where the matrix effect is much more pronounced. Spectrochemical techniques for the determination of low concentrations of volatile elements have been developed by Adamson *et al.*[32] They used IR absorption methods for mineralogical identification of residual melt beads. Interferences due to iron are minimized and volatile-element intensities are enhanced when the sample composition and the additions to the sample result in residue beads containing olivine- and pyroxene-group minerals. The method is applicable to copper, to iron and manganese oxides, to copper and iron sulfides, and to silica and silicate minerals in general. Recommendations of sample sizes and additives are summarized in Table I.

Various methods exist for presenting geochemical materials to the discharge. The original samples are preferably ground to −325 mesh and homogenized for uniformity. After blending with additives they are added to electrodes made of graphite or carbon such as ASTM shapes S-4 and

Table I

Composition of Electrode Charge for Volatile-Element Determination[a]

Sample	Sample weight, mg	Additions, mg				
		SiO_2	Fe_2O_3	MgO	MnO_2	K_2CO_3
Iron oxides, hydrates, or sulfides	75	15		8		2
Manganese oxides or hydrates	75	15		8		2
Silica or silicates	65				35	2
Copper oxides or sulfides	40	50	40	8		2
Copper–iron sulfide, Cu_5FeS_4	50	50	30	8		2
Copper–iron sulfide, $CuFeS_2$	80	50	5	8		2

[a] Reproduced from *Anal. Chem.* **39** (1967).

S-12.[33] These electrodes possess a cavity to receive the powdered sample. Other types are large-size carrier-distillation electrodes with boiler caps for obtaining improved sensitivities for volatile elements. Rotating disk, rotating horizontal platform, and vacuum-distillation electrodes are used in solution work. Powder samples are sometimes pressed into pellets prior to presentation to the discharge. Controlled atmospheres have come into wide spread use in the trace-element geochemical work of solid materials. For solutions the so-called plasma-arc or plasma-jet modifications of the dc-arc technique are of value. The sample solution is introduced into the arc discharge by an atomizing gas. At the same time a tangentially flowing auxiliary gas around the atomizer stabilizes and constricts the arc.

Certain elemental substances will not readily reduce in size on grinding. Malleable native silver is an example. If this occurs, 5- or 10-mg analytical samples, such as ordinarily used in dc-arc analysis, can produce erratic results. Fortunately, spectrographic silver determinations on thousands of rock pulps, soils, and other geologic materials in the writers' laboratories have shown that these errors are often less serious than may be expected. Errors in excess of the data in Table II are seldom encountered. In this table assay values of silver and copper in soils are compared by atomic

Table II

Comparison of Semiquantitative Spectrochemical with Atomic Absorption Data(ppm)

Sample	Spectrochemistry		Atomic absorption	
	Copper	Silver	Copper	Silver[a]
1	40	0.3	45	0.5
2	30	0.3	33	0.7
3	400	4.0	475	2.8
4	40	0.6	56	1.0

[a] Preconcentration by solvent extraction.

absorption and by dc-arc spectrography. The spectrochemical data were obtained without emulsion calibration or internal standards, but with controlled atmosphere and densitometric measurements.

The problem is more serious in the case of gold or platinum. These elements mostly occur in the metallic form as discrete particles of malleable metal. Nevertheless, results obtained with 5-mg samples have proved useful when fairly large numbers of samples from each presumed occurrence were analyzed.

The accuracy of spectrochemical data depends greatly on reliable values for available standards. The development of such standards can be time consuming and costly. In many instances poor agreement is obtained where several laboratories are involved. This fact is illustrated in Table III. It has also been shown by cooperative investigations conducted by the U. S. Geological Survey and others.[34,35] As a result of these studies, standards such as Granite G-1, Diabase W-1, Sulfide Ore 1, Syenite Rock 1, Granite G-2, Granodiorite GSP-1, Andesite AGV-1, Peridotite PCC-1, Dunite DTS-1, Basalt BCR-1, and others have become available and recommended concentration values have been reported for many elements.[36–38]

A typical dc-arc spectrographic procedure for geochemical analyses is shown in Table IV. It can be applied readily to other instrumentation. In Table V data on wavelengths and detection limits for 51 elements are presented. These elements are regularly determined at the writers' laboratories. Excitation potentials are calculated mainly from wave numbers reported by Meggers, Corliss, and Scribner.[39] First ionization potentials are from the *Handbook of Chemistry and Physics*.[40] The most sensitive lines of seven ele-

Table III

Development of Spectrochemical Standards; Nonspectrographic Assay Data (ppm)

Element	Sample No.	Number of laboratories	Range reported
Antimony	5	2	0.3–0.5
	12	2	0.6–1.1
	11	2	1.5–2.6
Arsenic	4	2	1.5–2.3
	1	6	3.0–12.0
	12	5	19.5–25.0
	11	4	100–161
Cadmium	5	2	<1.0–1.1
	11	3	11.0–12.0
Cobalt	1	2	0.5–2.0
	7	3	1.0–3.0
	5	3	17.0–25.0
Copper	2	7	5–11
	6	7	13–26
	15	4	110–180
	1	10	210–361
	12	10	710–952
	16	4	3000–3300
	14	8	4856–7424
Lead	1	6	2–6
	15	5	10–52
	16	5	96–235
	6	5	597–788
Manganese	1	2	10–12
	12	3	40–90
	3	3	490–730
Molybdenum	15	4	<0.5–3.0
Tin	5	2	2.0–3.0
	12	2	5.0–6.0
Tungsten	5	3	0.5–3.0
	4	3	1.0–5.0
	12	2	10–19
Zinc	1	5	13–14
	15	4	35–75
	3	4	105–148
	8	5	712–1016
	16	4	3970–4600
	11	4	8290–10700

Table IV

Example of Semiquantitative dc-Arc Spectrochemical Procedure

Instrumentation	
Excitation source	Multisource with arc–spark stand
Atmosphere control	Enclosed Stallwood jet with gas supply
Spectrograph	Three-meter, concave grating, 30,000 lines/in.; reciprocal dispersion 2.75 Å/mm, first order
Recording equipment	Plate holder with 4 × 10-in. photographic plates
Developing equipment	Mechanical photo processor with timer, automatic agitation, and temperature control
Microdensitometer	Projection type, 15× and 20× magnification
Calculating equipment	Programmable desk calculator
Procedure	
Sample preparation	Powders, −325 mesh; 5-mg samples, 10-mg additives (87.5% graphite, 12.5% GeO_2); natural rock samples and synthetic materials as primary standards; secondary standards for most routine work; 10 sec preroasting in burner flame
Electrode system	Upper electrode (cathode), high-purity graphite, $\frac{1}{8}$-in. diameter, 120° included angle tip; lower (sample) electrode (anode), high-purity graphite, shallow crater, $\frac{1}{4}$-in. diameter (ASTM S-4); 3.5-mm analytical gap; 70% argon–30% oxygen atmosphere, flow rate 5 l/min
Excitation conditions	dc-Arc discharge; 12 A, 230 V
Exposure conditions	Spectral region 2200–3600 Å, first order; slit-width 25 μ; three-step rotating sector (50, $12\frac{1}{2}$, and $3\frac{1}{8}$% transmissions); exposure time 90 sec, complete burn, no prearc
Photographic processing	Emulsion, Kodak SA-1 plates; developing, Kodak D-19, 4 min at 71° ± 0.3° F; short stop, 5 sec, 5% acetic acid; fixing, 5 min; washing, 5 min, running water; drying, warm air, about 2 min

ments listed at the end of Table V are ordinarily detected only with vacuum equipment.

Molecular band spectra are of interest for the indirect determination of nonmetals. Examples are the use of CaF bands for fluorine and CN bands for carbon. The determination of water in minerals and rocks through use of the 3064 Å OH-band system has been reported by Quesada and Dennen.[41]

Table V

Selected Data for Use in Geochemical Analyses

Element	Symbol	Wavelength	Spectrum	Detection limits for various elements, %		Excitation potential, V	First ionization potential, V
				Normal work	Special procedure		
Aluminum	Al	3082.2	I	0.001		0.0–4.0	5.96
		2378.4	I			0.0–5.2	
Antimony	Sb	2877.9	I	0.002	0.001	1.1–5.4	8.5
		3267.5	I			2.0–5.8	
Arsenic	As	2860.5	I	0.02	0.002	2.3–6.6	10.5
		2288.1	I			1.4–6.8	
Barium	Ba	4554.0	II	0.0005		0.0–2.7	5.19
		2335.3	II			0.7–6.0	
Beryllium	Be	3130.4	II	0.0001		0.0–3.9	9.28
		2348.6	I			0.1–5.4	
Bismuth	Bi	3067.7	I[a]	0.0003	0.0001	0.0–4.0	8.0
		2898.0	I			1.3–5.6	
Boron	B	2496.8	I	0.002		0.0–4.9	8.26
		2497.7	I			0.0–4.9	
Cadmium	Cd	2288.0	I	0.003	0.0001	0.0–5.4	8.96
		3261.1	I			0.0–3.8	
Calcium	Ca	4226.7	I	0.001		0.0–2.9	6.09
		3181.3	II			3.2–7.0	
Cerium	Ce	3201.7	II	0.05		0.9–4.7	6.54
		4222.6	II			0.4–3.3	

Chromium	Cr	2843.3	II	0.0005		1.5–5.9	6.74
		4274.8	I			0.0–2.9	
Cobalt	Co	3453.5	I	0.0002		0.4–4.0	7.81
		3449.2	I			0.6–4.2	
Copper	Cu	3274.0	I	0.0005		0.0–3.8	7.68
		2824.4	I			1.4–6.8	
Gallium	Ga	2943.6	I	0.0002		0.1–4.3	5.97
Germanium	Ge	3039.1	I	0.001	0.00001	0.9–4.9	8.09
		2651.2	I			0.1–4.8	
Gold	Au	2676.0	I	0.0003	0.00004	0.0–4.6	9.18
		2428.0	I			0.0–5.1	
Hafnium	Hf	3134.7	II	0.01		0.4–4.3	
		2866.4	I			0.0–4.3	
Indium	In	3039.4	I	0.001	0.00002	0.0–4.1	5.76
		3256.1	I			0.3–3.0	
Iron	Fe	3020.6	I	0.001		0.0–4.1	7.83
		2844.0	I			1.0–5.4	
		2599.4	II			0.0–4.8	
Lanthanum	La	3337.5	II	0.003		0.4–4.1	5.6
Lead	Pb	2833.1	I	0.0003	0.00005	0.0–4.4	7.38
		2663.2	I			1.3–6.0	
Lithium	Li	4602.9[b]	I	0.02		1.8–4.5	5.36
		3232.6	I			0.0–3.8	
Magnesium	Mg	2852.1	I	0.001		0.0–4.3	7.61
		2776.7	I			2.7–7.2	
Manganese	Mn	2798.3	I	0.001		0.0–4.4	7.41
		3070.3	I			2.2–6.2	
Mercury	Hg	2536.5	I	0.1	0.01	0.0–4.9	10.39
Molybdenum	Mo	3170.4	I	0.0005		0.0–3.9	7.35
		3194.0	I			0.0–3.9	

Table V (*continued*)

Element	Symbol	Wavelength	Spectrum	Detection limits for various elements, %		Excitation potential, V	First ionization potential, V
				Normal work	Special procedure		
Nickel	Ni	3050.8	I	0.0003		0.0–4.1	7.61
		3414.8	I			0.0–3.6	
Niobium	Nb	3195.0	II	0.005		0.3–4.2	
		3094.2	II			0.5–4.5	
Palladium	Pd	3404.6	I	0.001	0.0001	0.8–4.4	8.3
		3421.2	I			1.0–4.6	
Phosphorus	P[c]	2554.9	I	0.03		2.3–7.1	10.9
Platinum	Pt	3064.7	I	0.001		0.0–4.0	8.88
		2659.5	I			0.0–4.6	
Potassium	K	4044.1	I	0.5	0.1	0.0–3.0	4.32
		3446.7	I			–3.6	
Rhenium	Re	3460.5	I	0.005[d]	0.0001[d]	0.0–3.6	
		3464.7	I			0.0–3.6	
Scandium	Sc	3353.7	II	0.0005		0.3–4.0	6.7
Silicon	Si	2881.6	I	0.001		0.8–5.1	8.12
		2568.6	I			–6.7	
		2435.2	I			–5.9	
Silver	Ag	3280.7[e]	I	0.00002		0.0–3.8	7.54
		3382.9	I			0.0–3.6	
Sodium	Na	3302.3	I	0.01		0.0–3.8	5.12
		3303.0	I			0.0–3.8	

Strontium	Sr	4607.3	I	0.001		0.0–2.7	5.67
		3464.5	II			3.0–6.6	
Tantalum	Ta	3311.2	I	0.3		0.7–4.4	
		2685.2	II			0.5–5.1	
Tellurium	Te	2385.8	I	0.02	0.0006	0.6–5.8	8.96
Thallium	Tl	2767.9	I	0.01	0.0001	0.0–4.5	6.07
Thorium	Th	2870.4	II	0.05		0.2–4.6	
		2837.3	II				
		3325.1	II			0.5–4.2	
Tin	Sn	3175.1	I	0.001	0.0001[f]	0.4–4.3	7.30
		2840.0	I			0.4–4.8	
Titanium	Ti	3168.5	II	0.001		0.2–4.1	6.81
		3372.8	II			0.0–3.7	
		3349.4	II			0.0–3.7	
Tungsten	W	2947.0	I	0.005	0.0001[g]	0.4–4.6	8.1
		2896.4	I			0.4–4.6	
Uranium	U	4244.4	II	0.05		0.0–2.9	
		2860.5[h]					
		2882.6					
		2882.7					
		2882.9	II				
		3270.1	II			0.0–3.8	
Vanadium	V	3184.0	I	0.001		0.0–3.9	6.71
		3185.4	I			0.1–3.9	
Ytterbium	Yb	3464.4	I	0.001		0.0–3.3	7.1
		3289.4	II			0.0–3.8	
Yttrium	Y	3327.9	II	0.002		0.4–4.1	6.5
		3242.3	II			0.2–4.0	

Table V (*continued*)

Element	Symbol	Wavelength	Spectrum	Detection limits for various elements, %		Excitation potential, V	First ionization potential, V
				Normal work	Special procedure		
Zinc	Zn	3345.0	I	0.005	0.001	4.1–7.8	9.36
		3345.6	I			4.1–7.8	
Zirconium	Zr	3273.1	II	0.002		0.2–4.0	6.92
		3392.0	II			0.2–3.8	

The most sensitive lines of the following elements are below the range of wavelengths ordinarily used:

As	1890.5
Hg	1849.7
P	1774.9
S	1807.4
Sb	2068.4
Se	1960.9
Zn	2138.6

[a] Actually two lines of about equal intensity 0.09 Å apart.
[b] Iron interference is serious on Li 4602.9.
[c] Molecular bandheads at 3255.3 and 3270.5 can also be used.
[d] Sensitivities in siliceous matrix only; not in MoS_2 matrix.
[e] Mn 3280.8 may interfere.
[f] Sensitivity in siliceous matrix only.
[g] Sensitivity in siliceous matrix only; interelement effects are severe.
[h] Arsenic interference.
Note: The term "special procedure" in most cases refers to volatile-element techniques; I refers to atom lines, II refers to ion lines.

3.3. Quantitative Analyses

The transition from semiquantitative to quantitative spectrographic techniques is accomplished through emulsion calibration and the addition of one or more internal standards. Their purpose is the compensation of variables that are beyond satisfactory control by the spectrochemist. Compensation is accomplished through the use of an intensity ratio of the analytical line to the internal standard line. Internal standards are added directly to the sample material and are selected to have volatility, excitation properties, and line positions similar to the elements to be determined. The internal standard must be an element not of analytical interest. The choice of internal standards has been well discussed by Ahrens and Taylor.[19] A list of recommended internal standards as related to the volatilization characteristics of many elements is given in Table VI.

Table VI

Volatilization Behavior of 51 Elements and Recommended Internal Standards

Volatile elements			
For the best sensitivity volatile-element methods are required (unmixed powders, low currents, recording of early arcing periods only)			
Group IA	— Li, Na, K	Group VIA	— Te
Group IIIA	— Ga,[a] In, Tl	Group IB	— Cu,[a] Ag,[a] Au
Group IVA	— Ge, Sn, Pb	Group IIB	— Zn, Cd, Hg
Group VA	— P, As, Sb, Bi		
Recommended internal standars[b]: Ge, In, Li (Group IA)			
Medium-volatile and non-volatile elements			
Conventional methods are satisfactory (usually total consumption, carbon admixtures, high currents)			
Group IIA	— Be, Mg, Ca, Sr, Ba	Group VB	— V, Nb, Ta
Group IIIA	— B, Al	Group VIB	— Cr, Mo, W[c]
Group IVA	— Si	Group VIIB	— Mn, Re[c]
Group IIIB	— Sc, Y, La	Group VIII	— Fe, Co, Ni, Pd[c], Pt
Group IVB	— Ti, Zr, Hf	Lanthanides	— Ce, Yb
		Actinides	— Th, U
Recommended internal standards[d]: Pd, Ge, Be (Group IVA)			

[a] This is somewhat more sensitive by volatile-element techniques, but ordinarily determined by nonvolatile element procedures.
[b] Others used occasionally are Na and K.
[c] Sensitivity in siliceous materials can be improved by use of volatile-element techniques.
[d] Others used occasionally are Ca, Fe, Sr, and Ti.

Most of the semiquantitative and quantitative dc-arc methods follow the "total energy" technique reported by Slavin in 1938 in which samples are burned to completion.[42] During this process elements volatilize selectively. This has led to valuable modifications of the technique in which certain groups of elements are favored over others. For example, the detection limits of volatile elements can be lowered by recording only the early excitation periods and by omitting powdered graphite from mixtures with diluents. Graphite suppresses selective volatilization.

When photographic methods are used in quantitative work, calibration of the photographic emulsion becomes necessary. Many methods of calibration have been discussed in the literature.[43,44] Variation of the two-step method is illustrated in Fig. 8. Steps of different densities are produced on the photographic record through the use of filters or rotating step sectors near the slit. The transmissions of two line steps of known relation are measured. This is repeated for several lines of different densities within the spectral region of interest. In this way data are accumulated for plotting the so-called preliminary curve. This curve is utilized in the preparation of the emulsion calibration curve. In a specific example successive readings are taken by starting with a transmission value of one on the ordinate. The corresponding abscissa value is 2.5. If 2.5 is now used as the new ordinate value, a second abscissa value is obtained, and so on. This procedure results in a sequence of numbers as shown in Table VII.

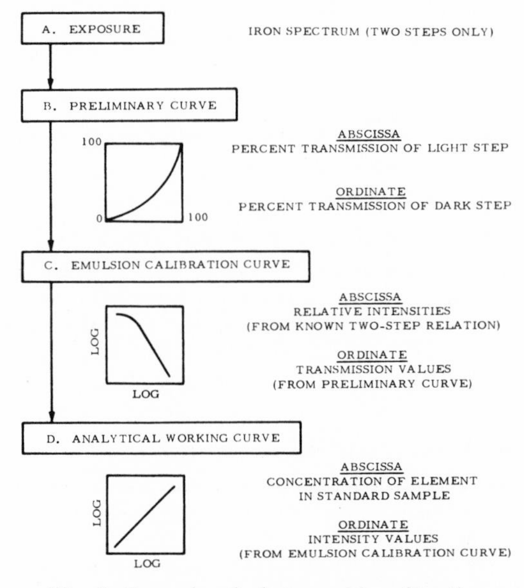

Fig. 8. Example of photographic calibration.

Table VII

Step No.	Relative intensity	Transmission, %
1	1.66	96.4
2	1.66^2	89.0
3	1.66^3	69.3
4	1.66^4	36.7
5	1.66^5	16.0
6	1.66^6	6.5
7	1.66^7	2.5
8	1.66^8	1.0

These numbers represent transmissions that would be obtained if eight steps were used on one line instead of two steps on a group of lines. Each number is related to the others by the relative transmissions of the two steps. In the table one step is $1.66\times$ darker than the previous step. When the corresponding relative intensity and transmission values are plotted logarithmically against each other, the emulsion-calibration curve results. It permits the analyst to relate photographic response to the intensity of light emitted by a sample. The so-called multistep method is similar, but calibration is done on several steps of a single line. Such methods have the disadvantage of requiring an optical alignment that will evenly illuminate the entire length of the spectral line. Emulsion-calibration curves are influenced by factors such as wavelength, emulsion characteristics and history, photographic processing, exposure time, slit width, and intermittency effects.

The homologous-line method uses pairs of lines of known intensity ratios obtained from the literature. This method does not require spectral lines to be uniformly illuminated throughout their vertical extent. It depends, however, upon intensity values which are not always known correctly. A recent improvement in the two-step method involves a beam splitter which separates the two steps horizontally instead of vertically so that uniform illumination is not necessary.[45] Emulsion calibration has been studied by Parodi and Burch.[46]

4. GEOCHEMICAL APPLICATIONS

Geochemical applications of optical emission spectroscopy are as diverse as the various fields of geochemistry. They can, however, be divided into three principal categories: (a) element abundance and distribution, (b) environmental studies, and (c) mineral exploration.

4.1. Element Abundance and Distribution

Spectrochemistry is well suited for studies of element abundance or distribution and has been used for this purpose for several decades. Early work by Clarke[47] and Clarke and Washington[48] was completed prior to the full development of spectrochemical techniques. It preceded Goldschmidt's[49] significant contributions to both geochemistry and spectrochemistry. More recently Green,[50] Turekian and Wedepohl,[51] and Vinogradov[52,53] have added to the Goldschmidt data. Computer treatment of spectrochemical results has come into widespread practice. Horn and Adams[54] applied spectrochemical abundance data to theoretical studies of the origin of the earth, its continents and sedimentary basins, and smaller areas.

4.2. Environmental Studies

Spectroscopy finds application wherever it is necessary to develop data on the chemical composition of materials. One example of environmental analysis involves the study of trace elements in soils or waters at potential industrial construction sites. Prior to construction it is desirable to establish base levels of environmental conditions so that subsequent changes caused by the industrial activity can be determined. Such studies have legal implications, especially in relation to pollution and health. For example, the construction of a lead–zinc mill or smelter may be unwise in a populated area with preexisting high natural levels of lead in soils and waters. A second category of environmental studies involves the monitoring of trace elements in food supplies, water supplies, and soils where food crops are grown. The important elements in such studies are toxic elements such as arsenic, cadmium, lead, and selenium. Others, such as cobalt, iron, manganese, and zinc, are also of interest. Contributions by Warren *et al.*[55] to this field are noteworthy.

4.3. Mineral Exploration

The use of spectrochemistry in mineral exploration is limited only by the ability and imagination of the scientist conducting the survey. Rock-chip samples are frequently taken as character samples for a prospect or for the various rock types in the area under investigation. In detailed work samples are typically taken 250 or 1000 ft apart.

Soil surveys are conducted in a similar way. In this case various horizons are sampled. In a common system of identification these horizons or layers of the well-developed soil profile are labelled A, B, and C. In simple terms, the surface soil, including the organic-rich layer, is called the A horizon. The B horizon is the intermediate layer just below A. The parent material from which the soil was derived is the C horizon. These major designations are further subdivided, sometimes in considerable detail. In general, the organic-rich A horizon, if present, is high in trace elements. The resulting data, however, are often erratic since they depend on localized concentrations of organic materials. When it is well developed the B horizon is used more frequently. Metal enrichment makes it highly suitable for geochemical prospecting. The C horizon is also useful. In comparison with the other horizons it is more uniform and more closely related to the composition of the underlying rock. Its trace composition is frequently lower. This type of work can extend from the collection of a few characteristic samples in each area to detailed grid surveys. A pioneering text on soil and plant analysis was published by Mitchell in 1948. It was most recently updated in 1964.[56]

Surface-water samples are frequently taken from streams or less frequently from lakes or swamps in an effort to relate the trace-metal content of these waters to the terrain from which they have been derived. Special-purpose studies can be made of the runoff from an area during different seasons or under changing conditions. Ground water and formation water (connate water) can be analyzed to determine the source material for sediments in a given formation. Spectrochemical methods are well suited to these applications, especially when information on several elements is desired. If only one or two elements are to be studied simpler methods are often more practical.

Stream-sediment sampling is an extremely important part of geochemical prospecting and is probably used more than any other approach, especially in the early reconnaissance stages of a survey. It is also applied to the follow-up of anomalies found during the reconnaissance stage. When the investigation proceeds to more detailed work, soil sampling or rock-chip sampling would ordinarily be utilized. The proper interpretation of anomalously high metal content in stream sediments is a complicated subject and one that may receive increased emphasis during the next few years. Emission spectroscopy can be used to provide more complete information than is gained by the usual geochemical determination of only one or two elements. A preliminary spectrochemical analysis will produce data which can be followed up by more precise methods for selected elements. Cruft[57] has published on trace-element determinations in soils and stream sediments by an internal-standard spectrographic procedure.

The analysis of drill cores or drill cuttings is very similar to work on rock chips. The analyses may be on composite samples made up, for example, of 50- or 100-ft intervals, or may be on samples selected on the basis of their apparent composition.

In biogeochemical prospecting samples usually analyzed consist of ashed plant material from deep-rooted plants such as mesquite or coniferous trees growing in the area. Recently, Bedrosian *et al.*[5] have reported on the direct-emission spectrographic analysis of trace elements in biological substances. They were able to achieve detection limits of 1 ppm or less on 26 elements without ashing or preconcentration. Prior to the analysis the samples were merely dried, mixed, and pelletized.

In an unusual application of spectrochemistry Hendricks *et al.*[59] have applied spectrographic techniques in conjunction with wet chemical methods to the investigation of metalrich oozes occurring in closed deeps in the median trough of the Red Sea. Young[60] has reported spectrographic data on cores from the Pacific Ocean and the Gulf of Mexico. Degens *et al.*[61] have reported criteria for differentiating marine and fresh-water shales based on spectrochemical data. Their results indicate that in the region studied, marine and fresh-water shales can be differentiated by means of quantitative spectrographic determinations of boron, rubidium, and gallium. Joensuu[62] has discussed spectrochemical methods in geochemistry in general with emphasis on the investigation of marine sediments. He reported interesting and significant details related to technique, including buffers, internal standards, automated sample presentation, direct readers, double-arc and rotating-disk procedures, the cathode-layer method, etc.

The application of specialized apparatus and techniques to geochemical problems has been reported in a number of interesting papers. Maxwell[63] and Snetsinger and Keil[64] discussed the use of lasers in mineral analysis, Cruft and Giles[65] the use of the direct reader as a geochemical tool, and Shaw[66] and Joensuu and Suhr[67] the Stallwood jet in silicate and general rock, mineral and related work, respectively.

5. LITERATURE

One or more textbooks and reference volumes are necessary in any geochemical laboratory with optical emission equipment. Among the more prominent publications are *Spectrochemical Analysis* by Ahrens and Taylor,[19] *Spectrochemical Procedures* and *Semiquantitative Spectrochemistry* by Harvey,[68,69] *Principles and Practice of Spectrochemical Analysis* by Nachtrieb,[70] and, with respect to spectroscopy in general, *Practical Spectroscopy* by Harrison, Lord, and Loofbourow[2]. Of equal importance are Harrison's *M.I.T. Wavelength Tables*,[17] Moore's[71] publications of atomic energy levels, and tables of spectrum lines by Zaider *et al.*[72] The theory of spectrochemical excitation is discussed in depth by Boumans.[73]

Among periodicals three distinct groups stand out. The first one is analytically oriented; it includes *Analytical Chemistry*, *Applied Spectroscopy*, and *Spectrochimica Acta, Part B, Atomic Spectroscopy*. A second group covers geological and mineralogical publications. These journals occasionally contain papers based on spectrochemical data or articles written to cover specific techniques. Among these publications are *Economic Geology*, *Canadian Mineralogist*, and *Geochimica et Cosmochimica Acta*. Related to this group are periodicals in the field of engineering, mining, and mineral industries. A third group is bibliographic in character. *Analytical Chemistry*, for example, carries biennial analytical reviews. In its April 1968 review on emission spectrometry 676 references were cited. In comparison, the April 1966 issue contained 351 references. *Spectrochemical Abstracts* has been published by Hilger and Watts for the years 1933 to 1951. The American Society for Testing Materials in its *Index to the Literature on Spectrochemical Analysis* (four parts) covered the period 1920 to 1955. In the first year, 1920, five articles were cited. By 1955 this number had risen to 579. Additional volumes of this valuable publication cannot be expected in the very near future, but the biennial review mentioned earlier should help to bridge the gap from 1955 to the present. *Chemical Titles*, published semimonthly by the American Chemical Society in a key word arrangement, allows rapid selection of significant articles from the current literature. *Chemical Abstracts* furnishes more detailed information on a delayed basis.

Among government publications related to spectrochemistry are monographs by the U.S. Bureau of Standards and bulletins by the U.S. Geological Survey.[74,75]

Other valuable sources of information include programs of scientific meetings and conferences which quite often contain abstracts of all papers presented, and literature obtainable from instrument manufacturers.

6. EVALUATION OF THE METHOD

As evidenced by its widespread use and multiplicity of applications, optical emission spectroscopy is a powerful analytical technique. In the writers' laboratories the spectrograph was found nearly indispensable in a particular comprehensive geochemical program. The technique has the advantage of rapidity and simplicity when large numbers of determinations are to be made on one sample, or when large numbers of samples must be processed. When photographic methods are used, permanent records of analyses become available for later checking or for the further determination of additional elements. Sample sizes are small. A reasonably complete semiquantitative analysis for 30–40 elements can be done rapidly on a 1-mg sample using routine methods. Even smaller samples can be handled by special procedures. High absolute sensitivities and low detection limits can be achieved by this technique.

One of the most significant advantages of optical emission work in geochemical applications is the certainty with which elemental identifications and estimates of abundance can be made. The possibility of spectral interference in elemental identification is often stated as a disadvantage of the method relative to chemical methods or other techniques. There are many possibilities of spectral interference, but in actual practice there is no danger of incorrect identification at the levels normally encountered if the work is performed carefully. For example, the identification of 10 ppm beryllium or silver in a rock sample is a completely unambiguous determination. The examination of several spectral lines of each element eliminates any possibility that an identification may be incorrect because of the coincidence of lines.

On the negative side, spectrochemistry is a destructive process. The sample is not available for further investigation. This, however, may not be important since the amount of sample used is small, and often large amounts of sample are kept in storage.

Since spectrochemical equipment is available throughout a wide price range, the question of high-vs-moderate cost may arise. The main advantage of the more expensive instrumentation lies in convenience and speed. With regard to "good results" the skill and patience of the operator are generally of greater importance than the price of the unit. The equipment and techniques are complex and do require trained personnel, especially with respect to method development and spectrum interpretation.

Examples of the analytical precision obtained in routine semiquantitative work are presented in Table VIII. These data stem from intralaboratory analyses performed at minimal two month intervals and reflect a procedure such as outlined in Table IV.

7. FUTURE DEVELOPMENTS

In spite of the relative age of spectrochemistry in comparison with other instrumental techniques, there is today considerable room and need for progress. Some areas of potential development are pointed out below.

TABLE VIII

Examples of Long-Term Reproducibility by a Semiquantitative DC Arc Procedure (in Percent)

Element	Number of analyses	Range reported	Mean	Standard deviation	Coefficient of variation
Aluminum	6	5–8.6	6.3	1.40	22
	8	5–9	6.6	1.3	20
Barium	6	0.08–0.20	0.14	0.042	30
Beryllium	6	0.0003–0.0004	0.00032	0.000041	13
Cobalt	6	0.001–0.004	0.0024	0.0012	50
Copper	8	0.06–0.10	0.079	0.013	16
Chromium	6	0.004–0.020	0.010	0.0057	57
Gallium	6	0.001–0.002	0.0018	0.0004	22
Lanthanum	4	0.010–0.015	0.011	0.0022	20
Lead	5	0.0028–0.0050	0.0038	0.0009	24
	8	0.0007–0.0010	0.00086	0.00011	13
Lithium	3	0.0010–0.0025	0.0018	0.00076	42
Magnesium	6	1.0–3.0	1.8	0.72	40
Manganese	5	0.008–0.03	0.014	0.008	57
Molybdenum	5	0.02–0.04	0.029	0.007	24
	9	0.0010–0.0020	0.0015	0.00025	17
Nickel	6	0.0023–0.008	0.0052	0.0022	42
Niobium	4	0.001–0.002	0.0012	0.0005	42
Silicon	6	26–34	31	2.7	9
Silver	6	0.00023–0.0006	0.00043	0.00016	37
Sodium	6	1.5–2.0	1.8	0.19	11
Tin	6	0.0002–0.003	0.0012	0.0010	83
Titanium	6	0.21–0.60	0.38	0.13	34
Vanadium	5	0.009–0.020	0.014	0.0056	40
Yttrium	6	0.002–0.003	0.0023	0.0005	22
Zinc	8	0.04–0.07	0.056	0.013	23
Zirconium	6	0.009–0.03	0.018	0.0078	43

As was mentioned earlier, the value of mobile spectrochemical laboratories has been established. The impetus for their use has come largely from the programs of the U.S. Geological Survey and from foreign efforts, notably in Australia, Canada, and the USSR. In the future such truck-mounted or trailer-mounted facilities will become more common in mineral exploration, also in environmental studies. Current trends are toward the installation of direct-reading spectrometers. Eventually, spectrograph–spectrometer combinations might be installed.

Direct readers interfaced with computers have been introduced into permanent laboratory facilities. Future trends may see these installations refined so that the data output would be in the form of contoured geochemical maps. Matrix effects will be corrected through programmed interelement compensation and arc-temperature measurements. Thompson *et al.*[76] and Decker and Eve[77] have reported on this topic.

A significant development in direct-reader instrumentation was reported by Margoshes.[78] This particular spectrometer–computer system, still in the product-development stage, reportedly can measure 2048 individual wavelengths simultaneously and will print out analytical results in numerical form.

Improvements are being made in the stability of the excitation discharge, especially the dc arc, as reported by Gordon.[79] Servocontrolled discharges combined with better focusing elements, perhaps interferometerically or electronically controlled, can be expected to improve the performance of emission equipment. Such advances may better the precision of the semiquantitative type of multielement analysis by as much as an order of magnitude and could open up new areas of application.

There will still be a need for photographic recording to obtain maximum sensitivity and reliability in the case of unusual element occurrences. The photographic process could be speeded up through the use of self-developing dry-process film, at least for qualitative determinations. An instrument for producing a photographic record simultaneously with direct reading of about 20 common elements would be of considerable interest. A system for computer analysis of photographic optical emission spectra has been reported by Helz *et al.*[80]

Much work remains to be done on systems and techniques for atmospheres other than air and pressures other than atmospheric. The use of controlled static atmospheres as contrasted to flow systems offers a promise of improved sensitivity.

We can expect further miniaturization of equipment, at least of electronic components, as advances continue in this area. The use of laser vaporization in microprobe analysis has already been introduced. Its applications can be expected to expand rapidly in the years ahead.

Many of these anticipated advances can be achieved with existing technology. The workers active in this field must provide the stimulus to bring about these developments within the limits of the economics involved in each application.

REFERENCES

1. Weise, E. K., History and Origin, in *The Encyclopedia of Spectroscopy*, G. L. Clark, ed., Reinhold, New York, 1960.
2. Harrison, G. R., Lord, R. C., and Loofbourow, J. R., *Practical Spectroscopy*, Prentice-Hall, Englewood Cliffs, N. J., 1963.
3. Kirchhoff, G., and Bunsen, R., *Chemical Analysis by Means of Spectral Observations* (in German), *Pogg. Ann. Physik u. Chem.* **110**, 161 (1860).
4. Gerlach, W., "The Correct Execution and Interpretation of Quantitative Spectrum Analysis" (in German), *Z. Anorg. Chem.* **142**, 383 (1925).
5. Goldschmidt, V. M., and Peters, C. (in German), *Nachr. Ges. Wiss. Göttingen, Math. Phys. Klasse*, 165, 257 (1931).
6. Goldschmidt, V. M., and Peters, C. (in German), *Nachr. Ges. Wiss. Göttingen, Math. Phys. Klasse*, **2**, 360, 528 (1931).
7. Goldschmidt, V. M., Berman, H., Hauptmann, H., and Peters, C. (in German), *Nach. Ges. Wiss. Göttingen, Math. Phys. Klasse* **3**, 141, 235, 278, 371 (1933).
8. Goldschmidt, V. M., Hauptmann, H., and Peters, C., Rare Elements in Rock Analysis (in German), *Naturwiss.* **21**, 363 (1933).
9. Goldschmidt, V. M., and Peters, C., The Geochemistry of Arsenic (in German), *Nachr. Ges. Wiss. Göttingen, Math. Phys. Klasse, N. F. Fachgr. IV* **1**, 11 (1934).
10. Goldschmidt, V. M., Bauer, H., and Witte, H., Geochemistry of the Alkali Metals. II. (in German), *Nachr. Ges. Wiss. Göttingen, Math. Phys. Klasse, N. F. Fachgr. IV* **1**, 39 (1935).
11. Fassel, V., *History of Flame Emission Spectrochemical Methods*, paper presented at the Seventh National Meeting, SAS, Chicago, Ill., May 14, 1968.
12. Walters, J. P., *Thirty Years of Emission Spectrochemical Development. The Spark Discharge*, paper presented at the Seventh National Meeting, SAS, Chicago, Ill., May 14, 1968.
13. Strock, L. W., *Thirty Years of Development of D. C. Arc Spectrochemical Analysis Methods*, paper presented at the Seventh National Meeting, SAS, Chicago, Ill., May 14, 1968.
14. Herzberg, G., *Atomic Spectra and Atomic Structure*, Dover, New York, 1944.
15. Richtmyer, F. K., Kennard, E. H., and Lauritsen, T., *Introduction to Modern Physics*, McGraw-Hill, New York, 1955.
16. McNally, J. R., Atomic Spectra, in *Handbook of Physics*, E. U. Condon and H. Odishaw, eds., McGraw-Hill, New York, 1967, pp. 7–38–53.
17. Harrison, G. R., *M.I.T. Wavelength Tables*, John Wiley & Sons, New York, 1939.
18. Grotrian, W., *Graphical Presentation of the Spectra of Atoms and Ions with One, Two, and Three Valence Electrons. I, II*, J. Springer, Berlin, 1928.
19. Ahrens, L. H., and Taylor, S. R., *Spectrochemical Analysis*, 2nd Ed., Addison-Wesley, Reading, Mass., 1961.
20. Guide to Scientific Instruments, *Science*, p. 158A, Nov. 28, 1967.
21. Buyers' Guide, *Industrial Research*, May 15, 1967.
22. Scribner, B. F., and Margoshes, M., Emission Spectroscopy, in *Treatise on Analytical Chemistry*, Part I, I. M. Kolthoff and P. J. Elving, eds., Vol. 6, pp. 3372–3379.
23. Gold Rush, 1967-Style, *Spectrum Scanner*, Jarrell-Ash Company, **22**, 2 (1967).
24. Stallwood, B. J., Air Cooled Electrodes for the Spectrochemical Analysis of Powders, *J. Opt. Soc. Am.* **44**, 171 (1954).

25. Boyd, B. R., and Goldblatt, A., *Atmosphere Excitation in Non-Enclosed Spark Stands*, paper presented at the Pittsburgh Conference on Analytical Chemistry and Applied Spectroscopy, March 1963.
26. Danielsson, A., Lundgren, F., and Sundkvist, G., The Tape Machine, I, II, and III, *Spectrochim. Acta* **1959**, 122.
27. Strock, L. W., *Spectrum Analysis with the Carbon Arc Cathode Layer*, A. Hilger, London, 1936.
28. Barnett, P. R., An Evaluation of Whole-Order, $\frac{1}{2}$ Order, and $\frac{1}{3}$ Order Reporting in Semiquantitative Spectrochemical Analysis, *U.S. Geol. Survey Bull.* **1084**-H (1961).
29. Grimes, D. J., and Marranzino, A. P., *Six-Step Field Standards for Semiquantitative Analysis of Rocks and Soils*, U.S. Geol. Survey, Denver, Colo., paper presented at the 1968 Pittsburgh Conference on Analytical Chemistry and Applied Spectroscopy, March 1968.
30. Decker, R. J., and Eve, D. J., *Appl. Spectros.* **22**, 13, 263 (1968).
31. Leys, J. A., and Dehm, R. L., Semiquantitative Industrial Chemical Analysis, in *The Encyclopedia of Spectroscopy*, G. L. Clark, ed., Reinhold, New York, 1960, pp. 285–286.
32. Adamson, D. L., Stephens, J. D., and Tuddenham, W. M., Application of Mineralogical Principles and Infrared Spectra in Development of Spectrographic Techniques, *Anal. Chem.* **39**, 574 (1967).
33. Designation of Shapes and Sizes of Graphite Electrodes, in *General Test Methods, ASTM Standards Part* 30, ASTM, Philadelphia, Pa. 1967, p. 334.
34. Stevens, R. E., *et al.*, Second Report on a Cooperative Investigation of the Composition of Two Silicate Rocks, *U.S. Geol. Survey Bull.* **1113**, 1 (1960).
35. Report of Nonmetallic Standards Committee of the Canadian Association for Applied Spectroscopy, *Appl. Spectros.* **15**, 159 (1961).
36. Webber, G. R., Second Report of Analytical Data for CAAS Syenite and Sulphide Standards, *Geochim. Cosmochim. Acta* **29**, 229 (1964).
37. Ingamells, C. O., and Suhr, N. H., Chemical and Spectrochemical Analysis of Standard Silicate Samples, *Geochim. Cosmo Chim. Acta* **27**, 897 (1963).
38. Flanagan, F. J., U.S. Geological Survey Silicate Rock Standards, *Geochim. Cosmochim. Acta* **31**, 289 (1966).
39. Meggers, W. F., Corliss, C. H., and Scribner, B. F., *Tables of Spectral-Line Intensities*, NBS Mon. 32, Parts I and II, 1961.
40. Weast, R. C., ed., *Handbook of Chemistry and Physics*, 45th Ed., The Chemical Rubber Co., Cleveland, Ohio, 1964, p. E-41.
41. Quesada, A., and Dennen, W. H., Spectrochemical Determination of Water in Minerals and Rocks, *Appl. Spectros.* **21**, 155 (1967).
42. Slavin, M., Quantitative Analysis Based on Spectral Energy, *Ind. Eng. Chem. Anal. Ed.* **10**, 407 (1938).
43. Churchill, J. R., Techniques of Quantitative Spectrographic Analysis, *Ind. Eng. Chem. Anal. Ed.* **16**, 653 (1944).
44. Photographic Photometry in Spectrochemical Analysis, in *ASTM Methods Chemical Analysis of Metals*, ASTM, Philadelphia, Pa., 1956, pp. 570–593.
45. Feldman, C., *A Beam-Splitter for Use in Calibrating Spectrographic Emulsions*, paper presented at the Seventh National Meeting, SAS, Chicago, Ill., May 16, 1968.

46. Parodi, J. A., and Burch, W. G., Jr., *A Study of Photographic Emulsion Calibration Techniques*, G.E., Hanford Atomic Products Operation, Richland, Wash., HW-28803, July 27, 1953.
47. Clarke, F. W., The Data of Geochemistry, *U.S. Geol. Survey Bull.*, 770 (1924).
48. Clarke, F. W., and Washington, H. S., The Composition of the Earth's Crust, *U.S. Geol. Survey PP 127*, 1924.
49. Goldschmidt, V. M., *Geochemistry*, Clarenden Press, Oxford, 1954.
50. Green, J., Geochemical Table of the Elements for 1959, *Geol. Soc. Am. Bull.* **70**, 1127 (1959).
51. Turekian, K. K., and Wedepohl, K. H., Distribution of the Elements in Some Major Units of the Earth's Crust, *Geol. Soc. Am. Bull.* **72**, 175 (1961).
52. Vinogradov, A. P., The Regularity of Distribution of Chemical Elements in the Earth's Crust, *Geokyhimiya* **1**, 6 (1956) (Engl. Transl. by Atomic Energy Research Establishment, Harwell, Berkshire, 1957).
53. Vinogradov, A. P., *The Geochemistry of Rare and Dispersed Chemical Elements in Soils*, 2nd Ed. (Engl. Transl. from the Russian by Consultants Bureau, New York, 1959).
54. Horn, M. K., and Adams, J. A. S., Computer-Derived Geochemical Balances and Element Abundances, *Geochim. Cosmochim. Acta* **30**, 279 (1966).
55. Warren, H. V., and Delavault, R. E., A History of Biogeochemical Investigation in British Columbia, *Transactions of the Canadian Institute of Mining and Metallurgy* **53**, 236 (1950).
56. Mitchell, R. L., *The Spectrographic Analysis of Soils, Plants, and Related Materials*, Commonwealth Bur. Soil Sci. Tec. Common. No. 44, 1948; reprinted with addendum, 1964.
57. Cruft, E. F., Trace Element Determinations in Soils and Stream Sediments by an Internal Standard Spectrographic Procedure, *Econ. Geol.* **59**, 458 (1964).
58. Bedrosian, A. J., Skogerboe, R. K., and Morrison, G. H., Direct Emission Spectrographic Method for Trace Elements in Biological Materials, *Anal. Chem.* **40**, 854 (1968).
59. Hendricks, R. L., Reisbick, F. B., Mahaffey, E. G., Roberts, D. B., and Peterson, M. N. A., Chemical Composition of Sediments and Intersticial Brines from the Atlantis II Discovery and Chain Deeps, in *Hot Brines and Recent Heavy Metal Deposits in the Red Sea*, E. T. Degens and D. A. Ross, eds., Springer Verlag, New York, in press.
60. Young, E. J., Spectrographic Data on Cores from the Pacific Ocean and Gulf of Mexico, *Geochim. Cosmochim. Acta* **32**, 466 (1968).
61. Degens, E. T., Williams, E. G., and Keith, M. L., Environmental Studies of Carboniferous Sediments, I. Geochemical Criteria for Differentiating Marine and Freshwater Shales, *Bull. Am. Assoc. Petrol. Geologists* **41**, 2427 (1957).
62. Joensuu, O. I., *Spectrochemical Methods in Geochemistry*, paper presented at the Seventh National Meeting, SAS, Chicago, Ill., May 15, 1968.
63. Maxwell, J. A., The Laser as a Tool in Mineral Identification, *Can. Mineral.* **5**, 727 (1963).
64. Snetsinger, K. G., and Keil, K., Microspectrochemical Analysis of Minerals with the Laser Microprobe, *Amer. Mineral.* **52**, 1842 (1967).
65. Cruft, E. F., and Giles, D. L., Direct Reading Emission Spectrometry as a Geochemical Tool, *Econ. Geol.* **62**, 406 (1967).

66. Shaw, D. M., Spectrochemical Analysis of Silicates Using the Stallwood Jet, *Can. Mineral.* **6**, 467 (1960).
67. Joensuu, O. I., and Suhr, N. A., Spectrochemical Analysis of Rocks, Minerals, and Related Materials, *Appl. Spectros.* **14**, 101 (1962).
68. Harvey, C. E., *Spectrochemical Procedures*, Applied Research Laboratories, Glendale, Calif., 1950.
69. Harvey, C. E., *Semiquantitative Spectrochemistry*, Applied Research Laboratories, Glendale, Calif., 1964.
70. Nachtrieb, N. H., *Principles and Practice of Spectrochemical Analysis*, McGraw-Hill, New York, 1950.
71. Moore, C. E., Atomic Energy Levels, *U. S. Bur. Standards Circ. 467*, Vol. I, 1949, Vol. II, 1952.
72. Zaidel, A. N., Prokof'ev, V. K., and Raiskll, S. M., *Tables of Spectrum Lines*, Pergamon Press, New York, 1961.
73. Boumans, P. W. J. M., *Theory of Spectrochemical Excitation*, Plenum Press, New York, 1966.
74. Bastron, H., Barnett, P. R., and Murata, K. K., Method for the Quantitative Spectrochemical Analysis of Rocks, Minerals, Ores, and Other Materials by a Powder dc Arc Technique, *U. S. Geol. Survey Bull.* **1084**-G (1960).
75. Myers, A. T., Havens, R. G., and Dunton, P. J., A Spectrochemical Method for the Semiquantitative Analysis of Rocks, Minerals and Ores, *U. S. Geol. Survey Bull.* **1084**-I (1961).
76. Thompson, G., Paine, K., and Manheim, F., A Flexible Computer Program for Evaluation of Emission Spectrometric Data, *Appl. Spectros.* **23**, 264 (1969).
77. Decker, R. S., and Eve, D. J., dc Arc in Emission Spectrography, IV. Correction for Matrix Effects, *Appl. Spectros.* **23**, 497 (1969).
78. Margoshes, M., *The Digilab* 2048-*Channel TVS Spectrometer*, paper presented at the Pittsburgh Conference on Analytical Chemistry and Applied Spectroscopy, March 1970.
79. Gordon, W. A., *Improvement of Analytical Precision Using a Servocontrolled dc Arc with Current Feedback*, paper presented at the Fifth National Meeting, SAS, Chicago, Ill., June 1966.
80. Helz, A. W., Walthall, F. G., and Berman, S., Computer Analysis of Photographed Optical Emission Spectra, *Appl. Spectros.* **23**, 508 (1969).

Chapter 8

ATOMIC ABSORPTION

L. R. P. Butler and M. L. Kokot

National Physical Research Laboratory
Pretoria, South Africa

1. INTRODUCTION

The relatively new method of atomic absorption spectrometry[1] has efficiently filled a gap that has existed for many years, by providing a simple, inexpensive technique whereby a wide range of elements in many different types of materials may be determined accurately and precisely. The large number of atomic absorption instruments now in use in geochemical and other laboratories bears testimony to the usefulness and popularity of the technique.

Many papers have been published describing applications of atomic absorption in geochemistry. These include the analysis of rocks,[2–4,22] ores,[5] waters,[6] brines,[7] etc. Several books have also been published on atomic absorption,[8–10] although only one[11] describes its use exclusively for the analysis of geologic materials.

The purpose of this chapter is to describe the method and to give a brief outline of its use for the analysis of silicates. It is not intended to give a general literature survey. The reader is referred to the papers and books mentioned for further and more detailed references. In particular the articles and bibliographies which appear in the *Atomic Absorption Newsletter* published by Perkin-Elmer (Norwalk, Conn., USA) provide an invaluable source of information.

Because of the high sensitivity of atomic absorption spectrometry, it has been applied in the majority of cases for the determination of minor and trace elements, but it has been shown that the technique is also adequately precise and accurate for the determination of major elements.

The requirement that the sample must be brought into solution is sometimes time consuming, but sample dissolution of silicate-base materials is generally not considered difficult. Inherent interference effects present with the determination of certain elements, notably some of the alkaline earths, and other elements that form refractory oxides, may create difficulties. The suggestion of using the nitrous oxide–acetylene flame[12] and a better understanding of the processes taking place in flames has enabled many of these difficulties to be overcome.

Atomic absorption techniques today offer very distinct advantages in respect of capital outlay, simplicity, accuracy, precision, speed, sensitivity, and cost of analysis. Most of the instruments offered by manufacturers, especially those in the medium and upper cost brackets, are reliable and well suited for the application to geochemical problems.

2. PRINCIPLES OF ATOMIC ABSORPTION

Atomic absorption depends on the phenomenon whereby atoms of an element are able to absorb electromagnetic radiation. This occurs when the atoms are unionized and unbonded to other or similar atoms. The wavelength of light at which atomic absorption proceeds is specific for the type of atom and occurs at the resonant frequencies. These correspond to the electronic transitions between the first allowable energy level and the ground state of the atoms. Atomic absorption also occurs for some elements with spectral lines other than those originating in the ground state, but the absorption probability, and thus the analytical sensitivity, is correspondingly lower. The reader who is interested in the theoretical explanation of the phenomena of atomic emission and absorption is referred to the literature dealing with spectral theory.[1,13,14]

A schematic diagram of the essential components of an atomic absorption instrument is given in Fig. 1.

In practice the sample solution is vaporized by means of a nebulizer and the aerosol is passed into a flame. The temperature of the flame is sufficiently high to further reduce the droplets by dehydration and dissociation into an atomic form so that the analytical elements in the vapor exist mostly, and preferably, as neutral unbound atoms in the ground state. If the flame temperature is high atoms may be raised to excited or even to ionized states. If this happens the population density of the ground state is decreased and the lower limit of detection is accordingly decreased.

The atomic vapor is illuminated by a light source radiating the char-

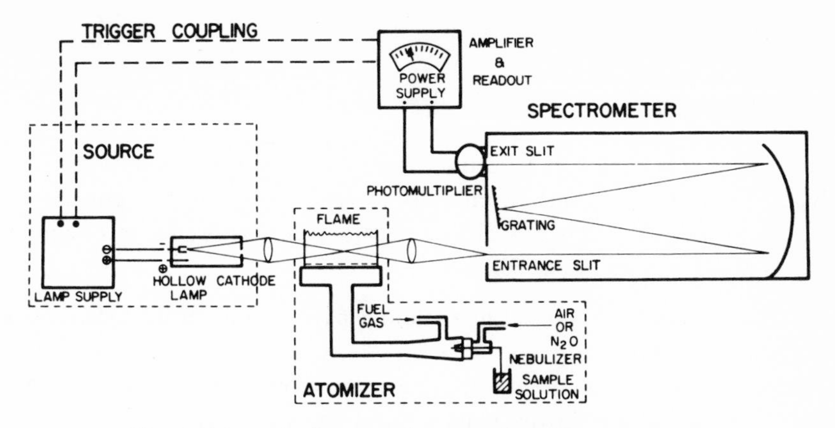

Fig. 1. Schematic diagram of atomic absorption apparatus.

acteristic light of the analytical element. A spectrometer is usually used to select the resonance spectral line of the element and to measure its intensity. When a solution is sprayed the atoms of the analytical element present absorb a portion of the light at the selected wavelength; the decrease in the intensity of the analytical line is measured by the spectrometer.

The usual procedure for determining the relationship between absorption and atomic concentration in the solution is by measuring the absorbance of a number of solutions containing known concentrations of the analytical element and then to draw an analytical graph or working curve by plotting absorbance against concentration. This graph is linear at low concentrations but may deviate from linearity at higher concentrations. It is preferable to measure absorbance rather than absorption. Atomic absorption follows Beer's law namely,

$$\frac{I}{I_0} = \log KCl$$

where I_0 and I are the intensities before and after passing through the flame, respectively, K is the absorption coefficient, C is the concentration, and l is the flame path length. If absorbance i.e., $\log(I/I_0)$, is plotted against concentration on linear paper, the graph is nearly linear, but if absorption I/I_0 is plotted, the analytical graph is curved.

There are a number of different steps in the procedure, each with variables which can contribute towards the final result. Whenever possible these variables are kept constant to improve the precision and accuracy of the method.

3. APPARATUS

3.1. Source

The source should radiate the characteristic spectral lines of the analytical element with sufficient intensity to enable the lines to be measured with ease on a relatively low-dispersion monochromator. Further requirements of the source are as follows.

(a) The emitted spectral lines should have narrow profiles, free from self-absorption.
(b) The spectrum should be pure with no interfering lines from other elements within the band pass of the spectrometer.
(c) The emission of light should be constant.
(d) The lifetime of the lamp should be long and its cost should be low.

The most successful source for atomic absorption has been the sealed-off hollow-cathode lamp. These lamps, manufactured for a wide variety of elements, are obtainable at reasonable prices. Other sources which have been used with some success for atomic absorption purposes are: electrodeless discharge tubes,[15,16,30] vapor-discharge lamps,[17] continuous emission sources,[18] flames,[19] and dc arcs.[20] Although a considerable amount of research is being conducted to develop new and more efficient sources, it is likely that the hollow-cathode lamp will continue to be used in atomic absorption instruments for analytical purposes for some time to come. In many instances lamps radiating the characteristic light of more than one element may be obtained.

Hollow-cathode lamps should be used according to manufacturers' instructions and the quoted maximum current should not be exceeded. The lifetimes of these lamps have been extended considerably in recent years by applying modern manufacturing techniques and it is not unusual to find lamps operating for considerably longer than 1000 h. The main reason for lamp failure is "cleanup" of the carrier gas. By this process sputtered particles from the metallic cathode trap gas atoms and thus reduce the pressure. Other reasons for lamp failure are out-gassing of metal components in the lamp and diffusion of molecular gases into the tubes. Most lamps are provided with "getters" and may be reconditioned by heating the getters. If the getter has been placed on the anode it may be heated by reversing the current through the lamp. This effectively extends the lifetime of the lamp.

Where vapor-discharge lamps are used as sources it is important that the current through the lamp should be reduced to increase sensitivity. The sources used for atomic absorption spectrometry have been exhaustively discussed in several books and publications.[8,21,45]

3.2. Absorption Flame

The sensitivity, precision, and accuracy of atomic absorption depends to a large extent on the type of flame used. This is especially important to the geochemist who is essentially working with solutions containing large quantities of various metal salts, as interferences are therefore most likely to be encountered.

The premixed acetylene–air flame is used for elements that are easily atomized, while the acetylene–nitrous oxide flame is used for refractory elements or those prone to interferences. By using a slit-type burner the flame can be made long and thin, giving the long path length suitable for atomic absorption. The sensitivity of the method depends on the number of atoms in the flame as well as the path length through which the light must pass. Most commercial instruments supply premixed, laminar-flow burners. With the latter the premixed gases issue from the burner orifice in a nonturbulent stream. The manner in which the flame burns plays a role in the precision with which absorption measurements can be made. When the ratio of fuel gas to oxydizing gas is high, the flame is termed fuel rich and is "softer" and burns at a lower velocity. Flames of this type are more easily affected by movement of the ambient air and may be less stable than more oxydizing flames which are "stiffer" and burn at a higher velocity.

Forced-diffusion burner–nebulizers, often called total consumption burners, are sometimes used for viscous solutions. These burner–nebulizers were actually developed for flame-emission spectrometry. Their short path length and wide spectrum of particle sizes make them unsuited for atomic absorption where high sensitivity and accuracy are required, as in the case of geochemical analyses.

The fuel gases used for the premixed burner are very important as they play an important role in controlling the temperature and the combustion properties of the flame. The gas mixtures which have been found most suitable are summarized in Table I, together with their maximum measured temperatures. For geochemical analyses the most popular gas mixtures are acetylene–air, acetylene–nitrous oxide, and in cases where selenium and arsenic are to be determined, hydrogen–air or hydrogen–argon with diffused air maintaining combustion. The latter flames are more transparent to UV

Table I

Maximum Temperatures of Atomic Absorption Flames[a]

Flame type	Maximum temperature, °C	Height above burner edge, mm
1. Acetylene–air, fuel-rich	2060	8–12
Acetylene–air, stoichiometric	2150	2.0
Acetylene–air, lean	2145	0.5
2. Propane–butane–nitrous oxide, rich	2480	10–14
Propane–butane–nitrous oxide, stoichiometric	2550	4
Propane–butane–nitrous oxide, lean	2535	0.5
3. Acetylene–nitrous oxide, stoichiometric	2795	4
Acetylene–nitrous oxide, lean	2740	0.5

[a] As measured by Butler and Fulton.[23]

light in the Schumann or vacuum spectral region of wavelengths shorter than 2000 Å.

An important part of the burner is the nebulizer. The nebulizer must convert the sample solution into a cloud of small droplets, the finest of which are conducted into the flame. There are many different types of nebulizers, but the most popular is the concentric pneumatic type. It has two anuli concentric about a common axis. Air, or the support gas, forced through the outer anulus creates sufficient suction in the central tube to lift a solution from the reservoir. The solution is drawn through the fine central tube and is broken up by the action of the fast-flowing gas. It is important that the droplets which are formed are small; particles less than 5 μ in diameter are preferred. The size of the droplets reaching the flame significantly influences the measurements. If a large spectrum of droplet sizes reaches the flame the process of atomization may be erratic because of the time required by the droplets to reach the final stages of dissociation. Not only will this cause absorption readings to fluctuate, but chemical interference effects (see Section 7.1) may be significantly greater. Larger droplets are prevented from reaching the flame by using spray chambers which throw them onto the walls, or by using spoiler tubes or glass bulbs to cause them to coalesce. Droplets smaller than 15 μ do not coalesce easily and may reach the flame without adhering to the walls of the spray chamber of the burner.

The greater the amount of solution that is fed into the flame, and the larger the number of atoms present, the better the sensitivity will be. Unfortunately, there is a limit to the amount of aerosol that can be fed into the flame, as too much aerosol may cool or even quench it.

A well-designed nebulizer will have a sample uptake of 1–3 ml per min and an efficiency, i.e., percentage of droplets less than 15 μ, of between 8 and 12% using water at 20°C as the test solution. The efficiency of a nebulizer may be increased by spraying organic solvents.[24] Well-designed nebulizer systems do not, however, give a very large increase in efficiency when organic solvents are sprayed.

3.3. Spectrometer

The role of the spectrometer is to separate the resonant spectral line from other lines being radiated by the lamp source. If as many elements as possible are to be determined the monochromator should transmit light from about 1900 to 10,000 Å. The monochromator's dispersion and resolution should be sufficient to separate the resonant line from other lines which do not show atomic absorption characteristics. In some cases, such as nickel, for example, where this is not possible with the low-dispersion monochromators, the sensitivity may be limited. If light sources such as the high-intensity lamp which boosts the intensity of resonance lines are used, this difficulty may be overcome.[25]

There are two dispersive media used in monochromators, namely, the prism and the grating. The former gives approximately uniform *intensity* across the spectrum, but the dispersion and resolving power change with wavelength. The grating gives (nearly) *linear dispersion* of the spectrum, but the intensity of reflectance varies with wavelength.

Prisms, with their poor dispersion at longer wavelengths, may give poorer sensitivity for elements having their resonance lines in the visible portions of the spectrum. The sodium doublet, for example, at 5890 and 5895 Å should be resolved for maximum sensitivity.

Modern gratings have what is known as "broad-blaze" angles, indicating that their reflectance is more uniform over a wider spectral range. Instruments with gratings may give better sensitivities for some elements with resonance lines at longer wavelengths. In some cases, however, sensitivity is lost, because higher lamp currents must be used to obtain sufficient spectral intensity for measurement.

It is imperative that the monochromator has reproducible slit openings. Because wavelength setting is critical for atomic absorption work, the wavelength adjustment should be precise and free from backlash. A coarse and fine adjustment is desirable, the coarse setting being used for rapid changeover from one element's wavelength to another and the fine setting for adjusting the wavelength critically so that the peak of the resonance line is measured. If only one setting is provided a compromise must be made.

3.4. Detector

It is desirable to use low source-lamp currents to reduce source self-absorption for maximum analytical sensitivity. Although the resonance lines are usually the most intense at currents as low as 5–15 mA, the actual intensity of these lines is very low, and it is desirable to use highly sensitive detectors. Photomultiplier tubes are extremely well suited for this purpose. The sensitivity of most light detectors, including photomultipliers, depends largely on wavelength. For the spectral range used in atomic absorption spectroscopy a single photomultiplier that is selected to give low noise is often sufficient. With a monochromator having a narrow-blazed grating it may be necessary to have available a second photomultiplier tube, sensitive to the wavelength region which is poorly reflected by the grating.

3.5. Amplifier and Measuring System

Photomultipliers seldom have sufficient photocurrent to enable direct measurement; therefore an electronic amplifier is required. The system proposed by Russell, Shelton, and Walsh[17] makes use of an ac amplifier, able to amplify only a very specific frequency. By introducing a mechanical modulator between the source and the flame or by switching the lamp (see Fig. 1) the light is modulated to the frequency to which the amplifier is tuned. This system has the advantage that any light emitted by the flame, at the wavelength of the resonance line, will be continuous, or at least not at the frequency of the amplifier. This light will give rise to a dc signal in the detector and will thus not be amplified. Many modern electronic systems make use of "lock-in" amplifiers which have a very narrow frequency acceptance and very low noise factor.

Some instruments make use of a double-beam system where the light from the hollow-cathode lamp is cyclically deflected around and through the flame. An advantage here is that fluctuations caused by the light source are reduced.[26]

Readout may be by a direct-indicating meter or galvanometer, or by digital voltmeters. It is advantageous to have some means for damping meter fluctuations or averaging results, as with some types of analysis fluctuations may be severe.

Scale expansion of the indicator meter is also a decided advantage, especially if very low concentrations are to be determined.

Digital concentration readout systems have proved useful, especially where rapid analyses are a requirement.[27] Small function generators are used to "linearize" calibration curves. This readout is then calibrated to read concentrations directly. This obviates the need for drawing a calibration graph. Function generators are more commonly used with automatic systems, where samples and standards are automatically nebulized on a conveyer system and the results typed out directly in units of concentration.[28]

4. SENSITIVITY, ACCURACY, AND INTERFERENCES

These factors play an important role in an analytical method, and as such must be discussed. As with many spectrochemical techniques a good deal depends on the care with which the instruments are operated and the analyses carried out.

4.1. Sensitivity

Gatehouse and Willis[29] suggested that a concentration giving 1.0% absorption or 0.004 absorbance should be accepted as the limit of detection. This is a very practical quantity, well suited to the measurement methods of atomic absorption, although not according to the accepted practices in spectrochemistry. A concentration giving a signal of intensity $2\times$ that of the statistical fluctuation or standard deviation of the background signal is usually defined as the detection limit. The latter definition is more correct scientifically, but unless a recorder is used to enable the background and signal fluctuations to be recorded it is difficult to determine in atomic absorption under normal working conditions. With a very stable measuring system of low noise, the latter definition will yield a somewhat lower detection limit than what is really the case. Many atomic absorption spectroscopists, therefore, prefer the more practical detection-limit figures according to the 0.004 absorbance criterion.

Sensitivity is defined as the slope of the analytical curve at any point. Most atomic absorption working curves are linear up to absorbance values of about 0.5. Gatehouse and Willis quoted concentrations giving 50% absorption, which, together with their lower limit 1% figures, lead to a good estimate of the sensitivity of atomic absorption for the various elements.

Analytical range is of equal importance to the analyst, although it may vary considerably from one person to another, depending on the requirements. It is felt that the lower realistic analytical range should commence where the signal is clearly distinguishable above background, and correspond to an absorbance reading of 0.01–0.02 A without scale expansion, i.e., about 3× the detection limit.

The upper analytical range figures given in Tables III–XI are based on the criterion that the limit is selected where the graph curves to approximately half its normal linear sensitivity. This varies from element to element. On an average the analytical range for an element by atomic absorption spreads over two orders of magnitude, e.g., 1–100 ppm.

4.2. Accuracy

The accuracy depends largely on the reliability of the standards and on the inherent interference effects. As with other comparative analytical methods great care must be taken to ensure that freshly standardized reference materials and standards are used. Should interference effects be present a standard material should be selected to ensure the same interferences as are present in the sample.

The accuracy which may be achieved will also depend on the precision with which measurements can be made. Atomic absorption results can usually be repeated, within a 95% confidence level, to between 0.5 and 2.5%. When the precision is poorer than this it is usually an indication of either instrumental failure or faulty technique.

4.3. Interferences

Interference effects are the main causes of inaccurate atomic absorption results. Several workers who in the past experienced the severe effects that occur with some elements and element groups made use of analyzed rock standards, which, hopefully, had the same interferences as the samples. This procedure is very useful, providing such standards are available.

The advent of the acetylene–nitrous oxide flame with its high temperature and special chemical properties has changed the situation. Many elements that are prone to severe interferences, such as calcium, may now be determined without difficulty. Interferences which have been recognized may be divided into the following categories: chemical, molecular, and spectral.

4.3.1. Chemical Interferences

Complex chemical reactions occur in the various regions of the flame with the result that a metallic species may appear in them from the aspirated sample solutions.

When a pure aqueous solution containing the analytical element is atomized some of the atoms will become excited, some ionized, and some may become chemically involved with other species or remain bound as molecules. If A is the total possible number of atoms, then

$$A = A' + XA + A^* + [A]$$

where A' are ions, A^* are excited atoms, XA are chemical combinations involving A where X may be either anions or cations, and $[A]$ are free atoms available for atomic absorption.

If an element P is present in the solution, the equilibrium between the various states may also lead to interference.

Enhancement occurs if $[A]$ increases:

(i) $P + XA \rightarrow PX + [A]$ — i.e., the interfering element P reacts preferentially with X to release atoms of element A.

(ii) $P \rightarrow P' + e$
then $A' + P' + e \rightarrow P' + [A]$ — ionization enhancement, but the interfering element has a low ionization potential; equilibrium shifts and the ionized atoms of the analytical element A become neutralized.

(iii) $P + A^* \rightarrow P^* + [A]$ — excitation enhancement. The same effect is achieved if the flame temperature decreases as a result of the combustion properties changing.

Depression will, on the other hand, occur if $[A]$ decreases:

(iv) $P + [A] \rightarrow PA$ — straight chemical combination.

(v) $P + XA + [A] \rightarrow PA + XA$ — chemical replacement.

(vi) $P^* + [A] \rightarrow A^* + P$ — excitation transfer.

(vii) $[A] \rightarrow A' + e$ — high flame temperature causes ionization.

(viii) $[A] \rightarrow A^*$ — excitation depletion of ground-state atoms.

The following methods may be used for overcoming interference.

For depression:

(a) High-temperature flames may be employed to break down molecular bonds. Although A' will be higher, $[A]$ will be constant even if P is present.

(b) Flames may be used with combustion properties that will prevent the formation of those molecules which do not dissociate easily such as refractory oxides, of which Al_2O_3 is a good example.

(c) Add an excess of suppressing agent S which will preferentially react with the interfering element P, i.e., where

$$P + [A] + S \rightarrow PS + [A]$$

and P and S have a higher affinity for each other than P and $[A]$.

For enhancement:

(a) Add an excess of an easily ionized element R so that the ionization potential of R is lower than S. This will effectively flood the flame with free electrons to prevent ionization of the analytical element:

$$R \rightarrow R' + e$$

$$R' + e + A' \rightarrow R' + [A]$$

(b) Combinations of ionization suppression and releasing agents.

The suggested methods for the different elements discussed in Section 5 have been found to be successful.

4.3.2. Molecular Interference

Angino and Billings[11] have reported an interference caused by non-atomic absorption effects. When a complex rock solution is measured against aqueous standards absorption caused by bands of molecular species may occur. This may lead to results that are too high. The effect is worse for elements that have resonance lines at shorter wavelengths or in bandheads.

4.3.3. Spectral Interference

Spectral interference is caused by the light source radiating spectral lines which are not resolved by the monochromator. The lines may be non-resonant lines of the analytical element, such as nickel, or they may originate from the carrier gas of the lamps. Neon and argon are usually used, and although they have relatively "clean" spectra there are cases where their lines are nearly coincident with analytical lines, as in the case of argon (2175 Å) with lead (2170 Å).[8]

When a highly luminous acetylene–nitrous oxide flame is used with a spectrometer having a wide entrance slit the high-intensity continuous radiation may saturate the photomultiplier detector, or cause the first amplifier stage to be overloaded. This can lead to severe fluctuations or erroneous readings. It is wise to use as narrow a slit as possible when a highly luminous flame is used.

Provided the correct procedures as recommended by the instrument manufacturers are followed, and, provided good lamps are used, spectral interference is seldom encountered. It does not necessarily lead to inaccuracies, but rather to loss of sensitivity and poor precision.

Once proper precautions have been taken to prevent interference the analyst may proceed to analyze samples with confidence. These measures include the addition of the necessary buffer or releasing elements and the use of the correct flame.

5. ANALYSIS OF GEOLOGIC SAMPLES BY ATOMIC ABSORPTION

In Chap. 3 a number of methods are given by which rock samples may be brought into solution. A scheme that has been found very useful for atomic absorption is outlined in Fig. 2. The following three sections are devoted to the complete analysis of rocks by atomic absorption.

5.1. Major Elements

The conditions listed below for the determinations of the various major elements may differ for different instruments. Analytical limits quoted are for the final sample solution, under normal conditions, such as the burner at full length, etc. A reduction in sensitivity may be achieved by using less sensitive lines or turning the burner through an angle to present a shorter flame path. This method has been found very useful and is recommended.

5.1.1. Silicon

The development of the acetylene–nitrous oxide flame has enabled silicon and other refractory elements to be determined. Amos and Willis[12] used a nitrogen–oxygen mixture with acetylene, but found the nitrous oxide flame to be superior and safer.

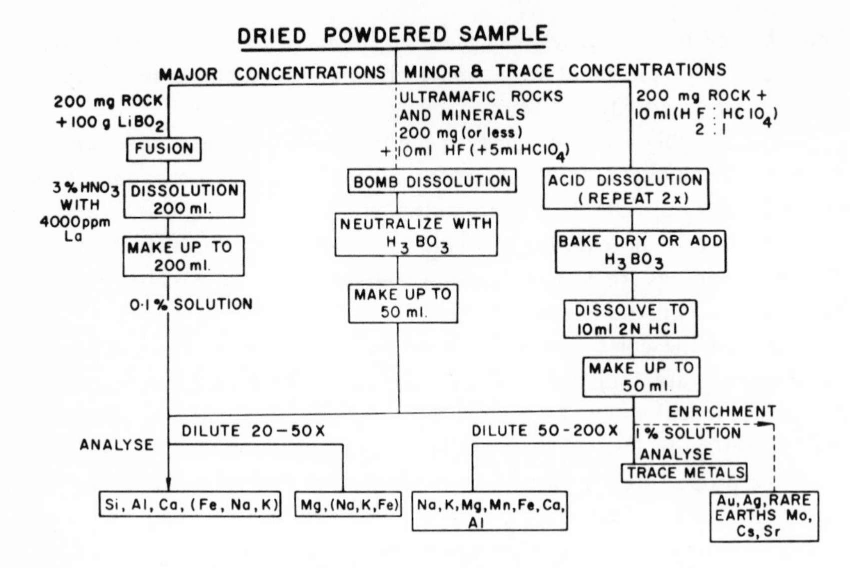

Fig. 2. Scheme for the dissolution of silicate rocks.

Sample dissolution is best done by lithium metaborate or other fusions, and provided the silicon is not too concentrated the solutions will keep for several days. It is advisable, however, to spray solutions as soon as possible after preparation. Standards should be prepared by the fusion of oven-dried silicon dioxide with lithium metaborate, and then standardized chemically.

For low concentrations of silicon, high-pressure techniques (see Chapter 3) may be employed to obtain more concentrated solutions. Standards prepared from waterglass or by dissolving silicon dioxide in hydrofluoric acid (see Table II) are satisfactory.

A fuel-rich acetylene–nitrous oxide flame gives the highest sensitivity. The region of the flame where maximum absorbance occurs is in the "red-feather" region. The authors have noticed that if the burner becomes very hot, a "memory" effect appears, i.e., absorbance values keep changing. The effect is aggrevated if the solution contains a high salt content. Cooling of the burner improves the situation, since partial clogging of the burner slot may result.

Interferences from other major elements in silicate rocks have not been reported to date.

The most sensitive line for silicon is the 2516-Å line. A normal hollow-cathode lamp will provide sufficient intensity with most spectrometers, even if only a narrow entrance slit is used to reduce flame radiation. Table III lists the conditions for silicon. The detection limits and analytical ranges for

Table II

Preparation of Standards

Element	Stock	Procedure	Resultant compound	Stock g/liter	Stock value, ppm
Li	Li_2CO_3	Dry in oven at 110°C; dissolve in dilute HCl[a]	LiCl	1.656	640
	Li metal	Dissolve in excess HCl; evaporate to dryness; weigh out chloride; standardize	LiCl	3.910 (chloride)	640
Na	NaCl[b]	Dissolve in freshly distilled water and store in polyethylene containers	NaCl	1.627	640
K	KCl	Dissolve in distilled water, after drying at 110°C	KCl	1.2210	640
	KNO_3	Ditto	KNO_3	1.655	640
Rb	RbCl	Dissolve in water; standardization is recommended; pipette 25 ml to Pt dish; add 1 drop H_2SO_4, evaporate to dryness, and heat to constant weight	RbCl	0.9055	640
Mg	Mg metal	Dissolve metal in minimum amount appropriate acid[a]	$MgCl_2$ (if HCl)	0.640	640
Ca	$CaCO_3$	Dry at 110°C; dissolve in minimum amount dilute HCl	$CaCl_2$	1.598	640
Sr	$SrCO_3$	Dissolve 1.078 g $SrCO_3$ in 300 ml water plus 2 ml conc. HCl[a] and dilute to volume	$SrCl_2$	1.078	640
Ba	$BaCl_2$	Dissolve 0.9747 g anhydrous $BaCl_2$ dried at 250°C for 2 h; dilute to volume	$BaCl_2$	0.9747	640
	$BaCO_3$	Dry at 110°C; dissolve in dilute HCl[a]	$BaCl_2$	0.9196	640
Al	Al metal	Dissolve in HCl with one small drop of Mg added as catalyst; filter off the mercury after the Al has dissolved	$AlCl_3$ (Al_2Cl_6)	0.640	

Table II (*continued*)

Element	Stock	Procedure	Resultant compound	Stock g/liter	Stock value, ppm
Si	SiO_2	Dissolve in HF + HNO_3 without heating; use Teflon or polypropelene containers; standardize	SiF_4	1.369	640
	SiO_2	Melt with 5 g of $LiBO_2$ at 1000°C or $Na_2B_4O_7$ at 500°C in carbon crucible; pour redhot melt into polypropelene beaker containing 3% HNO_3 solution while stirring (cold); standardize	SiF_4	1.369	640
	Na_2SiO_3	Dissolve in cold water; filter and standardize	Na_2SiO_3	2.7814	640
Cr	$K_2Cr_2O_7$	Dissolve in water, dilute to volume	$K_2Cr_2O_7$	1.4611	640
	Cr metal	Dissolve in HCl, or dilute H_2SO_4[a]	$CrCl_2$	0.640	640
Mo	$(NH_4)_6Mo_7O_{24}$, $4H_2O$	Dissolve in water and add few drops ammonia solution to make alkaline and standardize by gravimetric method	$(NH_4)_6Mo_7O_{24}$	1.1777	640
Mn	MnO_2	Dissolve in concentrated HCl[a]	$MnCl_2$	1.013	640
Fe	Fe metal	Dissolve fresh iron filings (or pure iron wire) in 10 ml 6*N* HCl and dilute to volume	$FeCl_2$	0.640	640
Ni	Ni metal	Dissolve fresh turnings in minimum amount nitric acid and dilute to volume	$Ni(NO_3)_2$	0.640	640
Pd	Pd metal	Dissolve fresh metal filings in aqua regia[a]	$PdCl_2$ and $Pd(NO_3)_2$	0.640	640
Pt	Pt metal	Ditto	$PtCl_2$ and $Pt(NO_3)_2$	0.640	640
Cu	$CuSO_4$, $5H_2O$	Dissolve fresh crystals in water; standardize	$CuSO_4$	2.523	640
	Cu metal	Dissolve fresh filings in dilute HNO_3[a] plus a few drops HCl	$Cu(NO_3)_2$	0.640	640

[a] Use as little acid as possible.
[b] All salts should be dried.

Table III

Conditions for the Determination of Silicon

	Wavelength, Å	Approx. relative sensitivity
Analytical lines	2516	1.0
	2507	0.36
	2529	0.32
	2217	0.26
	2211	0.15
	2208	0.06
Lamp	Hollow-cathode (10–20 mA), high-intensity lamps are used with some instruments	
Band pass	2–4 Å (see text)	
Flame	C_2H_2–N_2O fuel rich	
Burner	5 cm × 0.4 mm slot	
Burner height	Light should pass through "red feather" of flame	
Analytical range, ppm,	2–300	
Detection limit, ppm[a]	0.9	

[a] The detection limit is usually defined as a signal of magnitude twice the standard deviation of the background fluctuations.

this and all subsequent tables apply to the most sensitive conditions of the line and flame settings. More precise readings may be obtained for high concentrations by turning the burner.

5.1.2. Aluminum

Aluminum was the first element to be determined using the acetylene–nitrous oxide flame.[30] Since then several papers have been published on the determination of aluminum. Its determination in cement together with titanium and silicon[31] has been an important development. Aluminum has been determined in soils[32] and in high-silicon materials.[33] In all cases good agreement with accepted chemical results is reported.

Absorbance is found to be strongly dependent on the amount of acetylene in the flame. This is at a maximum with a strongly luminous fuel-rich flame. The region of the flame at which maximum absorbance occurs is critical and occurs in the region at the base of the flame. For maximum sensitivity it was found that by limiting the size of the cathode in the flame higher absorbance readings may be obtained.

Most elements cause enhancement, although calcium, zinc, copper, lead, magnesium, sodium, phosphate, and sulfate ions may cause depression. This is considerably reduced or eliminated by the addition of a releasing agent such as lanthanum and an element with a low ionization potential such as potassium or caesium.

For the rapid determination of aluminum in concentrations of over 5%, the fusion method of rock dissolution may be used, but for lower concentrations the acid-dissolution technique has been found to give more satisfactory results. In Table IV conditions are summarized for the determination of aluminum.

5.1.3. *Iron*

Iron is a major element in most rocks and exists in a variety of forms, namely, free metallic iron, primary oxides in several valency states, sulfide minerals, primary silicate minerals, and secondary compounds. Atomic absorption spectrometry provides a rapid and sensitive method for the

Table IV

Conditions for the Determination of Aluminum

	Wavelength, Å	Sensitivity factor
Analytical lines	3092.7	1.0
	3961.5	0.78
	3082.2	0.70
	3944.0	0.50
	2373.4	0.30
	2367.1	0.25
	2575.4	0.12
Lamp	Hollow cathode, 10 mA	
Band pass	2–4 Å	
Flame	C_2H_2–N_2O fuel rich (luminous)	
Burner	5 cm × 0.4 mm slot	
Burner height (below optical axes)	20–25 mm, adjust for max. absorbance	
	Without La	With La (1000 ppm)
Analytical range, ppm	3.0–500	1.0–200
Detection limit, ppm	0.90	0.6

determination of this important element. Several authors describe the use of atomic absorption in routine assays for iron, nickel, copper, and cobalt in ores. Price and Ragland[34] have determined iron in pure quartz and Galle[35] has determined iron, manganese, etc., in several rock types. Iron has also been determined in scleractinian corals and other skeletal carbonate material in recent sediments.[36] Silicate rocks have also been analyzed for iron by several other analysts.[3,37–40]

Interference with iron is mostly enhancement and it is usually small. A fuel-rich acetylene–air flame shows less enhancement and gives a higher absorbance signal than an acetylene–nitrous-oxide flame.

Both fusion and acid-digestion solution methods are suitable for the determination of iron. For basic rocks, where concentrations are high, considerable dilution is necessary, but for acid rocks, a dilution factor of 100 or 1000 usually suffices. Turning the burner and using the less sensitive lines listed in Table V reduces the sensitivity and hence the need for high dilution.

Table V also shows the conditions required for the determination of iron in rocks.

Table V

Conditions for the Determination of Iron

	Wavelength, Å	Sensitivity factor
Analytical lines	2483.3	1
	2522.8	0.5
	3020.6	0.2
	3719.9	0.1
	3589.9	0.05
Lamp	Hollow cathode 10–15 mA	
Band pass	1–3 Å	
Flame	C_2H_2–air (slightly fuel rich)	
Burner	10 cm × 0.5 mm slot	
Burner height[a]	10 mm	
Analytical range, ppm	0.2–25	
Detection limit, ppm	0.1	

[a] Distance from cathode image to burner top.

5.1.4. Calcium

Calcium has become one of the elements most frequently determined by atomic absorption spectrometry. Geochemically, calcium is considered an important element and is included in most rock analyses. Its determination is of importance for rock classification as well as for geochemical studies of minerals and rocks.

Several interference effects have been reported in the literature, and many of these are associated with the types of flames and gases used. Figure 3 shows the interference of phosphorus on calcium in various flames—hence a strong dependence on the flame type.

Aluminum also interferes with calcium, but this can be eliminated by using the acetylene–nitrous oxide flame and a releasing agent. Figure 4 shows how various amounts of lanthanum will reduce and eliminate this interference. The hotter acetylene–nitrous oxide flame requires much less lanthanum and also gives higher sensitivity when lanthanum is added.

Calcium absorbance depends markedly on the region of the flame through which the light passes as well as on the fuel flow.

If calcium concentrations are sufficiently high, sample fusions are more rapid and can be used for rapid dissolution, provided lanthanum is added to overcome interference from silicon. Acid digestion can also be used for

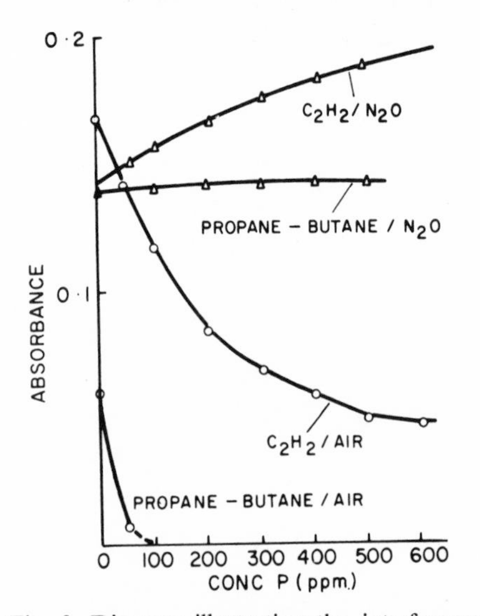

Fig. 3. Diagram illustrating the interference of phosphorus on calcium in various flames. (No releasing agent is present.)

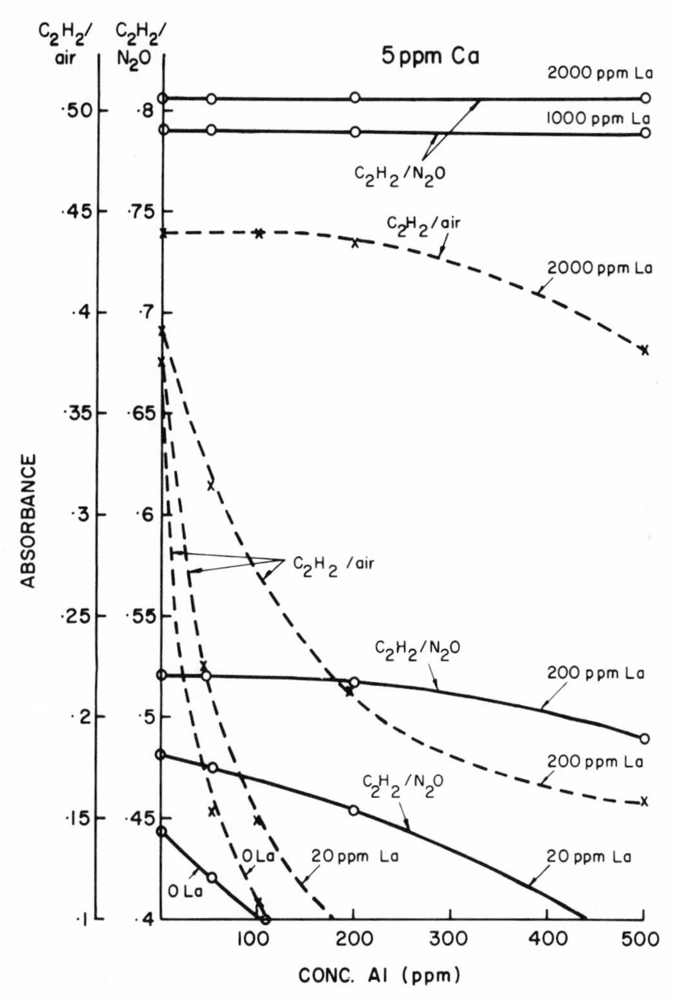

Fig. 4. Influence of aluminum on calcium in various flames, with various amounts of lanthanum added as releasing agent.

dissolution. The addition of caesium or potassium to standards is also recommended. The conditions for the determination of calcium are shown in Table VI.

5.1.5. Magnesium

The first published application of atomic absorption was the determination of magnesium by Allan.[41] Magnesium is subject to interferences similar to calcium but to a lesser degree, and analytical conditions are also similar.

Table VI

Conditions for the Determination of Magnesium and Calcium in Silicate Rocks

	Magnesium		Calcium	
Wavelength	2852 Å		4220 Å	
Lamp	Hollow cathode 5–10 mA		Hollow cathode 5–10 mA	
Band pass	2–4 Å		1–3 Å	
Flame	Max. sensitivity	Max. freedom from interference	Normal condition	Max. freedom from interference
	C_2H_2–air	C_2H_2–N_2O	C_2H_2–air	C_2H_2–N_2O
Burner	10 cm × 0.5 mm slot	5 cm × 0.4 mm slot add 200–1000 ppm La	10 cm × 0.5 mm slot	5 cm × 0.4 mm slot add 1000 ppm La
Flame setting	Lean or stoichiometric	Slightly fuel-rich (short red feather)	Fuel-rich luminous	Fuel-rich bright red feather
Burner height (below opt. axis)	5 mm	10 mm	10 mm	12 mm (light passes above red feather)
Analytical range (in solution), ppm	0.02–5.0	0.2–20	0.1–20	0.03–15
Detection limit, ppm	0.005	0.04	0.03	0.01

Belcher and Bray[42] found interference from aluminum, but this was suppressed by the addition of 1500-ppm strontium. Calcium and iron cause a sharp enhancement in the air–acetylene flame up to a concentration of 100 ppm, after which the absorbance remains constant.

Unless releasing agents are used changes in the concentration of any of the matrix elements will cause serious errors in the analysis, if aqueous standards are used. Even if synthetic or natural rock standards are used, the composition would have to match that of the samples very closely if accurate results are to be obtained. It is for this reason that releasing agents must be used for all alkaline earth determinations in rocks and minerals.

Dissolution methods for magnesium are the same as those for calcium. Table VI gives the analytical conditions for the determination of magnesium.

The accuracy obtained with the atomic absorption method appears to be good. While the precision may not be quite as good as x-ray fluorescence and colorimetric methods, the precision is similar to that obtained by emission spark methods.[43,44] The fact that the same precision is maintained to a very low concentration of magnesium in rocks also makes the method attractive to geochemists.

5.1.6. Sodium and Potassium

Sodium and potassium are considered to be major constituents in most geologic matter. They occur relatively abundantly in large numbers of silicate minerals typically associated with igneous rocks, as well as in a large variety of dissolved or soluble salts in sea water and deposits. In ultramafic rocks the concentrations of the alkali metals may be very low so that a sensitive method is required for their determination.

The atomic absorption method has been used with considerable success for the determination of sodium and potassium for all types of rocks and minerals. Accurate results have been reported by Billings,[38] Trent and Slavin,[37] and Belt.[40] Various methods of sample preparation were used, and the results were similar regardless of whether fusion or acid-dissolution techniques were used.

Earlier publications[29] listed a relatively cool flame, a town gas–air flame, as the most suitable for the determination of the alkali metals, and although this flame gives high sensitivity, it also gives rise to considerable depressive interference. Figure 5a shows the interference of various major elements on sodium, using propane–butane–air and in Fig. 5b the acetylene–air flames. Although sensitivity is reduced with the hotter flame depressive interference is less. Enhancement is increased, however.

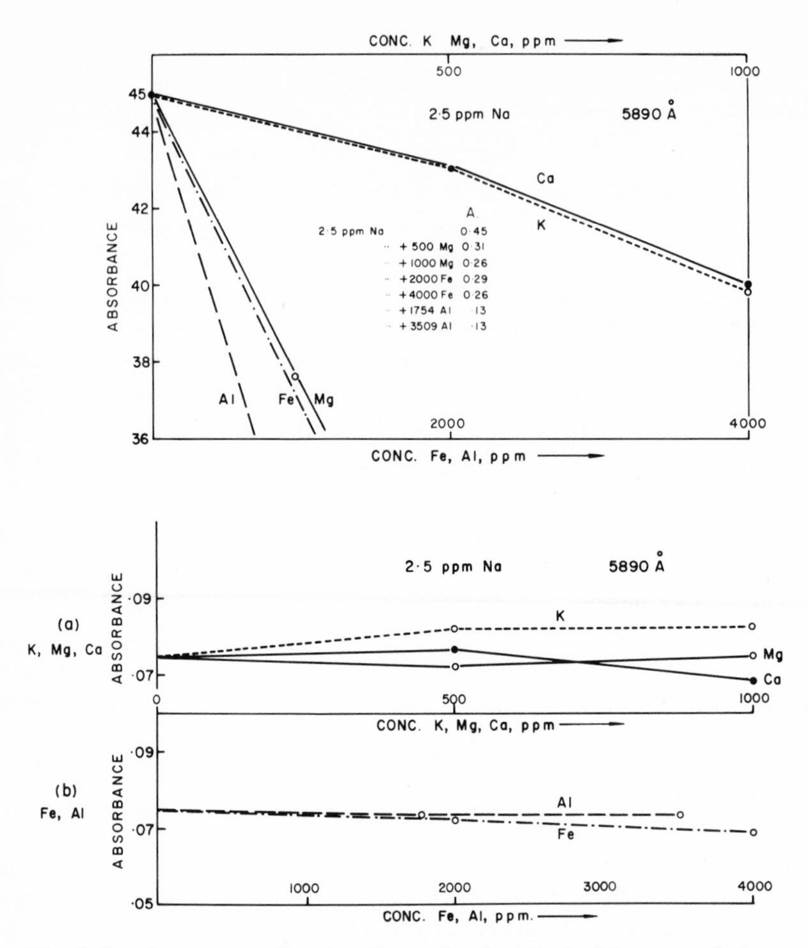

Fig. 5. Interference on sodium by other major elements in (a) propane–butane–air, and (b) acetylene–air flames.

A useful means for overcoming the enhancement interference is to add an excess of a low-ionization-potential element such as strontium or caesium to both standards and samples, similar to the technique described by Butler and Brink.[45]

When the concentration of sodium and potassium is high the potassium 4044-Å and sodium 3303-Å lines may be used.[40] Although these lines are less sensitive absorption is affected far less by the interfering elements. This is due to the relatively higher element concentrations.[46] It should be noted that when this is done the hollow-cathode lamps must be run at higher currents to provide sufficient intensity of these higher energy lines.

Table VII

Conditions for the Determination of Sodium and Potassium

Maximum sensitivity		
	Sodium	Potassium
Wavelength	5890 Å	7665 Å
Analytical range, ppm	0.01–2.0	0.03–5.0
Burner	Broad flame[a]	Broad flame
Flame	Prop.–but.–air	Prop.–but.–air
Band pass	2–4 Å	4–6 Å

Maximum freedom from interference				
	Sodium		Potassium	
Wavelength	5890 Å	3303 Å	7665 Å	4044 Å
Analytical range, ppm	0.1–10.0	10–500	0.1–10.0	20–1000
Burner	10 cm × 0.5 mm slot		10 cm × 0.5 mm slot	
Flame	C_2H_2–air stoichiometric		C_2H_2–air stoichiometric	
Band pass	4–6 Å		4–6 Å	

[a] From Butler, L. R. P., *Atomic Absorption Newsletter* **5**, 99 (1966).

For the same reason sensitivity with the acetylene–air flame is greater for these lines than with the propane–butane–air flame.

Table VII gives the most suitable conditions for the determination of sodium and potassium.

5.2. Minor and Trace Elements

Of the many elements that may be determined by atomic absorption only those elements considered to be of interest for geochemical prospecting will be discussed. Standard textbooks on atomic absorption[8,9] give details for the analysis of other elements.

5.2.1. Manganese

Manganese was determined by atomic absorption spectrometry by Allan[47] and David.[48] The sensitivity is good and few interference effects are encountered. It has been reported that silicon interferes with manganese.[6] However, when analyses of international rock standards were conducted by the authors using both fusion and acid-dissolution techniques, very little difference in manganese absorbance was noted. Lanthanum, which is generally used as a releasing agent, may eliminate this interference. The acid-dissolution technique has proved very successful for the determination of manganese (see Table VIII).

5.2.2. Copper and Zinc

The limit of detection for the determination of copper and zinc is low, and both elements are known to be free from interference effects.[5,11,49] The atomic absorption method has been used for the determination of these elements in a wide variety of materials, including geologic samples.[39,50]

Burrell[51] reported a slight enhancement on 5 ppm of zinc by iron above the 100-ppm concentration. He also reports severe depression by magnesium nitrate. Other authors[45,52,53,54] reported no interference from a number of elements on copper and zinc in an acetylene–air flame. Belt[55] reports no interference from other major elements in rocks.

Slight interference is noted when using the propane–butane flame. Although the sensitivity is poorer, the acetylene–air flame eliminates all interference and the results are correspondingly more reliable.

Table VIII

Conditions for the Determination of Manganese

	Wavelength, Å	Sensivity factor
Analytical lines	2795	1
	4041	0.1
Lamp	Hollow cathode 10 mA	
Band pass	2–4 Å	
Flame	C_2H_2–air (for maximum freedom for interference)	prop.–but.–air (for maximum sensitivity)
Burner	10 cm × 0.5 mm slot	
Burner height	15 mm	25 mm
Analytical range, ppm	0.1–10	0.05–5
Detection limit, ppm	0.05	0.02

The highest sensitivity is required for the determination of copper and zinc in common rocks, as their concentrations are low. For this reason acid digestion is recommended. In cases where all the silicate minerals are to be dissolved hydrofluoric–perchloric acid should be used. Slavin[50] describes a method whereby use is made of a hydrochloric–nitric acid mixture, followed by sulfuric acid. Farrar[49] describes the use of sodium hyposulfate followed by fusion. High-temperature fusion may, however, cause zinc to be lost through evaporation.

Certain minerals and elemental copper are dissolved more easily by dilute nitric acid. This characteristic is used by prospecting geochemists to obtain an indication of the mineral type.[56] The conditions for the determination of copper and zinc in rocks are given in Table IX. Other elements that require similar conditions to copper and zinc, are silver, gold, cadmium, lead, antimony, and bismuth.

5.2.3. Chromium

A wide variety of materials has been analyzed for chromium by atomic absorption spectrometry.[6,57]

Some interferences are experienced in an air–acetylene flame and iron and nickel suppress absorption in a fuel-rich air–acetylene flame.[8] When a hot oxidizing flame is used, however, the interference effects are reduced. The nitrous oxide–air flame is better suited for the determination of chromium in the various rock types, as interferences are reduced or even eliminated. The addition of 2% ammonium chloride to samples and standards is reported to have controlled interference effects of iron in chromium.[58] The addition of a low ionization element such as potassium or caesium will prevent ionization enhancement effects. Table X gives the conditions for the determination of chromium.

5.2.4. Lithium and Its Isotopes

Lithium has been the subject of some interest in the field of geochemistry. It may be determined comparatively easily by atomic absorption spectrometry using the strongly absorbent 6708-Å line. This technique has been employed by a number of researchers.[59–62]

The best flame for maximum sensitivity has been found to be a highly oxidizing fuel-lean acetylene–air flame with the light beam passing through as much of the primary combustion zone as possible (see Fig. 6).

Lithium has very few interference effects although sodium and potassium cause slight enhancement, and magnesium, calcium, and aluminum

Table IX

Conditions for the Determination of Copper and Zinc in Silicate Rocks

	Copper		Zinc	
	Wavelength, Å	Sensitivity factor	Wavelength, Å	Sensitivity factor
Analytical lines	3247 3274 2226	1.0 0.5 0.05	2138 3076	1.0 0.001 (approx.)
Lamp	Hollow cathode 5–10 mA		Hollow cathode 5–10 mA	
Band pass	2–4 Å		8 Å	
Flame	Max. sensitivity	Max. freedom from interference	Max. sensitivity	Max. freedom from interference
	prop.–but.–air	C_2H_2–air	prop.–but.–air	C_2H_2–air
Burner	Broad flame[a]	10 cm × 0.5 mm slot	Broad flame[a]	10 cm × 0.5 mm slot
Flame setting	stoichiometric	stoichiometric	stoichiometric (slightly fuel rich)	slightly fuel rich
Burner height	8 mm	4 mm	8 mm	10 mm
Analytical range, ppm	0.02–4.0	0.5–10.0	0.005–2.0	0.05–4.0
Detection limit, ppm	0.005	0.1	0.001	0.01

[a] From Butler, L. R. P., and Mathews, P. M., *Anal. Chim. Acta* **36**, 319 (1966).

Table X

Conditions for the Determination of Chromium in Silicate

	Wavelength, Å	Approx. relative sensitivity
Analytical lines	3594	1.0
	3579	0.73
	3605	0.62
	4254	0.40
	4275	0.29
	4290	0.22
Lamp	Hollow cathode 10–20 mA	
Band pass	2–4 Å	
Flame	C_2H_2–air (slightly fuel rich)	C_2H_2–N_2O (for freedom from interference)
Burner	10 cm × 0.5 mm slot	5 cm × 0.4 mm slot
Burner height	Light must pass through luminous zone above primary reaction region	
Analytical range, ppm	0.2–15	0.5–20.0
Detection limit, ppm	0.06	0.08

have a slight depressing effect. Iron alone has been found to cause depression.[63] However, a small amount of sodium present in the standards and samples virtually eliminates all interferences.

Fusion techniques are unsuitable for lithium determinations, as most alkali-fusion salts cannot be obtained in bulk in pure enough form. The most satisfactory method for dissolution when lithium is to be determined is a HF–$HClO_4$ digestion. Conditions for the determination of lithium are given in Table XI.

The isotopes of lithium were first determined by atomic absorption spectrometry by Zaidel and Korennoi.[64] Manning and Slavin[65] used a flame as source. When solutions of Li-6 and Li-7 were sprayed into this flame, absorbance could be measured in a normal absorption flame. A demountable Schüller–Golnow hollow-cathode lamp was used by others[66] as a means for producing an atomic cloud. Instrumental methods for determining the isotopic ratio of Li-6 to Li-7 have also been developed.[63,97]

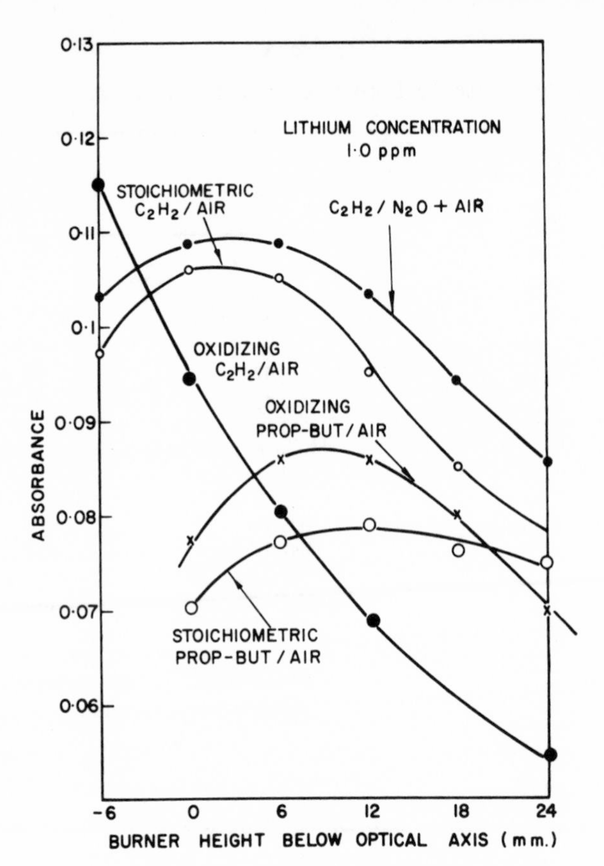

Fig. 6. The importance of flame-height setting and gas composition for lithium absorbance.

5.3. Microtrace Elements after Chemical Enrichment

Special methods must be applied for microtrace-element analysis, i.e., elements with concentrations below the normal detection limit when brought into solution. If interference effects are severe it may also be necessary to apply separation techniques to isolate, and concentrate, the analytical elements.

The most popular chemical method of extraction as applied to atomic absorption spectrometry is the extraction of metals from an aqueous solution into an organic solvent. Slavin[8] mentions four advantages when solvent extraction methods are used for atomic absorption.

(a) There is a 300–500% enhancement in sensitivity.

(b) The metal can be concentrated in the organic phase.

Table XI

Conditions for the Determination of Lithium

Analytical line	6708 Å
Lamp	Hollow cathode, 5 mA (currents above 10 mA result in self-absorption and decrease in sensitivity)
Band pass	2–4 Å
Flame	C_2H_2–air (fuel lean)[a]
Burner	10 cm × 0.5 mm slot
Analytical range	0.01–4.0 ppm
Detection limit	0.005 ppm

[a] The best sensitivity is obtained when the burner is adjusted without the flame burning to cause 10% of the light to be cut off.

(c) The salt content of a solution used for extraction can be far greater than that which can be handled in any burner.
(d) Interfering elements can often be left behind in the aqueous phase.

The most suitable organic solvents for use with atomic absorption have been found to be esters and ketones.[53] The use of ammonium pyrrolidine dithio-carbamate (APDC) as a complexing agent extracted into methyl-isobutyl ketone (MIBK) as a means for the concentration of many trace elements is well known, and is described in the literature for geologic applications.[11,37,67] Zinc, chromium, nickel, lead, cobalt, cadmium and many other metals can be extracted by this method. Wilson[52] determined copper, zinc, lead, and molybdenum at levels below 100 ppm in rock samples by precipitating the metals from acid digests with thioacetamide. Butler *et al.*[68] developed a completely automatic instrument to assay gold in cyanide mining solutions. Extraction of gold was effected by MIBK. A useful technique using normal amyl-methyl ketone for determining molybdenum in rocks was also developed.[69] Table XII shows extraction methods for various elements.

5.4. Water Analysis

Atomic absorption spectrometry has been applied with considerable success to water analyses. The main advantage of this method is that the sample can be sprayed into the flame directly. In the event that the metals sought are present below the instrumental detection limit they can easily be concentrated by partial evaporation or by organic solvent or ion-exchange extraction.

Table XII

Extraction Reagents for Different Elements

Element	Solvent	References
Antimony	APDC–MIBK	82, 83
Arsenic	APDC–MIBK	83
Bismuth	APDC–MIBK	82, 83
Cadmium	APDC–MIBK	84
Chromium	Diphenylthiocarbazone–MIBK APDC–MIBK	85, 67
Cobalt	APDC–MIBK	84, 86
Copper	APDC–ethyl amyl ketone (or MIBK)	53
Gallium	APDC–MIBK	83
Gold	MIBK	68
Indium	APDC–MIBK	83, 87
Iron	APDC–MIBK	83, 88, 89
Lead	APDC–MIBK	82, 88, 90
	Dithizone–MIBK	91
Magnesium	8-Hydroxyquinoline–MIBK	92
Manganese	APDS–MIBK	67
	Oxine in chloroform–methanol	67
	8-Hydroxyquinoline–chloroform	93
Molybdenum	Toluene-3,4-dithiol–MIBK	69
	APDC–*n*-methlamyl ketone	85
Nickel	APDC–MIBK	84
Palladium	Pyridine thiocyanate-hexone	94
Selenium	APDC–MIBK	83
Silver	Di-*n*-butylammonium salicylate–MIBK	42
	Dithiozonate complex–ethyl propionate	95
Vanadium	Cupferron–MIBK	96
Zinc	APDC–MIBK	53

Butler and Brink[45] describe the application of atomic absorption methods to the analysis of calcium, magnesium, iron, sodium, potassium, and copper in river, borehole, and well waters. All the samples were filtered to remove any suspended material. Formalin was added to prevent bacterial-induced changes in the pH from causing the precipitation of salts. In some cases the concentrations of sodium and magnesium were too high to be determined directly, and suitable dilutions had to be made. Some interferences were noted, particularly for calcium and magnesium, but these were suppressed by the addition of 1500-ppm strontium chloride.

A method for determining iron, chromium, nickel, cobalt, manganese, and copper in water at the parts-per-thousand-million level has been described by Wheat.[70] The samples were filtered and concentrated tenfold by evaporation. Burrel[71] determined nickel and cobalt by preconcentration and solvent extraction, and Platte and Marcy[6] described the atomic absorption method as a general tool for the water chemist. Price and Ragland[61] determined alkali metals and earths, and a few transition elements in ground water. The samples were filtered and treated with EDTA. Fishman and Downs[60] have analyzed natural waters by filtering the sample through 0.45-μ micropore-membrane filters. The elements manganese, zinc, copper, lithium, sodium, and potassium were determined by direct aspiration into the flame.

Little, if any, prior treatment is required for the study of most trace metals in the concentration range of 0.01–10 ppm in fresh water. Filtration is advisable, however, to ensure that only the dissolved portion is being examined.

6. GEOCHEMICAL PROSPECTING

One of the most successful applications of atomic absorption spectrometry has been in the field of geochemical prospecting. The underlying principles for the discovery of mineral deposits have been described in several books and articles.[72–74] Certain elements, such as copper, zinc, and lead, are considered to be good tracer elements. Mercury has also received considerable attention.[74] It is therefore not surprising that atomic absorption is widely practiced by many mining companies and geology laboratories throughout the world. The reasons for this are the relatively low cost of the equipment, the good precision and sensitivity obtainable, the versatility, and the low cost of sample analyses. The limitation of determining only one element at a time is offset by the advantages mentioned. When more than one element is being determined in a large number of samples, it is not unusual to find several atomic absorption spectrometers in use, each of which is set for determining one particular element only. This system is often less expensive than a large direct-reading spectrometer with its associated accessories to determine the elements simultaneously.

As the accent is on rapidity of analysis with reasonable precision, soil samples are not ground unless a large proportion of the sample is gravel. The samples may be taken a few inches below the surface, according to a grid pattern, or along chosen traverses or river beds. It is more usual to collect the portion passing through a 80–150-mesh sieve. After mixing, a portion of this fraction is then weighed and transferred to the dissolution vessel.

Dissolution may be carried out stepwise, i.e., a number of samples placed together in a basket or tray and then heated on a steam or sand bath, or they may be treated individually on a conveyor belt or continuous feeding mechanism.

The most important requirement is that samples should be treated in an identical way, as the absolute value of the element determined is of lesser importance than deviations from a norm. An unusually high value is termed an anomaly, and it is these anomalies which are followed. Figure 7 shows how an underground ore body shows up as an anomaly.

With uniform soil types it may be possible to forego the time-consuming weighing process with the use of a scoop of a constant volume. Surprisingly precise weights of soil may be measured in this way.

Many laboratories have automated the dissolution process, as this is the most time-consuming part of the analysis. Sample handling is kept to a minimum, and it is usual for the sample to be kept in the same vessel for the remainder of the procedure. Solutions are usually added with commercial or selfmade automatic dispensing flasks. After dissolution and making

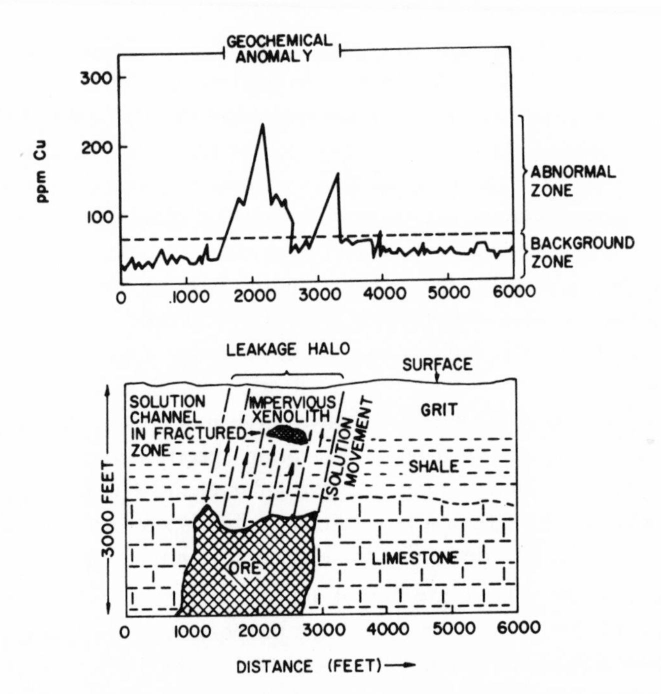

Fig. 7. Schematic diagram relating an anomaly of the concentration of copper along a traverse, with an underground ore body. Note the lower copper content above the impervious xenolith.

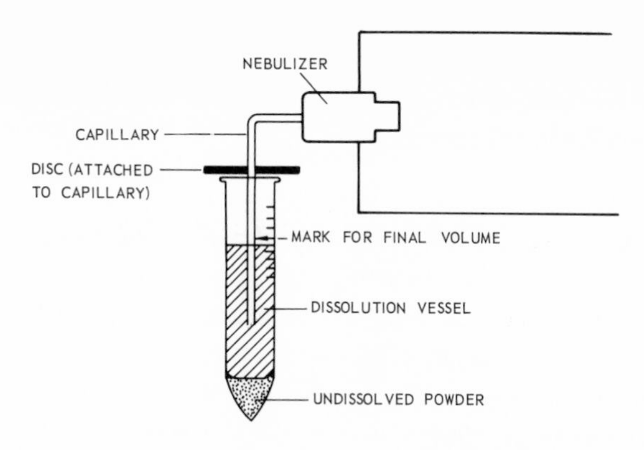

Fig. 8. An example of a typical dissolution–centrifugation vessel for the rapid dissolution of a prospecting sample.

up to volume, the sample vessel may be centrifuged or left to stand to enable undissolved material to settle. The solution above the powder is then presented to the nebulizer. Precautions are usually taken to prevent the capillary from inadvertently being dipped into the settled powder and becoming blocked. Figure 8 shows a typical dissolution flask and capillary with guard.

Other types of samples that may also be analyzed by atomic absorption include drill or water samples. Drill samples are usually taken where definite indications of mineralization have been found to obtain a three-dimensional picture of the deposit. In these cases, a higher degree of precision and accuracy is necessary than with grid samples, and greater care must be taken to ensure that these requirements are met.

Samples are dried, ground, and mixed, and several carefully weighed aliquots are taken to improve precision. It is also necessary that the dissolution procedure used is adequate for dissolving the minerals quantitatively (see Chap. 3, Sect. 3).

Atomic absorption is also well suited to the analysis of water samples. The presence of elements in subsurface water in dry regions may be an indication that the water has percolated or passed over mineralized strata. Many factors will influence the degree to which minerals will dissolve in such water, e.g., pH of the water, contact time or flow rate, rainfall pattern, temperature, etc. The analytical methods used must be very sensitive as concentrations may be relatively low. Simple concentration by evaporation or solvent extraction enables very low concentrations to be determined accurately.

Plants growing in mineralized soils may also have tell-tale high concentrations of indicator elements present in their leaves. By analyzing plant material it may thus be possible to detect anomalies. While this method of geochemical prospecting is relatively new, it may be very useful when used in conjunction with other methods.

7. RECENT AND FUTURE DEVELOPMENTS

The advances made in the technique and instrumentation of atomic absorption have been remarkable, and the question arises of whether this progress can be maintained. A trend which will prove useful to the geochemist will be the development of a satisfactory multielement analysis system, aimed at certain applications such as geochemical prospecting and routine water analysis. Multielement systems have been suggested.[75,76] Although "custom-built" instruments for specific applications may be developed, general multielement atomic absorption instruments appear to have inherent difficulties due to the problem of obtaining efficient multielement sources as well as difficulties in determining several elements with a single flame setting.

A selective modulator[77] with a solar-blind photomultiplier and interference-filter system for each element around a common flame was built by Walsh's group for geochemical prospecting. Selective modulation makes it possible to develop inexpensive instruments for each element and so do away with the most expensive item of instrumentation—the spectrometer. This modification may be useful for the geologist or geochemist with the possible extension to field instruments.

The resonant-detector atomic absorption spectrometer, suggested by Sullivan and Walsh,[25,78] has found some useful applications. The resonance fluorescence from a hollow-cathode or thermal-sputtering tube is measured using a photomultiplier and electronic set. As no wavelength adjustments are required, the spectrometer is insensitive to mechanical or thermal shocks and changes. These features are ideal for rough usage, and an interesting instrument for automatically measuring the gold content on gold-mine cyanide solutions using a resonance spectrometer was developed.[98] There are also several other applications of these spectrometers for geochemical purposes.

Considerable interest has been shown recently in the development of new atomization sources for converting solid, undissolved material into an atomic cloud to enable atomic absorption measurements to be made. Venghiattis[79] has suggested the use of solid-rocket-propellent powder mixed with an ore or concentrate sample to provide an atomic vapor of the sample.

This method may be very useful for geochemical prospecting. The recent L'vov furnace has been followed by other nonflame atomizing devices.[80,81] One of the drawbacks of nonflame atomizers is to efficiently vaporize the elements without fractional distillation and separation occurring.

REFERENCES

1. Walsh, A., *Spectrochim. Acta* **7**, 108 (1955).
2. Langmyhr, F. J., and Paus, P. E., *Anal. Chim. Acta* **43**, 397 (1968).
3. Medlin, J. H., Suhr, N. H., and Bodkin, J. B., *Atomic Absorption Newsletter* **8**, 25 (1969).
4. Yule, J. W., and Swanson, G. A., *Atomic Absorption Newsletter* **8**, 30 (1969).
5. Strasheim, A., Strelow, F. W. E., and Butler, L. R. P., *J. S. African Chem. Inst.* **13**, 73 (1960).
6. Platte, J. A., and Marcy, V. M., *Atomic Absorption Newsletter* **4**, 289 (1965).
7. Billings, G. K., and Angino, E. E., *Bull. Can. Petrol. Geol.* **13**, 529 (1965).
8. Slavin, W., *Atomic Absorption Spectroscopy*, Interscience, New York, 1968.
9. Ramírez-Muñoz, J., *Atomic Absorption Spectroscopy and Analysis by Atomic Absorption Flame Photometry*, Elsevier, New York, 1968.
10. *Atomic Absorption Spectroscopy*, A.S.T.M. STP 443.
11. Angino, E. E., and Billings, G. K., *Atomic Absorption Spectrometry in Geology*, Elsevier, Amsterdam, 1967.
12. Amos, M. P., and Willis, J. B., *Spectrochim. Acta* **22**, 1325 (1966).
13. Mitchell, A. C. G., and Zemansky, M. W., *Resonance Radiation and Excited Atoms*, Cambridge University Press, 1934, new Ed., London, 1961.
14. Alkemade, C. Th. J., *Appl. Optics* **7**, 1261 (1968).
15. Dagnall, R. M., and West, T. S. *Appl. Optics* **7**, 1287 (1968).
16. Winefordner, J. D., and Vickers, T. J., *Anal. Chem.* **36**, 161 (1964).
17. Russel, B. J., Shelton, J. P., and Walsh, A., *Spectrochim. Acta* **8**, 317 (1957).
18. Fassel, V. A., Mossotti, V. G., Grossman, W. E. L., and Kniseley, R. N., *Spectrochim. Acta* **22**, 347 (1966).
19. Rann, C. S., *Spectrochim. Acta* **23B**, 245 (1968).
20. Human, H. G. C., Butler, L. R. P., and Strasheim, A., *Analyst* **94**, 81 (1968).
21. Mavrodineanu, R., and Boiteux, H., *Flame Spectroscopy*, John Wiley & Sons, New York, 1965.
22. Butler, L. R. P. and Kokot, M. L., Geol. Soc. S. A. Special Publication No. 2, 449 (1969).
23. Butler, L. R. P., and Fulton, H. A., *Appl. Optics* **7**, 2131 (1968).
24. Robinson, J. W., *Anal. Chim. Acta*, **23**, 479 (1960).
25. Sullivan, J. V., and Walsh, A., *Spectrochim. Acta* **21**, 721 (1965).
26. Kahn, H. L. and Slavin, W., *Appl. Optics*, **2**, 931 (1963).
27. Slavin, W., and Trent, D. J., *Atomic Absorption Newsletter* **4**, 351 (1965).
28. Gaumer, M. W., Sprague, S., and Slavin, W., *Atomic Absorption Newsletter* **5**, 58 (1966).
29. Gatehouse, B. M., and Willis, J. B., *Spectrochim. Acta* **17**, 710 (1961).
30. Dagnall, R. M., Thompson, K. C., and West, T. S., *Talanta*, **14**, 1511 (1967).

31. Capacho-Delgade, L., and Manning, D. C., *Analyst.* **92**, 553 (1967).
32. Laflamme, Y., *Atomic Absorption Newsletter* **6**, 70 (1967).
33. Van Loon, J. C., *Atomic Absorption Newsletter* **7**, 3 (1968).
34. Price, V., and Ragland, P. C., *Southeastern Geol.* **7**, 93 (1966).
35. Galle, O. K., *Appl. Spectros.* **22**, 404 (1968).
36. Harris, R. C., and Almy, C. C., *Bull. Marine Sci. Gulf Caribbean* **14**, 418 (1964).
37. Trent, D. J., and Slavin, W., *Atomic Absorption Newsletter* **3**, 17, 118 (1964).
38. Billings, G. K., *Atomic Absorption Newsletter* **4**, 312 (1965).
39. Bowditch, D. C., Chalmers, A. H., and Powell, J. A., *Tech. Memo. S/3*, A.M.D.E.L., Adelaide, Australia, 1966.
40. Belt, C. B., Jr., *Anal. Chem.* **39**, 676 (1967).
41. Allan, J. E., *Analyst* **83**, 466 (1958).
42. Belcher, C. B., and Bray, H. M., *Anal. Chim. Acta* **26**, 322 (1963).
43. Suhr, N. H., and Ingamells, C. O., *Anal. Chem.* **38**, 730 (1966).
44. Govindaraju, K., *Publication G.A.M.S.* **3**, 217 (1963).
45. Butler, L. R. P., and Brink, D., *S. African Ind. Chemist* **17**, 152 (1963).
46. Hollander, T., Ph.D. thesis, University of Utrecht, Holland, 1964.
47. Allan, J. E., *Spectrochim. Acta* **10**, 800 (1959).
48. David, D. J., *Atomic Absorption Newsletter* No. 9 (1962).
49. Farrar, B., *Atomic Absorption Newsletter* **4**, 325 (1965).
50. Slavin, W., *Atomic Absorption Newsletter* **4**, 243 (1965).
51. Burrel, D. C., *Norsk. Geol. Tidssler* **45**, 21 (1965).
52. Wilson, L., Austr. Dept. Supply, Report ARL/MET 17, 1963.
53. Allan, J. E., *Spectrochim. Acta* **17**, 467 (1961).
54. David, D. J., *Analyst* **83**, 655 (1958).
55. Belt, C. B., Jr., *Econ. Geol.* **59**, 240 (1964).
56. Sampey, D., personal communication, 1968.
57. Kinson, K., Hodges, R. J., and Belcher, C. B., *Anal. Chim. Acta* **29**, 134 (1963).
58. Giammarise, A., *Atomic Absorption Newsletter* **5**, 113 (1966).
59. Angino, E. E., and Billings, G. K., *Geochim. Cosmochim. Acta* **30**, 153 (1966).
60. Fishman, M. J., and Downs, S. C., U.S. Geol. Surv. Water Supply Paper, 1540-C, U.S. Yort, Printing Office, Washington, 1966.
61. Price, V., and Ragland, P. C., *Water Resources Paper No. 18*, University N. Carolina, 1966, p. 8.
62. Slavin, W., *Atomic Absorption Newsletter* **7**, 11 (1967).
63. Butler, L. R. P., *Atomic Absorption Spectrometry and Its Application in Geochemistry*, Ph.D. thesis, University of Cape town, 1968.
64. Zaidel, A. N., and Korennoi, E. P., *Opt. and Spectroscopy* **10**, 299 (1961).
65. Manning, D. C., and Slavin, W., *Atomic Absorption Newsletter* **1**, 39 (1962).
66. Goleb, J. A., and Yoko Yuma, Y., *Anal. Chim. Acta* **30**, 213 (1964).
67. Mansell, R. E., and Emmel, H. W., *Atomic Absorption Newsletter* **4**, 365 (1965).
68. Butler, L. R. P., Strasheim, A., and Strelow, F. W. E., XII Coll. Spectr. Int., Exeter, England, 1965.
69. Butler, L. R. P., and Mathews, P. M., *Anal. Chim. Acta* **36**, 319 (1966).
70. Wheat, J. A., Report DP-879, U.S. Dept. Comm. Office Tech. Serv., also AEC Res. and Dev. Report TID-4500, 1964.
71. Burrel, D. C., *Atomic Absorption Newsletter* **4**, 328 (1965).

72. Hawkes, H. E., and Webb, J. S., *Geochemistry in Mineral Exploration*, Harper and Row, New York, 1962.
73. Kvalheim, A., *J.O.S.A.* **37**, 585 (1967).
74. Cameron, E. M., Ed. Proc. Symp. on Geochem. Prospecting, Ottawa, 1966.
75. Butler, L. R. P., and Strasheim, A., *Spectrochim. Acta* **21**, 1207 (1965).
76. Dawson, J. B., and Ellis, D. J., XII Coll. Spectr. Int., Exeter, England, 1965.
77. Bowman, J. A., Sullivan, J. U., and Walsh, A., *Spectrochim. Acta* **22**, 205 (1966).
78. Sullivan, J. V., and Walsh, A., *Spectrochim. Acta* **22**, 1843 (1966).
79. Venghiattis, A. A., *Atomic Absorption Newsletter* **6**, 19 (1967).
80. Massmann, H., *Spectrochim. Acta* **23B**, 215.
81. West, T. S., *S. African Minerals Science and Eng.* **1** (1969).
82. Willis, J. B., *Anal. Chem.* **34**, 614 (1962).
83. Mulford, E. C., *Atomic Absorption Newsletter* **6**, 28 (1966).
84. Sprague, S., and Slavin, W., *Atomic Absorption Newsletter* **3**, 37 (1964).
85. Delaughter, B., *Atomic Absorption Newsletter* **6**, 66 (1965).
86. Burrel, D. C., *Atomic Absorption Newsletter* **4**, 309 (1965).
87. Lakanen, E., *Ann. Agr. Finniae* **1**, 109 (1962).
88. Trent, D. J., and Slavin, W., *Atomic Absorption Newsletter* **3**, 118 (1964).
89. Menis, O., and Rains, T. C., *Anal. Chem.* **32**, 1837 (1960).
90. Berman, F., *Atomic Absorption Newsletter* **3**, 111 (1964).
91. Elwell, W. T., and Gidley, J. A. F., *Atomic Absorption Spectrophotometry*, Pergamon Press, London, 1966.
92. Suzuki, M., Yanagisawa, M., and Takeuchi, T., *Talanta* **12**, 989 (1965).
93. Calkins, R. C., *Appl. Spectros.* **20**, 146 (1966).
94. Erinc, G., and Magee, R. J., *Anal. Chem. Acta* **31**, 197 (1964).
95. West, P. W., and Ramakrishna, T. V., *Environ Sci. Technol.* **1**, 717 (1967).
96. Sachdev, S. L., Robinson, J. W., and West, P. W., *Anal. Chim. Acta* **37**, 12 (1967).
97. Butler, L. R. P. and Schroeder, W. W., *Int. Atomic Absorption Spectroscopy Conf., Sheffield, England, 1969.*
98. Rossouw, J., personal communication, 1968.

Chapter 9

X-RAY TECHNIQUES

H. A. Liebhafsky*

Texas A & M University
College Station, Texas

and

H. G. Pfeiffer*

General Electric Company
Schenectady, New York

1. INTRODUCTION

A chapter of this length can be at most a guide to x-ray methods in geochemical analysis. To see why this is true, consider, for example, that the 1968 review in *Analytical Chemistry* of x-ray absorption and emission by Campbell and Brown[1] lists 17 new books and 695 references, and that its counterpart by Merritt and Streib[2] on x-ray-diffraction records over a third as many items of each kind. These reviews, which appear every two years, and several books[3–6] are the main sources of information for this chapter. They are only a fraction of what should be consulted by anyone seriously interested in using x-ray methods in geochemistry.

This chapter aims at giving enough general information about x rays to convince the reader (if necessary) of their comparative simplicity and

* The authors are grateful for permission to use certain passages, tables, and figures from Liebhafsky, H. A., Pfeiffer, H. G., Winslow, E. H., and Zemany, P. D., *X-Ray Absorption and Emission in Analytical Chemistry*, John Wiley & Sons, New York, 1960.

outstanding usefulness in geochemistry, and to reinforce this general information by citing representative specific applications. The x-ray processes involved are absorption, emission, and diffraction. These processes will usually lead to determinations of elements in a sample or, in the case of diffraction, to a determination of its crystal structure. Chemical analysis, defined as the characterization and control of materials,[7] has a much broader assignment that x-ray methods alone usually cannot accomplish.

The explosive increase in the use of x-ray methods in analytical chemistry since World War II did not result from new knowledge about the rays themselves. It occurred primarily because methods of x-ray detection were improved until it was a matter of simple routine to measure x-ray intensity with high precision. Rapid progress in experimental nuclear physics accelerated this improvement, which soon led to advances, still continuing, in x-ray equipment other than detectors. Now, the computer is beginning to make itself felt, and the end is not in sight.

X rays are the most powerful radiant energy obtainable from an atom without entering its nucleus. Their wavelengths range roughly from 0.1 ("hard" x rays) to 100 Å ("ultrasoft" x rays), which means that they lie between γ rays and the UV rays of shortest wavelength. The important advantages of x rays for chemical analysis are the simplicity of their spectra; almost complete freedom (for most elements) from chemical influences; predictability of behavior, which often includes predictability of deviations from simple behavior; the opportunity of designing the desired precision into determinations by x-ray methods; and the nondestructive nature of these methods. Most of these weighty advantages are traceable ultimately to the high energy of x rays.

2. THEORETICAL CONSIDERATIONS

2.1. Nature and Excitation of X Rays

As x rays originate outside the atomic nucleus, they must be excited; that is, energy must be supplied so that atoms (or ions) can emit them. Electrons and absorbed x rays, both of sufficient energy, are the two most important means of excitation. The Coolidge tube is the most common source of x rays for analytical work. It is sealed and operates at high vacuum.

In the Coolidge tube electrons leave a coiled tungsten filament (cathode) at high temperature (upper center of Fig. 1) to strike a target (anode), from which the x rays leave the tube through a window, usually of beryllium,

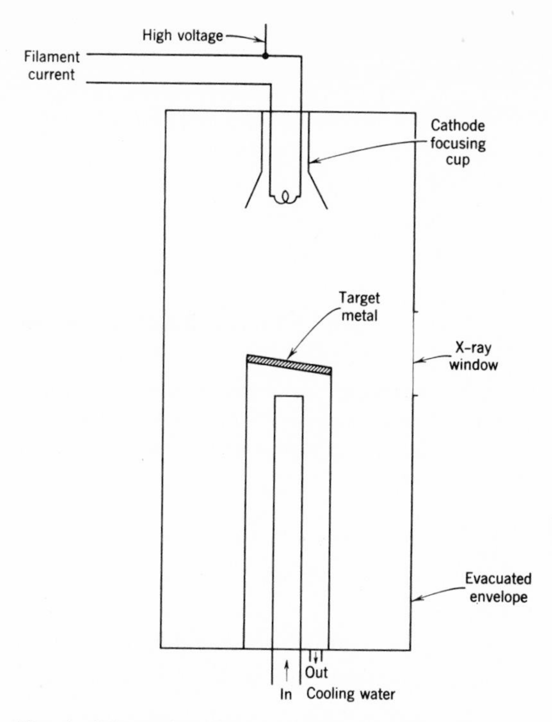

Fig. 1. Schematic diagram of the Coolidge (high-vacuum) x-ray tube. Coolidge tubes are widely used because they are stable and long-lived and because they permit tube current and voltage to be controlled independently.

chosen because this light metal does not absorb the rays unduly [Eq. (6)]. Tubes with interchangeable targets are available. Tungsten, molybdenum, chromium, copper, silver, nickel, iron, and cobalt are common target materials. Tungsten (for "hard" x rays) and chromium (for "soft" x rays) make a good pair of targets for a single tube.

Figure 2 shows the x-ray spectra observed to pass the window of a tungsten-target tube when the voltages accelerating the electrons are 20 and 50 kV. These are *continuous emitted* x-ray spectra and they are *polychromatic beams*. They have a short-wavelength limit λ_0, at which they intersect the abscissa. This shows that the excitation of x rays is a *quantum phenomenon.* The relation of λ_0 to the maximum potential difference V (in kV) across the tube is given by

$$\lambda_0 = 12.393/V \qquad (1)$$

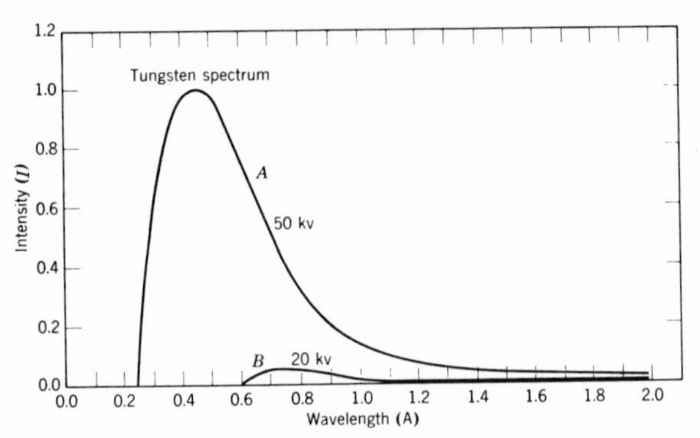

Fig. 2. The continuous x-ray spectrum. Note that the short-wavelength limit [Equation (1)] is 0.248 Å for 50 kV. (a) and 0.620 Å for 20 kV. (b) Such spectra are useless for the identification of elements.

Had V been increased above 70 kV the spectra in Fig. 2 would have undergone a startling change, which is shown in Fig. 3 for molybdenum excited by electrons and by x rays. The spikes are characteristic lines, or *monochromatic beams*, that can be used as analytical lines for the elements, just as the far more numerous lines in the visible or the UV region have

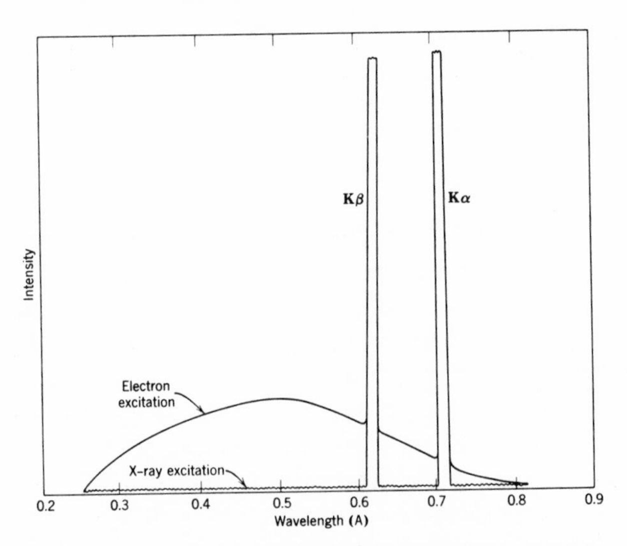

Fig. 3. An idealized drawing of the molybdenum spectrum excited by 35-kV electrons and by the polychromatic beam from a 35-kV x-ray tube. With x-ray excitation, most of the energy appears in the characteristic lines. With electron excitation, much of it is wasted in the continuous spectrum.

long been used in emission spectrography. Note that electron excitation of the molybdenum lines in Fig. 3 could have been accomplished with a molybdenum target in a Coolidge tube (Fig. 1), while x-ray excitation could have resulted had molybdenum as *element or as compound* been exposed to a suitable polychromatic beam generated by a Coolidge tube.

2.2. Interaction of X Rays with Matter

For our purposes the interaction of x rays with matter can be described under the headings absorption, scattering, and diffraction. We have already said that absorptive interaction (and electron bombardment of matter) can result in x-ray emission.

In the main, x-ray absorption is a physical process governed by Beer's law, which for a narrow, parallel, monochromatic x-ray beam incident perpendicularly upon an absorber of unit area and of uniform thickness and composition takes the form

$$\ln[I_1/I_2] = \mu(m_2 - m_1) = \mu\, \Delta m \tag{2}$$

where I_1 and I_2 are intensities transmitted by two samples of mass m_2 and m_1 of the same material, expressed in g/cm^2 of sample area for identical x-ray beams of initial intensity I_0. The proportionality constant is the *mass-absorption coefficient* μ.

The absorption of x rays principally results from photoelectric absorption and from scattering. In the simplest case of photoelectric absorption, all the energy of the absorbed x ray goes into the kinetic energy of a photoelectron and into the potential energy of an excited atom that can subsequently emit a characteristic line. The scattering of x rays is the basis of all x-ray-diffraction phenomena. Scattering may be unmodified [no change in wavelength; see Eq. (5)] or *modified* (wavelength increased; the Compton effect). Formally

$$\mu = \tau + \sigma \tag{3}$$

where τ, the coefficient for photoelectric absorption, generally makes the greater contribution to μ. Scattering increases in relative importance as atomic number and wavelength decrease.

Owing to their high energies the wavelengths of x rays often are comparable to spacings in crystals. Consequently, a crystal can act as an x-ray grating made up of equidistant parallel planes (Bragg planes) of atoms or ions from which unmodified scattering of x rays may occur in such a fashion

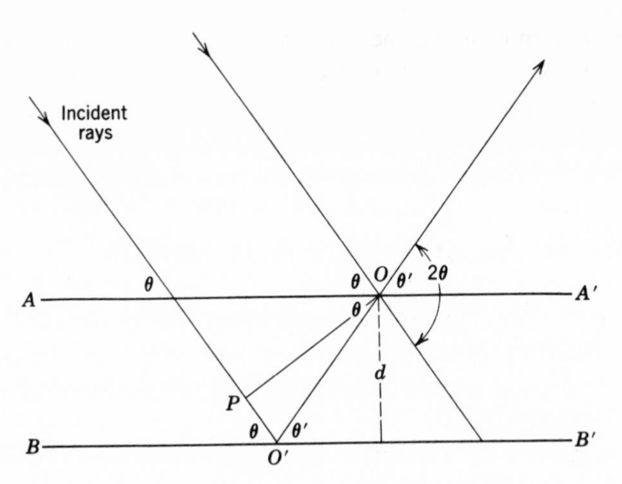

Fig. 4. Simplified "reflection" derivation of Bragg's law.

that the waves from different planes are in phase, and reinforce each other. When this happens the x rays are said to undergo *Bragg reflection* by the crystal, and a diffraction pattern results.

The conditions for Bragg reflection to occur are illustrated in Fig. 4, where AA' and BB' are the traces of successive Bragg planes, d is their distance apart, and θ is the glancing angle of the incident x-ray beam on these planes. Bragg reflection occurs when a wave scattered at O' can reinforce an identical wave scattered at O, these being the points at which the incident beam meets the Bragg planes.

Reinforcement will occur if the path length of a beam specularly reflected at O' exceeds by an integral number (n) of wavelengths the path length of a beam similarly reflected at O. This requires, therefore, that in Fig. 4

$$\theta = \theta' \tag{4}$$

and (Bragg's law)

$$n\lambda = 2d \sin \theta \tag{5}$$

Planes below BB' will contribute to the reflected beam to an exponentially decreasing extent. The relationship between λ and $\sin \theta$ determines the useful wavelength range of a particular crystal.

Equation (5) is obeyed so well that it is possible to use x-ray diffraction from crystals for highly precise determinations either of d or of λ. The former type of determination is basic in establishing crystal structure, whereas the latter is used in x-ray-emission spectrography. Both determinations are highly precise ultimately, because d is remarkably constant for the same Bragg planes in different crystals properly grown. When a polychromatic beam strikes such a crystal, only those wavelengths are detectable above background for which Eqs. (4) and (5) are obeyed. By positioning the detector to intercept a reflected beam that makes an angle of 2θ with the beam incident on the crystal, it is possible to measure the combined intensities of the wavelengths for which these equations are satisfied for a given crystal. By varying 2θ, and, if necessary, by changing crystals, the intensity-wavelength distribution of a polychromatic beam can be obtained, as in the case of Fig. 2. In this way, also, a "monochromatic" beam of desired wavelength can be selected from a polychromatic beam. If an even purer beam is needed two crystals in series (double monochromator) may be used. Clearly, the purer the beam, the lower will be its intensity.

2.3. X Rays and Atomic Structure

The short-wavelength limit (Fig. 2) and the appearance of characteristic lines (Fig. 3) both show that the interaction of x rays with matter involves quantum transitions, probably from one atomic energy level to another. X-ray-absorption spectra also bear this out.

Figure 5 shows that one may regard the absorption spectra roughly as parallel lines broken by the absorption edges, which are called $K, L, M, \ldots$ edges in order of increasing wavelength. Each edge has associated with it an emission spectrum, or a series containing relatively few lines, which are identified by the same letter. The wavelength of each such line exceeds that of the edge. Between edges the following approximation holds [see Eq. (3)]:

$$\tau = (CN/A)(Z^4\lambda^3) \simeq \mu \tag{6}$$

where C is a proportionality constant and Avogadro's number N divided by the atomic weight A gives the number of atoms per g. The approximation $\tau \approx \mu$ becomes more nearly true as Z and λ increase. In Fig. 5 note that (a) aluminum shows no absorption edges, since these occur at wavelengths beyond 10 Å, (b) that copper shows only the K edge, which occurs at a longer wavelength than the K edge for lead and at a wavelength comparable to those for the three L edges of lead, and (c) that the decrease of slope at short wavelengths, which is most noticeable for Al but undetectable for Pb, reflects the effect of σ on μ [see Eq. (3)].

Equation (6) shows why a polychromatic beam is strongly filtered as it passes through a sample. Filtering implies the preferential removal of longer

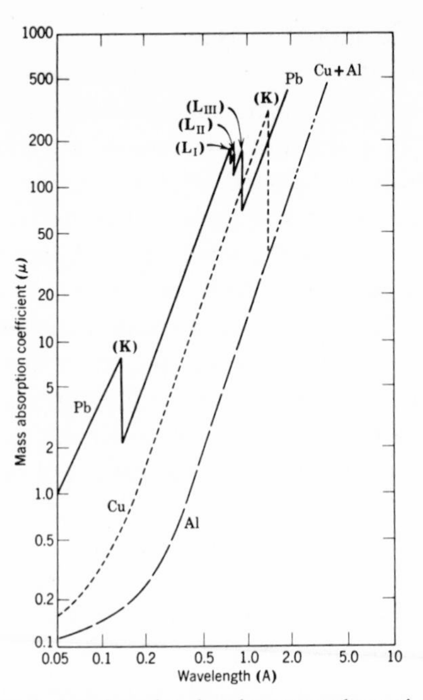

Fig. 5. Log-log plot showing mass-absorption coefficient as a function of wavelength for three common metals. Note that the discontinuities locate the absorption edges. After H. A. Liebhafsky, *Ann. N.Y. Acad. Sci.* **53**, 997 (1951).

wavelengths, which occurs because τ varies as λ^3. Another way to describe this change is by saying that the effective wavelength of the polychromatic beam has decreased.

Let us pass from absorption to emission. The K spectra—namely, those that are associated with the K absorption edge—will be described, although the others of longer wavelength are governed by the same fundamental considerations. Only the K and L spectra are presently important in analytical chemistry.

The simplicity of x-ray spectra, which derives from their high energy, is best illustrated by the fundamental significance of atomic number in determining the frequency of the characteristic lines. Moseley's celebrated relationship for the frequency ν of the K_α line of the element with atomic number Z is

$$\nu = 0.248(10^{16})(Z - 1)^2 \tag{7}$$

How is a K line generated? Every chemist knows that atomic nuclei are surrounded by shells of electrons that, when completed, contain 2, 8, 8 18,... electrons, this being the explanation of periodicity in chemical properties. For reasons that will become apparent later, these may be called K, L, M, N,... electrons, respectively, on the basis of conclusive x-ray evidence. Neither hydrogen nor helium has a K series, although each has K electrons. The reason is that the K series is generated only when the inner K shell contains a hole that is being filled by an electron from one of the outer L, M,... shells. That is, the generation of the K series requires (a) the absence of a K electron and (b) the presence of an outer-shell electron, whose transition to the K shell is permitted by the selection rules. This explains why—no matter what the method of excitation—all K lines, for a given element, have the same excitation threshold and appear together if they appear at all.

An atom (or ion) with a single hole (from which an electron has been removed) in the K shell is in the K state; the L, M,... states are similarly defined. The generation of the K lines may be accordingly represented as follows:

$$K \text{ state} \rightarrow L \text{ state} + K_\alpha \quad \text{and} \quad K \text{ state} \rightarrow M \text{ state} + K_\beta \tag{8}$$

The K spectrum is thus a series of lines originating from a single initial state, and the wavelengths of the characteristic lines in each series differ because each line represents a transition to a different final state. Because the energy of any M state is less than that of any L state, it follows that K_β will have a shorter wavelength than K_α.

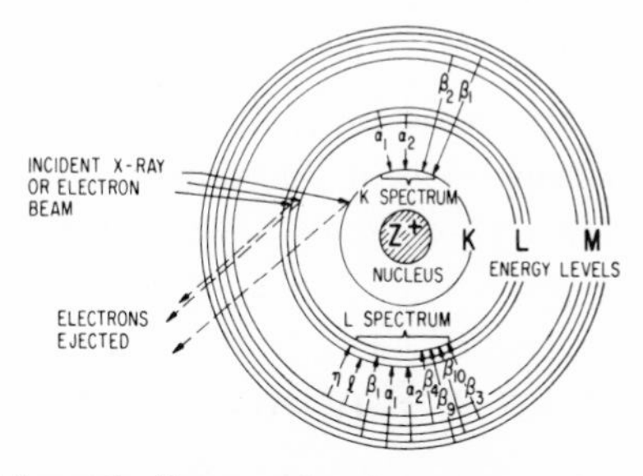

Fig. 6. Schematic diagram showing x-ray or electron excitation of analytical x-ray lines for any element with atomic number 29 (copper) or greater. The K spectrum results when x rays of wavelengths shorter than that of the K edge (Fig. 5) are absorbed to eject a K electron, and the vacancy is filled by an electron from an outer shell. Electrons may replace x rays to eject the electron. L spectra are produced in a similar manner.

In Fig. 6 the characteristic lines of possible importance in analytical chemistry are shown for a typical atom. The simplicity of x-ray emission is attested to not only by the small number of lines, but even more eloquently by the fact that such a simple drawing can represent the x-ray excitation of analytical x-ray lines for any element with atomic number 29 (copper) or greater. The L spectra are useful substitutes for the K spectra when, as with very heavy atoms, the energies required to excite the K spectra are beyond the capabilities of the equipment available.

3. INSTRUMENTATION

3.1. Analytical Systems

A *photometer* consists of a source, sample, and detector. The sample acts as absorber to filter and attenuate a polychromatic beam. Let us replace the sample with a crystal undergoing Bragg reflection. If the crystal is being used to produce a monochromatic beam of known wavelength, the instrument is a *spectrometer*. If there is in place of the crystal a sample whose lattice spacings are being determined, we have a *diffractometer*. Let us add a sample as fourth component to the spectrometer. If the sample undergoes excitation by x rays (or electrons), we have an *emission spectrograph*, with the crystal serving to analyze (resolve) the emitted spectrum. If the sample primarily attenuates an x-ray beam, we have a *spectrophotometer* (see Fig. 7).

In the spectrophotometer the crystal may be placed between source and sample, in which case it acts as a monochromator. Or, it may be inserted between sample and detector, where it acts as an analyzer of the transmitted beam.

This abbreviated summary is only an introduction, but no more is possible. An x-ray-emission electron microprobe will later be described more fully to give the reader a feeling for the complexities that are possible, and sometimes indispensable.

3.2. Measurement of X-Ray Intensities

X rays are detected by observing an effect of their interaction with matter. These effects often are translated into electric currents, pulsed or continuous, for which complex electronic circuitry is usually needed. Among the more important effects are (a) latent image formation on a photographic plate (chemical effect), (b) ionization in a gas (electrical effect), and (c) excitation of a phosphor to yield visible light (optical effect).

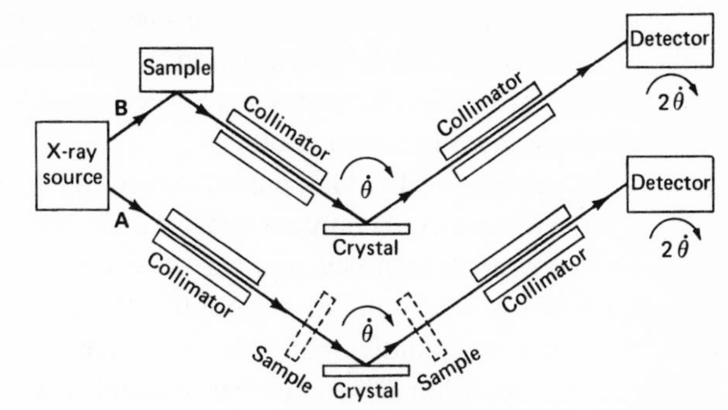

Fig. 7. Schematic diagrams of instruments for x-ray spectral analysis with x-ray excitation. To cover a spectral range, the crystal is moved with angular velocity $\dot{\theta}$, and the detectors at twice this rate, or $2\dot{\theta}$.

An *instantaneous detector* measures directly the intensity I, which is the number of x-ray quanta per sec that produce the effects mentioned above. When I is low, an *accumulative detector* may be required. The x-ray intensity is never constant, although the fluctuations are usually negligible in a good spectrograph. An instantaneous detector will indicate these fluctuations, whereas an accumulative (or integrating) detector will not. The absorption of x-ray quanta on their way from the sample is often a vital matter that is governed by Eq. (6). Absorption on the way to the detecting medium reduces intensity and is generally undesirable. Absorption within the detecting medium is desirable as it produces the effect on which the intensity measurement rests.

In most detectors the quantum effects resulting from the x-ray quanta will eventually yield electrons. Under the simplest conditions these electrons appear as separate, well-defined pulses or bundles, one pulse for each x-ray quantum. The pulses may then be counted individually, and each pulse, hence each x-ray quantum, will be registered as a unit. As the intensity increases it becomes increasingly difficult to maintain the individuality of the pulses.

The principal functions performed by electronic circuitry in the measurement of x-ray intensity are as follows:

(a) Stabilization of x-ray voltage and current.
(b) Stabilization of detector voltage.
(c) Amplification as required to give useful signal.
(d) Shaping of pulses, when required.
(e) Pulse-height selection. In some detectors the mean height of the pulse (roughly, the size of the electron bundle just mentioned) can

be made proportional to the energy of the x-ray quantum. By means of pulse-height selection, which is a sort of electronic filtering process, it is possible to select pulses of certain heights for counting and to reject others.

(f) Counting. Individual pulses are counted either by accumulating them for a fixed counting interval or by measuring the time required for a predetermined number of counts.

(g) Scaling. The number of counts often is so large that they cannot be recorded individually, as by a mechanical counter alone. Under this condition a fixed number of counts (say, 2 or 10) is accumulated into a larger unit. These units are similarly accumulated into a still larger unit, and so on. This process is known as scaling.

(h) Integration. A detector that yields discrete pulses can be used as an instantaneous detector if the pulses can be averaged to form a continuous electric current, as in a counting-rate meter.

The precision of x-ray-intensity measurements by counting techniques is limited ultimately by errors inherent in the statistics of counting. Instability in the electronic circuitry, which tends to become more serious as the counting circuits grow more complex, introduces errors in the intensity measurements that will combine with, and may overshadow, the irreducible statistical counting error.

Our main concern for the measurement of x-ray intensities is with the photographic plate, the Geiger counter, the proportional counter, and the scintillation counter.

3.2.1. Photographic Plate

The photographic plate, especially useful in x-ray diffraction, benefits as detector because, for an x-ray beam, but not for ordinary light, the same photographic density results from intensity A acting for time B as from intensity B acting for time A.

3.2.2. Geiger Counter

In the Geiger counter the absorbed x-ray quantum initially ionizes a gas, and the photoelectron thus produced is caused by an electric field to ionize other molecules in the gas-filled counter. Because the field is high the initial ionization is amplified enormously. The Geiger counter is trouble-free, easy to use, and particularly valuable for x rays of long wavelength, for which the energy produced in the initial ionization is low. The Geiger counter *cannot discriminate* among x rays of different wavelengths.

For identical geometry proportional counters operate with internal electric fields below those of Geiger counters; consequently, the mean pulse height (or total energy) can be made proportional to the energy (inverse wavelength) of the x-ray quantum. The detector can also handle intensities that give counts approaching a million per sec without serious losses. Neither advantage accrues to the Geiger counter. Proportional counters are particularly valuable in x-ray-emission spectrography in the following three ways: (a) as simple counters, in the manner of Geiger counters, especially for high counting rates; (b) with a pulse-height analyzer, as a means of resolving an x-ray beam when the wavelengths of interest have no close neighbors; and (c) with a pulse-height selector or discriminator to obtain reduced background.

It has always been difficult to do quantitative work with the characteristic x-ray lines of elements below titanium in atomic number. These spectra are not easy to obtain at high intensity, and the long wavelength of the lines makes attenuation by absorption a serious problem. The use of helium in the optical path has been very helpful. The design of special proportional counters, called gas-flow proportional counters, has made further progress possible. The windows in these gas-flow proportional counters are exceedingly thin, fragile, and unavoidably leaky. Such a window is satisfactory only if a steady flow of the filling gas is maintained at minimum pressure differential against the helium atmosphere in the optical path—hence the term gas-flow counter. The purging of impurities from the counter is an incidental benefit derived from the gas flow.

3.2.3. Scintillation Counter

In the scintillation counters usual in analytical chemistry, individual x-ray quanta can be absorbed by a single crystal which is highly transparent to light (for example, an alkali halide crystal with thallium as activator). The resultant visible scintillations can produce an output pulse of electrons from a multiplier phototube. The multiplier phototube internally amplifies up to a millionfold the photoelectric current produced when the light of these scintillations strikes the photocathode of the tube. The electrical pulses that result can be counted similarly to those from a proportional counter.

3.2.4. Proportional Counter

Proportional counters and scintillation counters can both give electrical pulses of a height, amplitude, or total energy proportional to that of the x-ray quantum absorbed. Because this energy is inversely proportional to the x-ray wavelength, pulse-height selection, which is done electronically, offers the possibility of spectrum analysis. Though inferior in this respect to Bragg reflection [Eq. (5)], pulse-height selection is useful for intensity measurements on analytical lines in special cases and for suppressing background and eliminating unwanted characteristic lines of an order [n in Eq. (5)] different from that of the analytical line. Pulse-height selection is less effective with scintillation than with proportional counters. New solid-state detectors based on germanium and silicon give promise of great future usefulness because they can give extremely narrow peaks for x-ray lines under favorable conditions.

4. X-RAY DIFFRACTION

4.1. Description

The invariant angles between crystal faces (or between faces obtained by cleavage) can be explained only in terms of an internally periodic crystal structure consisting of parallel planes that contain relatively large numbers of atoms or molecules and that are separated by a uniform distance d. Any crystal will have more than one kind of plane, as illustrated in Fig. 4, and hence more than one value of d, the *interplanar spacing* or *repeat distance.* It may therefore be assumed that

$$d \text{ (Bragg reflection)} = d_{hkl} \tag{9}$$

for every spacing in every crystal. Here d, the Bragg reflection, is either an experimental value obtained by x-ray diffraction or a theoretical value obtained by a mathematical analysis of the diffraction process; d_{hkl} is based on a geometrical analysis of crystal symmetry that makes use of the unit cell, where the subscripts are the "Miller indices." This oversimplified summary amounts to saying that crystals are useful three-dimensional diffraction gratings, and that determinations of *crystal structure by x-ray diffraction* may be relied upon in chemical analysis. For most applications of this kind, the statement just made can be narrowed further by substituting "determinations of interplanar spacings (d's) and of concomitant intensities".

Only diffraction from fine powders (Hull–Debye–Sherrer) need be considered here. A narrow beam of monochromatic x rays impinges upon the sample, which is usually rotated so as to produce the effect of completely random crystal orientation, with all possible values of θ for the same weight.

We have then an experiment at constant λ and variable θ. Bragg reflection will occur according to Eq. (5), which means for the simplest geometry (see Fig. 8) these reflections will form cones about the incident x-ray beam. The cones will form diffraction rings on a vertical photographic plate as shown. The numbers are the *hkl* values for each reflection, in this case, for a face-centered cubic lattice whose relation to the *d* spacings was indicated above. One ring can obviously be formed by reflections from more than one plane.

In practice the diffraction measurements are done differently, since, not only the spacings, but the intensities of the different reflections are usually needed. The measurements may be done in a diffraction camera, usually with a cylindrical photographic film around the x-ray beam as detector. Alternately, a diffractometer may be used, in which case a trace is obtained. With the photographic film as detector the entire diffraction pattern is recorded at once, for which one pays the price of slow response, need for developing film, and low precision in intensity measurements. The recording diffractometer has the advantage as regards the last three items. Complex qualitative determinations show photographic methods at their best. With some cameras samples need weigh only 0.1 mg.

An x-ray diffraction determination for an unknown substance consists of (a) sample preparation, (b) generation and interpretation of the diffraction pattern, and (c) comparison of the pattern with that of known substances. If no pattern is obtained, or if the pattern is so diffuse as to be useless, the method has shown only that the sample is not well crystallized and cannot be identified in this way. If the pattern matches a known pattern for a single substance completely, identification is almost always satisfactory. If the pattern is good but the match is incomplete, the sample is a crystalline

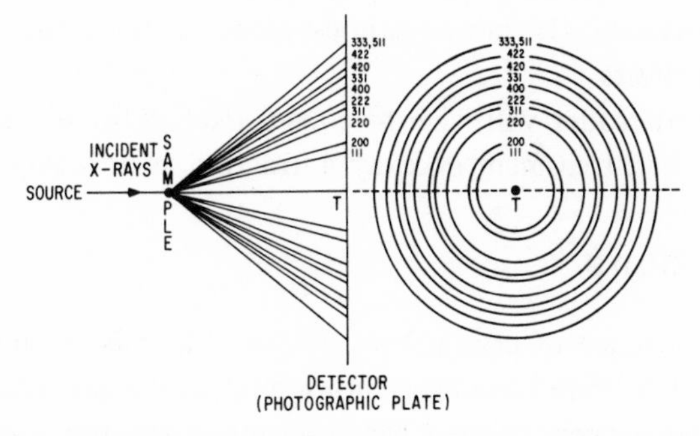

Fig. 8. Schematic representation of Hull-Debye-Sherrer powder method for simple geometry. The x-ray beam transmitted by the sample strikes the photographic plate at T. After W. L. Bragg, *The Crystalline State.*

mixture for which intensities must be known or estimated if quantitative or semiquantitative results are desired.

In complex cases the matching of diffraction patterns for sample and knowns can be a Herculean task. Such matching could proceed effectively only after Hanawalt, Rinn, and Frevel published their compiled patterns in 1938. The Joint Committee on Chemical Analysis (sic!) by Powder Diffraction, which includes ASTM, has made available a file of 8000 patterns and published several books. Frevel[8] devised a comprehensive computer program to carry out the entire searching and matching process, and describes the program in a paper that is an excellent critique of the diffraction method. He has gone one step further[9] by using a fully digitized precision microphotometer in the automated measurement of powder patterns to obtain a printout of interplanar spacings (d's) and of concomitant intensities.

We conclude by pointing out some relationships of x-ray diffraction to other x-ray methods. (a) Because diffraction can at best establish crystal structure, it must in most cases be supplemented by a method that identifies at least some of the elements in the sample. For this x-ray-emission spectrography is almost ideal for the heavier elements and gaining in value for the light elements as analytical lines of longer wavelength (e.g., K lines of F or Na) become easier to use. (b) In some cases these lines are generated when diffraction measurements are made, but they now disappear into the background: simultaneous recording of diffracted and emitted lines for analytical purposes may some day be possible. (c) Monochromatic K_α lines for diffraction work are obtained by filtering out the longer wavelength K_β lines according to Eq. (6); this relates diffractometry and absorptiometry. (d) In diffractometry there is absorption by the sample of incident and diffracted beams, or there is an *absorption effect* similar to those in x-ray-emission spectrography. In both methods internal standards are used to cope with these effects.

The interested reader will wish to consult Ref. 6 for a complete, and Ref. 10 for a briefer, treatment of x-ray diffraction as an analytical tool.

4.2. Applications

The high incidence of crystallinity among minerals invites x-ray diffraction studies, and there have been thousands. Two, both involving silica in some form, have been selected for description because they show the relation of x-ray-diffraction determinations to geochemical analysis.

Four forms of silica, namely, amorphous, quartz, cristobalite, and tridymite, need to be considered and distinguished in the examination of siliceous dusts that are silicosis hazards. Klug and Alexander (Ref. 6, pp. 417–433) give an excellent discussion of x-ray-diffraction determinations on siliceous samples. Noteworthy features are (a) emphasis on careful preparation of samples, (b) use of an internal standard (fluorite), (c) careful determination and analysis of the various errors, (d) test of method on known mixtures, (e) comparison of photographic and scanning methods, (f) early use of automatic recording, and (g) determinations of silica in three crystalline forms. Needless to say, this is an application in which conventional wet methods of analysis are at an enormous disadvantage.

The nature of a soil sample is largely fixed by its minerals, especially by those in the clay fraction, in which primary and secondary silicate minerals are of first importance. After pointing this out Whittig[11] claims the following: "x-ray diffraction has contributed more to mineralogical characterization of clay fractions of soils than has any other single method of analysis." The great care needed in pretreating soil samples shows the complexity of this kind of geochemical analysis. Kunze[12] states that removal during pretreatment of free iron oxide, of soluble salts, of carbonates, and of organic matter from a soil sample is either desirable or necessary for each of the constituents named if good x-ray-diffraction patterns are expected. A monograph by Brindley[13] is a guide to earlier work in this field.

5. X-RAY ABSORPTIOMETRY

Little need be said about absorptiometric methods because they are the least important of the x-ray methods, although an understanding of x-ray absorption is of prime importance. The methods use either polychromatic or monochromatic beams. Absorptiometry with polychromatic beams is closely related to radiography and is analogous to colorimetry with white light. The method is not specific, but it can give a fast response with simple equipment. An obvious application is, therefore, the control of industrial processes, or the rapid examination of ores to indicate, for example, the total content of the heavy metals. The most interesting use of monochromatic beams is in measuring absorption on two sides of an absorption edge as a way of both identifying an element and estimating its amount. Reference 5 describes both kinds of absorptiometry.

The advent of portable x-ray equipment (see Section 6.4) may make x-ray absorptiometry more popular as a tool for geologic or mineralogical exploration.

6. X-RAY-EMISSION SPECTROGRAPHY

6.1. Discussion

X-ray-emission spectrography consists of exciting, either with x rays or with electrons, a characteristic analytical line for each element sought in a sample. Each individual element is then identified by measuring the wavelength of the analytical line. The amount of each element present is then determined by measuring the intensity of the analytical line. From Fig. 9 the usefulness of the method may be deduced for identifying those elements that give a characteristic K or L line within the range of the spectrograph. The simplicity of the spectra, the speed with which determinations can be made, the nondestructive character of the method, its applicability either to an element or to most of its compounds—all these advantages recommend the method to every laboratory that is large enough to support an x-ray spectrograph, or that has x-ray-diffraction equipment that can be converted into such a spectrograph. X-ray-emission spectrography is already the most important of the x-ray methods under discussion. If progress continues toward removing difficulties in the determination of light elements, the method may become one of the most important for the determination of elements above atomic number 7 (nitrogen).

Moseley began x-ray-emission spectrography in 1913 with a detector in the form of a photographic plate, which has now been replaced mainly by proportional and scintillation counters. The broad scope of the method has led to a great diversity of equipment. The first spectrographs were often converted diffractometers. Gradually, emission spectrographs with

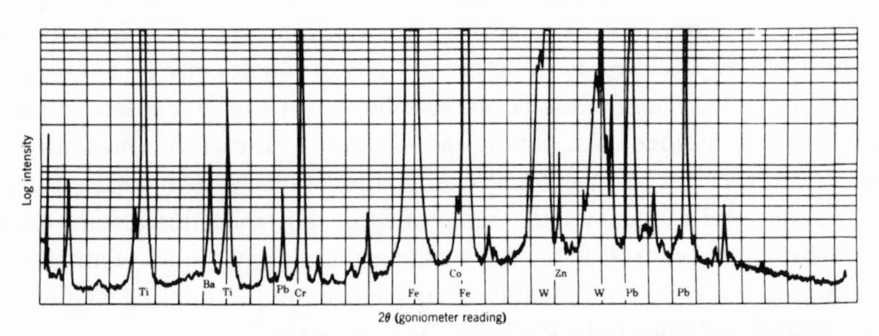

Fig. 9. Chart recording, made in minutes, of the emission spectrum from a genuine bank note. The goniometer reading for a peak identifies the element through Eq. (5) and the height of the peak is related to the amount of element present.

air paths were developed. Because air strongly absorbs x rays of longer wavelength, air came to be replaced by helium. Vacuum spectrographs are now increasingly available. Curved crystals to focus characteristic lines from small areas are becoming more popular. The range in size of spectrographs is illustrated by two instruments at the U.S. Geological Survey. The curved-crystal spectrograph of Adler and Axelrod[4] takes tiny crystals as samples and can be placed on a desk top. The other spectrograph is a one-ton General Electric device to be used on lunar samples. Automated spectrographs for industrial use are either multichannel or sequential. In the former, the beam from a sample travels simultaneously along as many as 22 channels, each with its own crystal and detector. In the sequential spectrograph, as many as 24 analytical lines in an unknown sample can each be compared at the proper goniometer setting with that of the same line in a standard. The corresponding angles are traversed sequentially.

The intensity of an analytical line (a) *may* depend upon the thickness of the sample, (b) *will* depend upon the amount present of the element sought, and (c) *will often* depend upon the matrix, namely, the other elements in the sample.

The variation of line intensity with thickness provides an introduction to absorption effects encountered whenever characteristic lines are emitted from samples of finite thickness. Consider individual atoms of an element deposited on a thin substrate highly transparent to x rays—say, atoms of molybdenum upon paper. Let a characteristic line (say molybdenum K_α) be excited by a polychromatic beam, with the x-ray source and detector both being located above the sample. As long as the number of molybdenum atoms is small, they will not noticeably attenuate the incident beam, nor will an x-ray quantum radiated by any molybdenum atom be absorbed by any other. Under these conditions, the intensity of the characteristic line will be proportional to the number of molybdenum atoms and hence to the thickness of the molybdenum film.

As the film of molybdenum grows thicker the metal will begin to filter and to attenuate the incident polychromatic beam, which will become shorter in wavelength and weaker as it penetrates. The intensity of molybdenum K_α will now increase with thickness at a continuously decreasing rate. If the metal is thick enough, even the shortest wavelength x rays will fail at a certain depth to excite K_α quanta at a rate high enough to reinforce measurably the emergent beam. Virtually all such "deep" quanta will be absorbed by the molybdenum on their way to the detector. The depth at which this first occurs is the *critical depth* for the experimental conditions. Unless the thickness of a sample exceeds this critical depth, the intensity of the analytical line will be too low for the amount of the element present in the sample.

To proceed, let us take for granted that it is logical to relate analytical-line intensity to weight fraction. Suppose the weight fraction $W_E{}^S$ of element E in sample S is to be determined by measuring the intensity $I_E{}^S$ of an analytical line. In the simplest case, we may assume that

$$I_E{}^S = W_E{}^S I_E{}^E \tag{10}$$

where $I_E{}^E$ is the line intensity for the pure element at "infinite" thickness.

Deviations from Eq. (10), called *deviations from proportionality*, fall into three principal classes: (a) absorption and enhancement effects, which are placed together because they both involve absorption; (b) effects traceable to heterogeneity in the samples, principally surface effects and segregation; and (c) instability, including drifts and fluctuations, in the spectrograph and associated equipment.

Comparison with similar deviations elsewhere in chemical analysis shows these to be easier to cope with, because they are to some extent predictable. Comparison of sample and standard, use of an internal standard, and dilution of the sample with a material such as water, cellulose, or alumina, which are all transparent to x rays, may be very helpful. References 3, 4, and 5 may be consulted in this regard.

Moseley's law [Eq. (7)] assures us that the K spectra of elements with low atomic numbers will lie in or near the ultrasoft-x-ray region of, say, 11 Å and up. Among these elements are several of great importance in geochemistry, such as sodium and oxygen. Though work in this x-ray region is difficult because of low line intensities and absorption effects it has proved most rewarding.[14,15,16]

6.2. Applications

The reader is urged to consult Adler's book[4] for an overview of what this method has already done in geochemistry. The determinations carried out cover the entire range of the method, including trace to major constituents of large samples, light to heavy elements, qualitative through semi-quantitative to quantitative results.

Carl and Campbell[17] point out that mineral analyses by x-ray-emission spectrography may be conveniently classified as follows:

Group I. Element of interest variable in a virtually constant matrix. Examples—Se in S, Th in monazite.

Group II. A pair of elements, each independently variable.
Examples—Nb and Ta, Hf and Zr.
Group III. Two or more elements of interest, each independently variable in a virtually constant matrix.
Examples—Fe and Mn in domestic ores.
Group IV. One or more elements of interest in a variable matrix.
Examples—too diverse to list.

Group I, the simplest, usually requires only that the analytical line be counted, that a background correction be applied if necessary, and that the content of the element sought be obtained by simple proportion from counts made on a suitable standard, which can usually be prepared by adding the element to a synthetic matrix. The group-II problems listed are of particular interest because they show the importance of achieving high resolution without undue loss of intensity. In the determinations of niobium and tantalum in the same ore, this was accomplished by a happy marriage of chemical and x-ray methods. The constant "matrix" in group III is conveniently defined as all the sample except the elements being determined, but a word of caution is necessary. If these elements are present at appreciable concentration, then absorption and enhancement effects to which they give rise must of course be considered, for these effects do not reside in the "matrix" alone. Group IV is by all odds the most diverse and the most interesting of the four categories. Because the matrix is variable and often unknown, an internal standard is almost obligatory for all but the crudest exploratory work, and Carl and Campbell[17] outline an interesting approach to the quantitative determination of an element in a sample from this group.

Along with the determinations of silica by x-ray diffraction and of light elements by x-ray emission, the quantitative analysis of rare-earth ores ranks as an outstanding opportunity for x-ray methods to prove themselves in analytical chemistry. Not only do these ores often contain quite a few elements to be analyzed for, but these elements are generally so similar that their determination by conventional methods is laborious and costly. In the hands of Lytle, Botsford, and Heller x-ray-emission spectrography has proved highly successful in giving quantitative results for a relatively large number of rare-earth elements in a sample. Their *Bureau of Mines Report of Investigations* 5378 repays close study. The analytical problems faced by these investigators were so complex that they could be solved satisfactorily only by comparing the unknowns with standards closely resembling them in composition.

The general discussion of applications will have to be concluded by citing only a few of the more important activities directly related to geochemistry. The *Annual Conferences on the Applications of X-Ray Analysis* have covered many of these activities, and the papers of the conferences since 1957 have been published.*

As the (present) climax of the x-ray-emission work at Pomona College[14–16] we cite a study[18] based on results for 949 samples of the Rattlesnake Mountain pluton—results obtained with a spectrograph that permitted determinations of Na, Mg, Al, Si, P, S, K, Ca, Ti, Mn, and Fe on a single preparation of a silicate rock. The work is noteworthy for, among other things, the extensive use of computers to automate the x-ray determinations and to expedite the extensive statistical treatment of the data. Reference 18 is a useful guide to related papers published from Pomona College.

The U.S. Geological Survey has published many applications of the kind under consideration here. For the earlier work, see Ref. 4. Interesting among the more recent work is the "solution-dilution" method, which can cope with small samples and reduce deviations from proportionality,[19] the determinations of individual rare earths in complex minerals,[20] and the use of chemical and electrochemical methods in conjunction with x-ray-emission spectrography.[21] The U.S. Geological Survey has acquired from the General Electric Co. a "1-ton, $6\frac{1}{2}$-ft" x-ray-emission spectrograph that can work in the ultrasoft range.[22] To be used on lunar samples, the spectrograph will give information about elements as light as boron and it can detect spectral shifts of 0.03 Å.

Finally, a few words about the remarkable work being done by M. L. Salmon and his associates in the Fluo-X-Spec Analytical Laboratory, Denver, Colorado. A distinctive feature of x-ray-emission spectrography is this: for many samples, especially samples of minerals, semiquantitative results of a reliability sufficient for most purposes cost little more than qualitative results. This laboratory wisely exploits this feature. In the period 1956–1959, it processed about 30,000 samples. A description of its earlier activities[23] is supplemented in later references.[4,24,25]

6.3. Electron Microprobe

The modern electron microprobe may well be the most sophisticated instrument used in analytical chemistry. We restrict ourselves here to its main function, which is that of an x-ray-emission spectrograph with electron

* Volume 5 of this series, entitled *Advances in X-Ray Analysis*, has been cited,[14,15] Volume 14 deals with the 1970 conference.

excitation of analytical lines from small parts of a sample. The diameter of the region studied may be in the vicinity of 1 μ or 10^{-4} cm, and the surface should be electrically and thermally conducting.

As compared with x-ray excitation, electron excitation will usually give an analytical line of higher intensity with less absorption and enhancement effects. Its comparative disadvantages are a greater background, a need for high vacuum, and a greater risk of changing the sample. Apart from the exploration of a small area, electron excitation has one other advantage, and that is that the electron beams can be focused to make a point source. This may give an analytical line from a small area at far higher intensity than would be possible with x rays. Electrons are generally less penetrating than x rays of comparable energy, and the region from which analytical lines reach a detector will therefore be severely restricted in depth as well as in area. Few samples will be homogeneous enough to give electron-

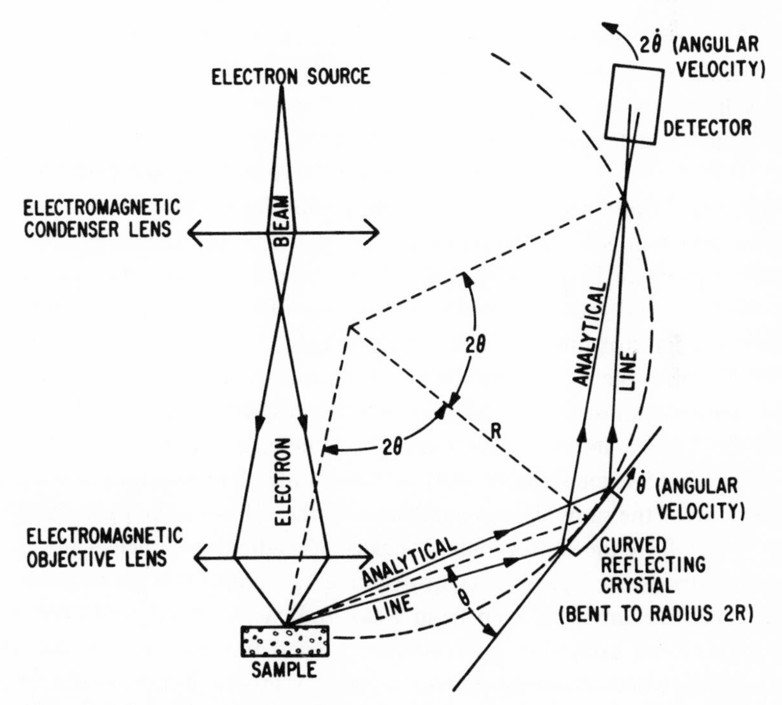

Fig. 10. Schematic operating diagram (not to scale) for an electron microprobe. Only one spectrograph channel is shown. Sample and curved reflecting (analyzing) crystal lie on the focusing (or Rowland) circle while the detector is near enough to the circle to intercept the focused analytical line at full intensity. Complex means are used to maintain good focusing when θ is changed. The curved crystal is needed because the x rays come from a point source.

microprobe results independent of location, and therefore both the strength and the weakness of the probe are self-evident.

The bare essentials of an electron microprobe appear in Fig. 10. One important missing member is an optical microscope to give visual information for the area being explored and to help locate the electron beam. Baird and Zenger[26] emphasize the importance of this matter in geochemistry. What geochemistry needs, and Baird and Zenger have provided, is a microscope akin to petrographic microscopes operating on transmitted light.

It should therefore be apparent that the x-ray-emission spectrograph for the determinations on all or much of a sample and the electron microprobe for the detailed examination of such a sample are really complementary instruments.

6.4. Portable X-Ray-Emission Spectrograph

The significant contribution x-ray-emission spectrography can make to geochemistry would become even greater if it could be made in the field as well as in the well-equipped laboratories that are needed today for geochemical determinations by this method. Portable spectrographs, such as that being developed on the basis of British work[27,28] by the Texas Nuclear Corp.,[29] are badly needed and highly welcome. The problem of this portable instrument reminds one of exploring for uranium with the Geiger counter, but the problem is in reality much more difficult. X-rays must be excited in the field, and more than one element will often have to be determined. Taking a detector into the field is not enough.

The portable x-ray-emission spectrograph uses radioactive isotopes as the ultimate source of the energy that excites the analytical lines in the sample. The excitation process may be simple, as in the case where the isotope Fe-55 excites the K lines of vanadium or chromium. Here, x-ray excitation of the sample is accomplished by the K lines of manganese, which are given off by the source mainly because its radioactive nuclei can capture its K electrons. In Fig. 11 a "central-source" geometry that gives reasonable analytical-line intensity even from weak radioactive sources is illustrated. More complex excitation processes are needed for the effective excitation of analytical lines for elements over a range of atomic numbers. If a "target element" is chosen for bombardment by electrons, γ rays, or α particles from the radioactive source, then the target will emit characteristic lines that effectively excite analytical lines in the sample.

Weak sources that yield intensities that are approximately 10^{-7} times that of an x-ray tube will require no special shielding. A complete instrument

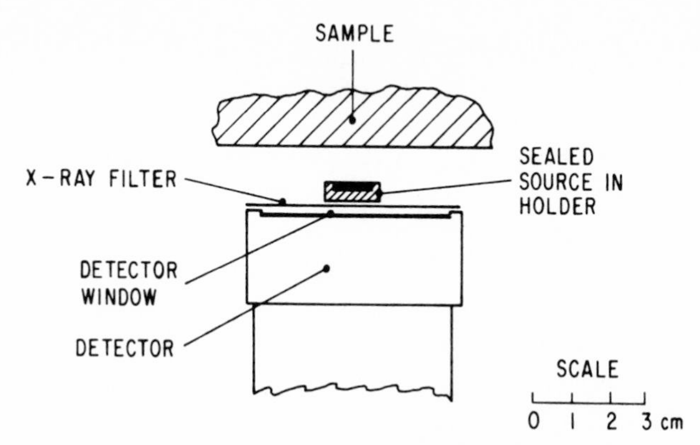

Fig. 11. Simple "central source" arrangement of source, sample, and detector in a portable x-ray emission spectrograph. The source could be ^{55}Fe and the sample, chromium. A scintillation counter is the preferred detector. With such an arrangement, the *K* lines of manganese would travel from source to sample, but *not* from source to detector. X rays from the sample would reach the detector after passing through two Ross balanced filters[30] operating sequentially.

weighing about 15 lb will give rise to analytical lines at intensities that give 10^3–10^5 counts per sec from pure elements—or, intensities high enough so that determinations can be made on minerals over counting times ranging from 10–100 sec.

Some portable spectrographs may be used as is on a flat surface. Samples of other kinds are placed above the source (see Fig. 11) and covered with a metal cap. Applications envisaged are the field assay of minerals, exploration of dry holes, and determinations on slurries of cements and ores, for example. A special version containing six filters has been used on board ship for determinations of the principal constituents (Mn, Ni, Fe, Co) in the manganese nodules present in enomous amounts on the floors of the Pacific and Indian Oceans.[31] This instrument will not only expedite exploration by enabling one to make determinations almost as fast as one can travel, but it will help further by giving a basis for quick, on-the-spot decisions.

REFERENCES

1. Campbell, W. J., and Brown, J. D., *Anal. Chem.* **40**, 346R (1968).
2. Merritt, L. L., Jr., and Streib, W. E., *Anal. Chem.* **40**, 429R (1968).
3. Jenkins, R., and De Vries, J. L., *Practical X-Ray Spectrometry*, Springer-Verlag, New York, 1967.

4. Adler, I., *X-Ray Emission Spectrography in Geology*, Elsevier, New York, 1966.
5. Liebhafsky, H. A., Pfeiffer, H. G., Winslow, E. H., and Zemany, P. D., *X-Ray Absorption and Emission in Analytical Chemistry*, John Wiley & Sons, New York, 1960.
6. Klug, H. P., and Alexander, L. E., *X-Ray Diffraction Procedures*, John Wiley & Sons, New York, 1954.
7. Liebhafsky, H. A., *Anal. Chem.* **34**, 23A (1962).
8. Frevel, L. K., *Anal. Chem.* **37**, 471 (1965).
9. Frevel, L. K., *Anal. Chem.* **38**, 1914 (1966).
10. Liebhafsky, H. A., Pfeiffer, H. G., and Winslow, E. H., in *Treatise on Analytical Chemistry*, I. M. Kolthoff and P. J. Elving, eds., Interscience, New York, 1964, Part I, Vol. 5, p. 3081.
11. Whittig, L. D., *Agronomy* **9**(1), 671 (1965).
12. Kunze, G. W., *Agronomy* **9**(1), 568 (1965).
13. Brindley, G. W., ed., *X-Ray Identification and Crystal Structure of Clay Minerals*, The Mineralogical Society, London, 1951, 345 pp.
14. Henke, B. L., in: *Advances in X-Ray Analysis*, Plenum Press, New York, 1962, Vol. 5, p. 285.
15. Baird, A. K., MacColl, R. S., and McIntyre, D. B., in: *Advances in X-Ray Analysis*, Plenum Press, New York, 1962, Vol. 5, p. 412.
16. Baird, A. K., and Henke, B. L., *Anal. Chem.* **37**, 727 (1965).
17. Carl, H. F., and Campbell, W. J., *Am. Soc. Testing Materials Spec. Tech. Publ.*, No. 157, 1954, p. 63.
18. Baird, A. K., McIntyre, D. B., and Welday, E. E., *Geol. Soc. America Bull.* **78**, 191 (1967).
19. Rose, H. J., Jr., Cuttitta, F., and Larson, R. R., *U.S. Geol. Survey Prof. Paper* 525-B, 1965.
20. Rose, H. J., Jr., and Cuttitta, F., private communication, June 18, 1968.
21. Rose, H. J., Jr., and Cuttitta, F., in: *Advances in X-Ray Analysis*, Plenum Press, New York, 1968, Vol. 11, p. 23.
22. U.S. Department of Interior, news release, May 11, 1968.
23. Liebhafsky, H. A., Winslow, E. H., and Pfeiffer, H. G., *Anal. Chem.* **32**, 240R (1960).
24. Salmon, M. L., in: *Advances in X-Ray Analysis*, Plenum Press, New York, 1961, Vol. 4, p. 433.
25. Salmon, M. L., in: *Advances in X-Ray Analysis*, Plenum Press, 1964, Vol. 7, p. 604.
26. Baird, A. K., and Zenger, D. H., in: *Advances in X-Ray Analysis*, Plenum Press, New York, 1966, Vol. 9, p. 487.
27. Bowie, S. H. U., Darnley, A. G., and Rhodes, J. R., *Trans. Instn. Min. Metall.* **74**, 361 (1964–65).
28. Rhodes, J. R., *Analyst* **91**, 683 (1966).
29. Rhodes, J. R., and Furuta, T., in: *Advances in X-Ray Analysis*, Plenum Press, New York, 1968, Vol. 11, p. 249.
30. Ross, P. A., *J. Opt. Soc. Am. Rev. Sci. Instr.* **16**, 433 (1928).
31. Rhodes, J. R., and Furuta, T., *Trans. Instn. Min. Metall.*, Nov. or Dec., 1968.

Chapter 10

RADIOMETRIC TECHNIQUES

L. Rybach

Institut für Kristallographie und Petrographie
Eidgenössische Technische Hochschule
Zurich, Switzerland

1. INTRODUCTION

The discovery of natural radioactivity radically affected developments in science and technology. The study of radioactive rocks and minerals yielded a vast amount of information on decay processes, half-lives, isotopic abundances, etc. Now that these physical constants have been evaluated, today's investigations aim towards the determination of abundance and distribution of the natural radioelements in the earth's crust, in its upper mantle, in the atmosphere, and even in space.

The most important natural radioelements—uranium, thorium, and potassium—are very inhomogeneously distributed; to establish their distribution patterns a great number of data is indispensable. Simple and rapid radiometric techniques have been developed to perform serial analyses. Statistical analysis of the results (see Chap. 2) is necessary for solution of geologic problems related to the origin of rock bodies and for both small- and large-scale assessments of geochemical balance among different rock units and other material in the earth's crust.[1]

In geologic processes such as magmatic differentiation, hydrothermal alteration, or geochemical cycles, the natural radioelements act as tracers; the Th/U ratio in particular is characteristic of the mechanism involved (see the excellent review of the geochemistry of uranium and thorium by Adams *et al.*[2]). Radiometric methods are applied in sedimentology to trace the ancient routes of sediments and to determine the rate of sedimentation.[3]

Geophysical investigations of the terrestrial heat flow are carried out to obtain information about the earth's internal structure. The main heat source is the radioactive heat generation in rocks; the generally accepted heat production constants are: 0.73 cal/yr/g U, 0.20 cal/yr/g Th, and 27 μcal/yr/g K.[4] From U, Th, and K abundances determined by radiometric methods, their heat production can be calculated; by combining radioactive heat production and heat-flow data in a given area one can construct possible geophysical–geological structure models.[5]

In common rocks U, Th, and ^{40}K occur only in trace amounts. To be economically significant, for instance, uranium contents of several hundred grams per ton must be present. After a temporary decline in activity during the early 1960's, the fast-growing demand for nuclear fuel gave impetus to the world-wide prospecting for uranium, and, to a certain extent, for thorium ores. In the course of this activity more man-hours were spent in the search for uranium than for all other metals together throughout history. At current prices ($ 8/lb U_3O_8) exploitation of a uranium deposit is economically feasible if more than 2000 tons of uranium metal are present at a concentration of at least 1000 ppm. To evaluate these limits two kinds of investigation must be carried out.

(a) Determination of the size of the ore body. Proceeding from the grade distribution at the surface (anomaly map) the depth-structure below radioactive outcrops is studied by exploratory trenches, drifts and crosscuts, and/or by drilling.

(b) Determination of the uranium concentration distribution. Besides semiquantitative determinations *in situ* this is done by analyzing a large number of samples taken at the surface, from the drift walls, in drill holes. Radiometric methods are applied in almost all phases of exporation.

Age dating is an independent method of solving such geologic problems as the origin of formations, metamorphic processes, volcanic activity, etc. Though modern age dating is carried out by mass spectrometry preliminary screening of accessory mineral concentrates by radiometric techniques simplifies the dating procedure considerably.

2. FUNDAMENTALS

2.1. Nuclear Data of Natural Radioisotopes

All elements having an atomic number greater than 83 are radioactive. They decay towards stable end products by emitting alpha and/or beta

Table I

Some Naturally Occurring Radioisotopes[a]

Radioactive isotope	Type of disintegration	Half-life, yr	Isotopic abundance, %	Abundance in lithosphere, ppm	Stable end product
$^{40}_{19}K$	β, EC	1.3×10^{9}	0.0118	3.0	^{40}Ca, ^{40}A
$^{87}_{37}Rb$	β	5.2×10^{10}	27.85	75.0	^{87}Sr
$^{147}_{62}Sm$	α	1.1×10^{11}	14.97	1.0	^{143}Nd
$^{176}_{71}Lu$	β, EC	2.2×10^{10}	2.59	0.01	^{176}Hf
$^{187}_{75}Re$	β	5.0×10^{10}	62.93	0.001	^{187}Os

[a] From *Chart of the Nuclides*, 9th Ed., USAEC, 1966.

particles (the latter often accompanied by gamma radiation with characteristic energy) or by electron capture (EC). Whereas naturally radioactive isotopes of several lighter elements (some of them are listed in Table I) reach their stable end products in most cases through a simple one-stage decay process the long-lived heavy radioactive elements decay into daughter products that are in themselves radioactive. They in turn decay and form multistage chains (see Table II) which end when finally a stable daughter isotope is formed. The uranium, thorium, and actinium series start with ^{238}U, ^{232}Th, and ^{235}U as parent members, the lead isotopes ^{206}Pb, ^{208}Pb, and ^{207}Pb, respectively, being the stable end products.

The decay of any type of radioactive nucleus can be described by the exponential law. Given N_0 radioactive atoms at a time $t = 0$, there are

$$N = N_0 \exp(-\lambda t) \tag{1}$$

atoms at any given later time t. The decay constant λ is the probability of decay per nucleus per unit time and is related to the half-life $T_{1/2}$ by

$$T_{1/2} = \frac{0.693}{\lambda} \tag{2}$$

For N nuclei present, the activity, or number of decays per second, is λN. The activity unit is 1 Ci which is defined as 3.7×10^{10} disintegrations/sec

Table II

Half-Lives, Energy Levels, and Abundances for Members of the Naturally Radioactive ^{238}U and ^{232}Th Series[a]

Isotope	Half-life	Radiation	Energy, MeV, abundance (in parentheses), %
$^{238}_{92}U$	4.5×10^9 yr	α	4.18(77), 4.13(23)
$^{234}_{90}Th$	24.1 day	β	0.19(65), 0.10(35)
		γ	0.09(15), 0.06(7), 0.03(7)
$^{234}_{91}Pa$	1.18 min	β	2.31(93), 1.45(6), 0.55(1)
		γ	1.01(2), 0.77(1), 0.04(3)
$^{234}_{92}U$	2.50×10^5 yr	α	4.77(72), 4.72(28)
		γ	0.05(28)
$^{230}_{90}Th$	8.0×10^4 yr	α	4.68(76), 4.62(24)
$^{226}_{88}Ra$	1622 yr	α	4.78(94), 4.59(6)
		γ	0.19(4)
$^{222}_{86}Rn$	3.82 day	α	5.48(100)
$^{218}_{74}Po$	3.05 min	α	6.00(100)
$^{214}_{82}Pb$	26.8 min	β	1.03(6), 0.66(40), 0.46(50), 0.40(4)
		γ	0.35(44), 0.29(24), 0.24(11), 0.05(2)
$^{214}_{83}Bi$	19.7 min	β	3.18(15), 2.56(4), 1.79(8), 1.33(33), 1.03(22), 0.74(20)
		γ	2.43(2), 2.20(6), 2.12(1), 1.85(3), 1.76(26), 1.73(2), 1.51(3), 1.42(4), 1.38(7), 1.28(2), 1.24(7), 1.16(2), 1.12(20), 0.94(5), 0.81(2), 0.77(7), 0.61(45), 0.29(26), 0.24(12), 0.035(45)
$^{214}_{84}Po$	160×10^{-6} sec	α	7.68(100)
$^{210}_{82}Pb$	19.4 yr	β	0.06(17), 0.02(83)
		γ	0.05(4)
$^{210}_{83}Bi$	5.0 day	β	1.16(100)
$^{210}_{84}Po$	138.4 day	α	5.30(100)
$^{206}_{82}Pb$	Stable		
$^{232}_{90}Th$	1.41×10^{10} yr	α	4.01(76), 3.95(24)
		γ	0.06(24)

Table II (*continued*)

Isotope	Half-life	Radiation	Energy, MeV, abundance (in parentheses), %
$^{228}_{88}$Ra	6.7 yr	β	0.05(100)
$^{228}_{89}$Ac	6.13 h	β	2.18(10), 1.85(9), 1.72(7), 1.13(53), 0.64(8), 0.45(13)
		γ	1.64(13), 1.59(12), 1.10, 1.04, 0.97(18), 0.91(25), 0.46(3), 0.41(2), 0.34(11), 0.23, 0.18(3), 0.13(6), 0.11, 0.10, 0.08
$^{228}_{90}$Th	1.91 yr	α	5.42(72), 5.34(28)
		γ	0.08(2)
$^{224}_{88}$Ra	3.64 day	α	5.68(95), 5.45(5)
		γ	0.24(5)
$^{220}_{86}$Rn	54.5 sec	α	6.28(99+)
$^{216}_{84}$Po	0.158 sec	α	6.78(100)
$^{212}_{82}$Pb	10.64 h	β	0.58(14), 0.34(80), 0.16(6)
		γ	0.30(5), 0.24(82), 0.18(1), 0.12(2)
$^{212}_{83}$Bi	60.5 min	α	6.09(10), 6.04(25)
		β	2.25(56), 1.52(4), 0.74(1), 0.63(2)
		γ	0.04(1), with α 2.20(2), 1.81(1), 1.61(3), 1.34(2), 1.04(2), 0.83(8), 0.73(10), with β
$^{212}_{84}$Po	0.30×10^{-6} sec	α	8.78(100)
$^{208}_{81}$Ti	3.1 min	β	2.37(2), 1.79(47), 1.52, 1.25
		γ	2.62(100), 0.86(14), 0.76(2), 0.58(83), 0.51(25), 0.28(9), 0.25(2)
$^{208}_{82}$Pb	stable		

[a] From Ref. 58.

(dps). The fractional units mCi (10^{-3} Ci) and μCi (10^{-6} Ci) are more frequently used. The activity of 1 kg ^{238}U is about 0.3 mCi.

For a number of radioactive elements (A, B, C; $T_{1/2A}$ very long) in a decay chain which are in conditions of natural disintegration in an isolated system for prolonged periods, a state of radioactive equilibrium is reached which is determined by the equation

$$\lambda_A N_A = \lambda_B N_B = \lambda_C N_C \tag{3}$$

This holds especially true for the uranium and thorium series where the parent half-life is very long as compared to the daughter half-lives (see Table II). Thus, when a radioactive decay chain is in equilibrium the activity of each radioactive daughter isotope, measured in Ci, is the same. That is, in a rock sample the daughter isotopes are present in amounts which are proportional to the respective half-lives (e.g., 0.3 mg Ra in 1 kg U). Geologic processes such as erosion, leaching, etc., can separate parent and daughter isotopes, since they are chemically different. The gas members of the ^{238}U, ^{235}U, and ^{232}Th series, ^{222}Rn, ^{219}Rn, and ^{220}Rn, could escape from the system which would seriously affect the degree of equilibrium.

The presence or absence of radioactive equilibrium within a rock or mineral is of fundamental importance in radiometric determinations. Uranium, for example, is usually determined by measuring the gamma radiation emitted by the radioisotope ^{214}Bi, a member of the ^{238}U decay chain. If Eq. (3) were not fulfilled for a particular sample, the uranium value found by this means would be erroneous. Section 5.4 describes methods to master this situation.

Several modes of decay may be possible for a radioactive isotope so that several α particles and γ rays of discrete energies may be emitted (the energy is usually given in MeV units, 1 MeV $= 1.6 \times 10^{-13}$ J). For a particular nucleus each mode of decay has a definite probability, so that the energy distribution of the emitted particles is characteristic of that isotope. Emission probabilities (or abundances) of the different members in the ^{238}U and ^{232}Th series are given in Table II; beta energies are the maximum values, since beta emission has a continuous energy distribution. The relatively simple decay of ^{40}K is shown in Fig. 1.

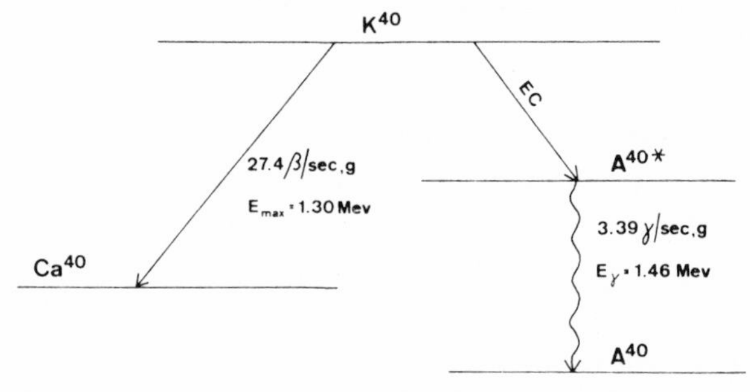

Fig. 1. Decay scheme of ^{40}K. Branching ratio (EC–gamma emission to beta emission) = 0.123.

2.2. Background Radiation

Radioactivity measured by a detector in the absence of a sample is called the background. This background arises from different sources: (a) cosmic radiation, (b) radioactivity in the environment, and (c) inherent radioactivity of the detector assembly.

(a) The primary flux of cosmic rays contains high-energy protons (0.19 p/cm$^2\cdot$sec$\cdot$sr), alpha particles (0.03 α/cm$^2\cdot$sec$\cdot$sr), and a fraction of lighter nuclei (Li, Be); the incident energy being of the order of 1.5 BeV/cm$^2\cdot$sec$\cdot$sr. Their reaction with nuclei of the earth's atmosphere produces a complex mixture of mesons, muons, hard and soft gamma rays, etc. (for more details see the excellent review by May and Marinelli[6]).
The intensity of this secondary radiation increases markedly with barometric altitude (roughly by a factor of two between 6000 and 12,000 ft). A much slighter variation with the geomagnetic latitude is also present. Considerable, but never total, absorption can be achieved by lead, iron, or mercury shields (see, also, Section 3.5).

(b) The environmental radiation itself is a sum of different contributions. Trace amounts of radioactive elements are present in any kind of material. The floor and walls surrounding a laboratory detector, even the shielding, contain minute quantities of potassium, uranium, and thorium. This contribution is constant, and thus can easily be considered.
A rather variable portion (usually less than 5% of the total background) represents the atmospheric radiation. Its main component is ^{222}Rn from the ^{238}U series diffusing from the earth's surface. The mean flux density of the exhaled Rn atoms is about $f \times 2 \times 10^{-10}$ μCi/cm$^2\cdot$sec, depending on the soil porosity f.[7] The ^{222}Rn concentration in the air is strongly dependent on meteorological factors like barometric pressure, precipitation, etc.; the daily variation can extend over more than one order of magnitude.[8]
Fission products from nuclear-weapon tests may also contribute to the atmospheric radioactivity. Washed out by fallout, these radionuclides can add substantially to the normal soil radioactivity (up to 30% of the total background[9]). The main product, ^{95}Zr–^{95}Nb decays with a half-life of about 60 days.

(c) Usually, special care is taken to select pure materials for detector construction. However, trace amounts of potassium, uranium, thorium, radium, etc., cannot be completely eliminated even with complicated procedures. Radioactivity traces are present in scintillation crystals, in the glass window of the multiplier phototube* (mainly potassium), etc. These contributions are constant for a given detector assembly. Besides these sources nonradioactive effects can also contribute to the background: the dark current of the phototube in a scintillation detector, and extremely small amounts of light penetrating to the cathode of the phototube. Some of the components of the total background radiation can be reduced considerably. For detailed discussion see Section 3.5.

2.3. Interaction of Radiation with Matter

Interactions of charged particles and gamma rays make their detection possible. Alpha particles interact with matter by removing the electrons from the target atom shells (ionization). The range or depth of penetration depends on the incident energy and the absorbing material. A 1-MeV alpha particle has a range in aluminum of about 0.0004 cm only. Electrons (beta radiation) can penetrate deeper: for 1-MeV electrons the range in aluminum is 0.15 cm. Besides ionization, electrons may lose energy by radiation as well (bremsstrahlung), depending on the atomic number Z of the absorber:

$$\frac{\text{Radiation loss}}{\text{Ionization loss}} = \frac{E_\beta Z}{800} \tag{4}$$

where E_β is the incident energy. In geologic materials (solid rocks or powdered samples) the maximum range of betas is a few millimeters only.

The production of secondary electrons gives the key to the detection of gamma rays. Gamma rays can be scattered and/or absorbed in three ways. In the first way, a gamma-ray photon strikes an orbital electron which carries away a part of the initial energy E_γ; the deflected gamma photon goes on with diminished energy, depending on the angle of the emerging electron. This type of interaction is called the *Compton effect*. There is no direct relationship between secondary electron energy and incident gamma energy. The electron energies are continuously distributed between zero and a maximum value, given by

$$E_{\text{max}} = \frac{E_\beta}{1 + \dfrac{m_0 c^2}{2E_\gamma}} \tag{5}$$

where m_0c^2 is the rest energy of the electron (0.511 MeV).

* For low-background application it is desirable to remove the mica-filled plastic base[10].

The second type of interaction is called the *photoeffect*. The whole incident energy of the gamma quantum is used up to eject a photoelectron. The binding energy E_b in a given electron shell is constant, thus a direct relationship exists between secondary electron energy E_e and incident gamma energy E_γ:

$$E_e = E_\gamma - E_b \tag{6}$$

or

$$E_e \propto E_\gamma \tag{7}$$

The third kind of interaction is the transformation of a gamma ray into a positron–electron pair. The positron immediately combines with an orbital electron, annihilating both and releasing, in opposite directions, two gamma-ray photons (the process is called, therefore, *pair production*) of 0.511 MeV each. Therefore, this type of interaction occurs only if $E_\gamma \geqq 1.022$ MeV. The secondary electron energy is

$$E_e \approx \frac{E_\gamma - 1.022}{2} \tag{8}$$

The probabilities for Compton scattering, photoelectric absorption, and pair production (cross sections τ, σ, and $\varkappa$, respectively) depend on the absorber material (atomic number Z) and incident gamma energy (E_γ):

$$\begin{aligned} \tau &= f(E_\gamma, Z) \propto Z^4 \\ \sigma &= f(E_\gamma, Z) \propto Z \\ \varkappa &= f(E_\gamma, Z) \propto Z^2 \end{aligned} \tag{9}$$

Gamma rays do not have definite ranges but are attenuated exponentially with increasing thickness of absorber. The intensity I after passing through x cm of material is

$$I = I_0 \exp(-\mu x) \tag{10}$$

where $\mu = \tau + \sigma + \varkappa$. Since μ is usually given in units of cm^2/g (mass absorption coefficient, a), in Eq. (10) the density ϱ (g/cm^3) must be introduced:

$$I = I_0 \exp(-a\varrho x) \tag{11}$$

Figures 2a and 2b show the energy dependence of a, τ, σ, and $\varkappa$ in the most frequently used shielding (Pb) and detector (NaI) materials. Photoelectric

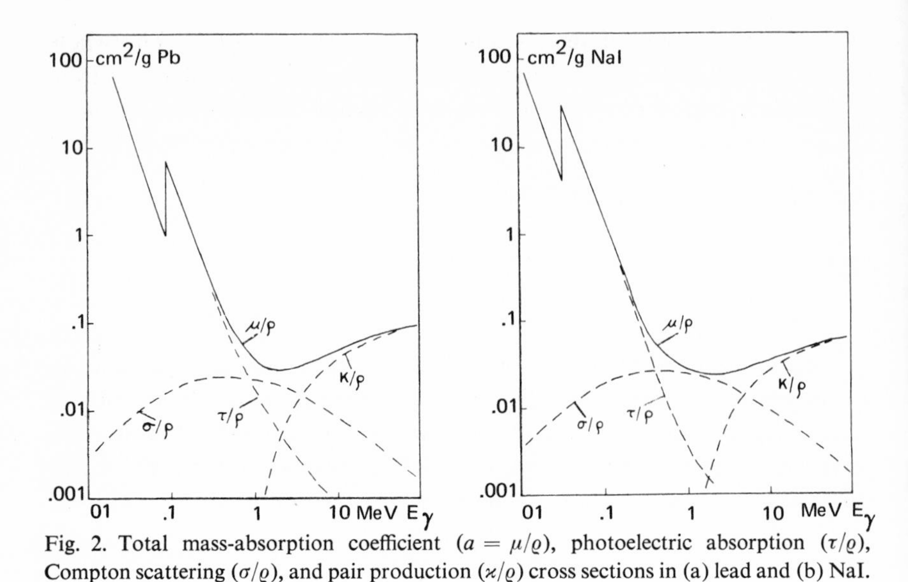

Fig. 2. Total mass-absorption coefficient ($a = \mu/\varrho$), photoelectric absorption (τ/ϱ), Compton scattering (σ/ϱ), and pair production ($\varkappa/\varrho$) cross sections in (a) lead and (b) NaI.

absorption is predominant at low, Compton scattering at medium, and pair production at high energies.

With increasing absorber thickness originally parallel gamma rays are more and more scattered diffusely and a deviation from the exponential law occurs (Fig. 3): scattered gammas can join the original ones and the effective absorption decreases.[11] To describe this phenomenon the buildup factor B is introduced:

$$I = I_0 B(ax') \exp(-ax') \tag{12}$$

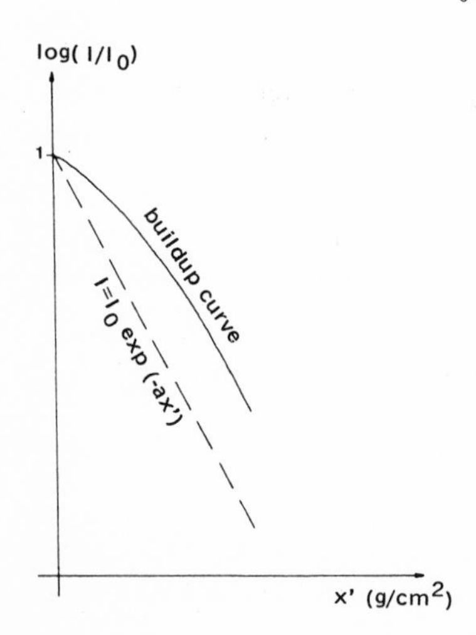

Fig. 3. Deviation from the ideal exponential law of gamma-ray absorption due to the buildup effect.

where $x' = x\varrho$ (g/cm^2). B itself is a function of the absorber material and thickness and of the incident gamma energy. In most cases one has to deal with this type of absorption.

3. RADIATION DETECTORS

3.1. Beta and Gamma Measurement*

The function of radiation detectors is to convert the energy of incident particles of gamma photons into pulses of electrical energy that can then be measured. The classical and still widely used radiation detector is the Geiger (GM) tube. Though alpha particles and, to a certain extent, gamma rays can be detected, its main application is the measurement of beta radiation. For laboratory measurements the end-window type is commonly used (the window consists of a very thin, <2 mg/cm^2 mica foil), while for field prospecting thin wall (30 mg Al/cm^2) types are more convenient. Figure 4 shows a circuit for adapting a longer cable between the detector and the pulse-registering counter.

The number of pulses produced in a GM tube (for a given radiation intensity) is relatively independent of the high voltage applied (if operating in the "Geiger-plateau" voltage range), which allows the use of simple, unstabilized power-supply units. Disadvantages are numerous:

(a) After each registered pulse the counter is blocked for a certain time. This time ("dead time") is in the order of 200 μsec/count. Especially at higher count rates (the usual units are cpm, counts per minute, or cps, counts per second) a correction for this effect must be made: the effective incident count rate $A_{\text{eff.}}$ can be calculated by

$$A_{\text{eff.}} = \frac{A_{\text{reg.}}}{1 - \tau A_{\text{reg.}}} \simeq A_{\text{reg.}}(1 + \tau A_{\text{reg.}}) \tag{13}$$

where $A_{\text{reg.}}$ is the count rate measured and τ is the dead time. τ can easily be determined by the two radioactive-source method.[12]

(b) The counting efficiency for gammas is very low (around 1%). Though a certain improvement can be achieved by coupling several

* For practical use a vast number of counting devices, electronic equipment, etc., exist on the market. It is far beyond the scope of this article to select or to compare different types and manufacturers. If reference is given to a particular type this is because the writer has special experiences with that model or unit.

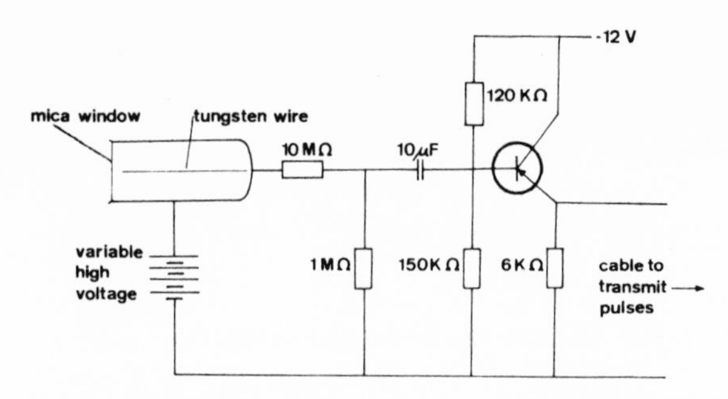

Fig. 4. Emitter–follower circuit with a low-impedance output for a longer cable between GM tube and registering unit.

GM tubes in parallel, scintillation counters are definitely better suited for gamma registration.

(c) There is no relation between the amplitude of the pulse generated by the incident particle and particle energy (all pulses have the same height). Thus, no spectral differentiation is possible.

The latter two shortcomings are eliminated with a scintillation detector, which is the most commonly used instrument for gamma measurements. Figure 5 shows the principle of the scintillation counter. The main parts, the crystal (usually sodium iodide containing a trace of thallium) and the multiplier phototube, are mounted together. Gamma rays entering the crystal produce secondary electrons by the three processes mentioned in Section 2.3. These electrons cause the crystal to scintillate. The scintillation light is converted to electric pulse and amplified in the attached phototube. The photocathode and dynodes of the multiplier phototube are made of material that is capable of producing secondary electrons and are maintained at successively higher positive potential.

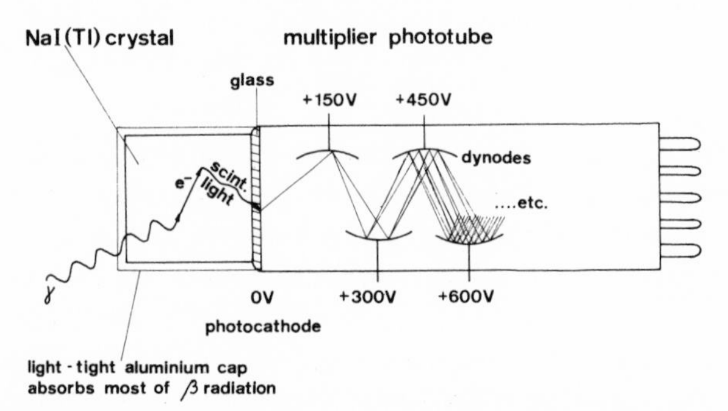

Fig. 5. Scintillation detector for gamma counting. (NaI is hygroscopic and must be hermetically sealed.)

The gain V obtainable from a multiplier phototube depends mainly on the voltage E_d between each of the n dynodes:

$$V = kE_d^n \tag{14}$$

where k is a constant, depending on the dynode material. Usually a resistor string supplies the individual potentials to the dynodes. If there is a small change ΔE_d in E_d, from Eq. (14) if follows that

$$\frac{V}{\Delta V} = n \frac{E_d}{\Delta E_d} \tag{15}$$

thus, stabilization of the supply voltage within close limits is very important. Different gain-stabilizing equipment is now commercially available (see also Section 5.3). The amplification of commercial phototubes is approximately 10^8. Even with such high amplification the output signal is relatively small (as compared to a GM tube) and special additional electronics are necessary. The dead time τ is, on the other hand, about $10^3 \times$ less than for a GM counter.

The amount of scintillation light (and thus the pulse height at the photomultiplier output) is proportional to the energy of the secondary electron. Depending upon the process involved a unique pulse height (photoelectric absorption) or a continuous distribution (Compton scattering) will appear at lower and medium incident energies. At high energies a positron–electron pair is produced. The positron is immediately annihilated, giving rise to two 0.51-MeV gamma rays. If both of these gammas are stopped in the crystal the pulse height will correspond to the incident gamma energy E_γ. If one or both gammas can escape the pulse height will be less, correspond-

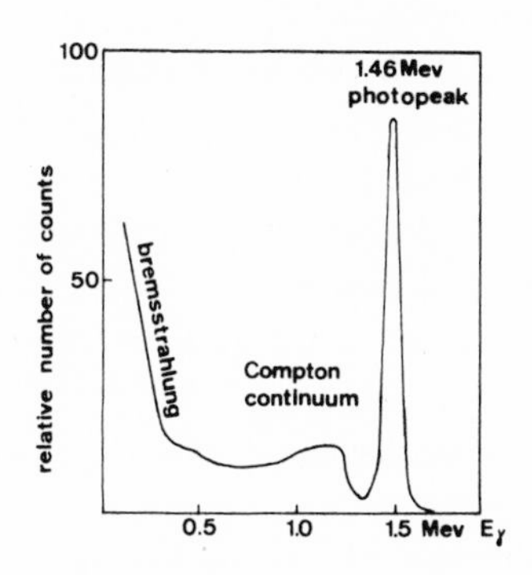

Fig. 6. Gamma-ray spectrum of ^{40}K measured by scintillation detector.

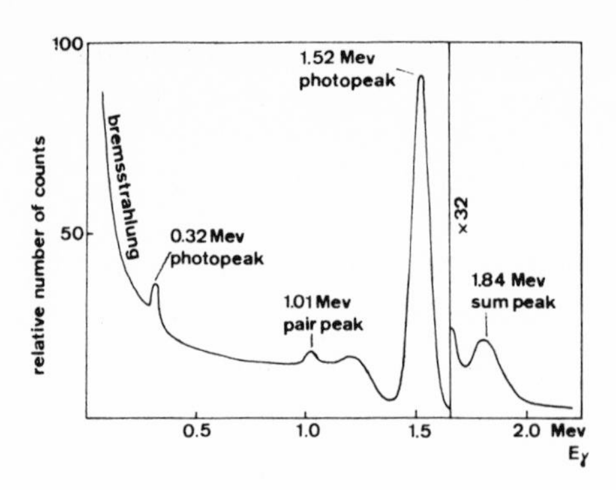

Fig. 7. Gamma-ray spectrum of ^{42}K showing pair and sum peaks.

ing to E_γ -0.51 MeV or E_γ -1.02 MeV. Thus, for a simple radioisotope like ^{40}K emitting monoenergetic gamma rays (see Fig. 1) a whole distribution of pulse heights occur (Fig. 6). A similar spectrum is shown by the artificial isotope ^{42}K (Fig. 7). Besides photopeaks, Compton continua, and pair production peaks, three additional features are visible in Figs. 6 and 7.

(a) The effect of instrumental resolution (a broad peak instead of a single line appears). With the half-width ΔE the resolution r is given by

$$r(\%) = 100 \frac{\Delta E}{E_\gamma} \tag{16}$$

NaI(Tl) crystals have a resolution* of about 7% in the region of 0.7 MeV.

(b) The bremsstrahlung increase. When beta particles are stopped in matter a certain fraction of their energy is radiated as bremsstrahlung (photons). A continuous energy distribution of photons results which predominates at lower energies. For radioisotopes with high beta branching directly to the ground state (branching ratio for ^{40}K is 8 : 1, see Fig. 1) the contribution of the bremsstrahlung to the gamma-ray spectrum is considerable. The probability of bremsstrahlung emission increases with Z^2 of the absorber. Thus, for beta absorbers low Z material (Be, polyethylene) must be used (see detailed discussion in Ref. 10, p. 19).

* Much better resolution can be achieved by solid-state detectors, e.g., Ge(Li). However, their sensitivity is insufficient for the activities encountered in normal radiometric practice. Figure 8 shows a spectrum of a relatively strong source ($\sim$0.2 μCi).

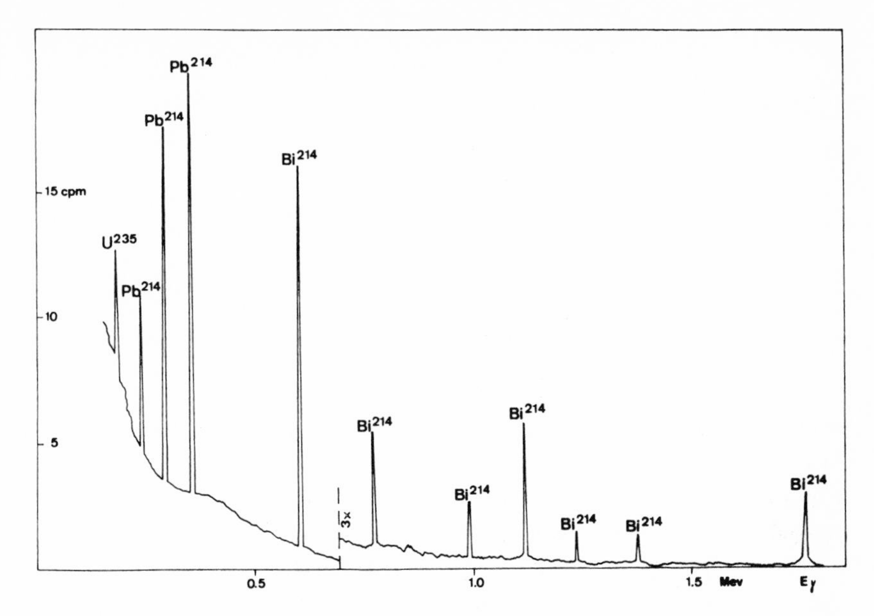

Fig. 8. Gamma-ray spectrum of the mineral brannerite (U, Th, Ca)(Ti, Fe)$_2$O$_6$. A 1-g substance at 6 cm distance from a 20-cm^3 Ge(Li) detector. Note the sharp lines of ^{214}Bi.

(c) Summation peak (at $0.32 + 1.52 = 1.84$ MeV) in the ^{42}K spectrum which occurs because of the finite resolving time of the scintillation detector. This effect is especially important at higher activities, since the probability p for the overlap of decay events (they will sum in amplitude) in time is given by

$$p \propto 2A^2\tau \tag{17}$$

where A is the incident count rate and τ the overall resolving time. For further details see Ref. 13, p. 18.

3.2. Alpha Measurement

Of the different possibilities of alpha-particle detection the two most popular detector types should be mentioned here. Zinc sulfide as a scintillator has been utilized since the early days of nuclear research. ZnS in contact with a multiplier phototube is still a very useful device for carrying out integral alpha counting (see Section 5.5). Besides gas-flow GM counters operating in the "proportional region" (see Section 3.1) and ionization chambers for special applications,[14] solid-state detectors are commonly used, especially for alpha spectrometry.

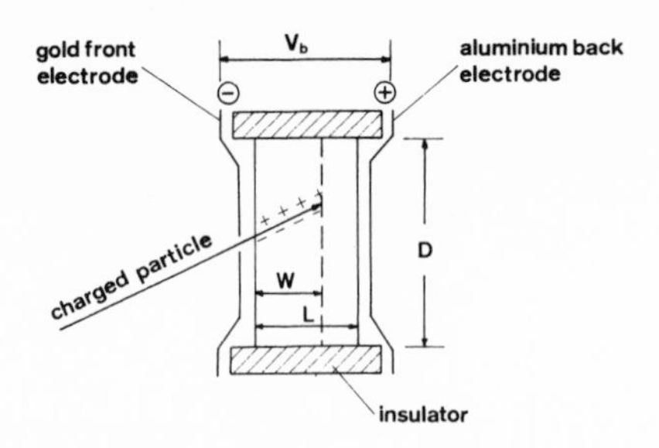

Fig. 9. Surface-barrier-type detector diode (schematic). D, diameter; L, total thickness; W, depth of the sensitive (depleted) region.

Solid-state detectors are usually of the surface-barrier type: these are basically semiconductor (silicon) wafers between deposited electrodes (see Fig. 9). An incident alpha particle releases ionization charge carriers which are separated and collected by the applied electric field V_b. The detector area $Q = D^2\pi/4$ determines the sensitivity (counting rate) of the device, and therefore large-area detectors should be used. On the other hand, the signal capacity of the diode, which mainly determines its noise, is

$$C_d \propto \frac{Q}{\sqrt{V_b}} \tag{18}$$

Thus, a compromise must be made in selecting an appropriate sensitive area.

A simple equivalent circuit is shown in Fig. 10, where R_s is the diode series resistance, R_L corresponds to the reverse leakage in the diode, R_b is the load resistor through which the bias voltage is applied, C_s represents all of the stray capacity in cables, etc., and C_A is the input capacitance of the associated preamplifier.

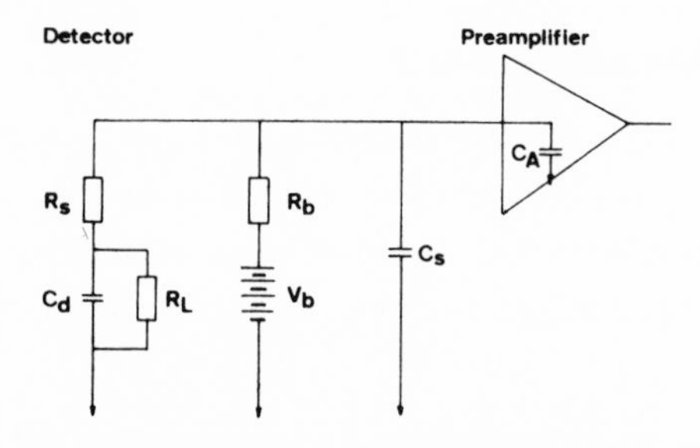

Fig. 10. Equivalent circuit for surface-barrier detector used in alpha spectrometry.

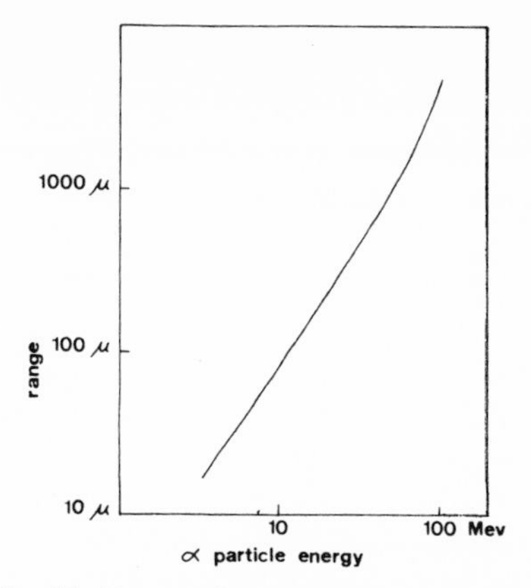

Fig. 11. Range of alpha particles in silicon.

If the total energy of the incident particle is deposited in the sensitive (depleted) region the output pulse height will be proportional to the incident energy. Thus, the maximum range of the particles to be analyzed establishes the minimum usable sensitive depth. Figure 11 gives the ranges for alpha particles with different energies in silicon.

Alpha sources emit particles of discrete energies ("line spectrum"). In this type of detector, again, a broadening occurs; the measured spectrum shows peaks, instead of lines (see Fig. 12). The instrumental resolution is usually given as the peak width at the half-maximum height (FWHM).

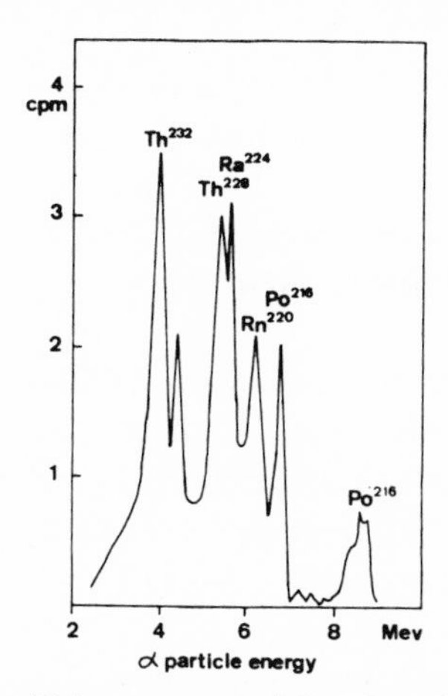

Fig. 12. Alpha spectrum of the mineral thorite ($ThSiO_4$) measured by a 1-cm² solid-state detector. Resolution *ca.* 5% FWHM (after Cherry *et al.*[53]).

Table III

Semiconductor Alpha-Detector Dates

Sensitive area, mm^2	Depletion depth, μ	Resolution FWHM, keV
25	100	18–25[a]
100	100	20–35
200	100	25–40
450	100	35–55
950	100	70–100

[a] This depends on the ORTEC model.

Table III lists typical FWHM values for different detector sizes measured at the 5.48-MeV ^{241}Am peak. Since no secondary processes take place in alpha absorption, measured alpha spectra show simple forms, making their interpretation simple.

It should be mentioned here that the introduction of portable x-ray-fluorescence analyzers to mineral exploration has been a success.[15] The surface of the rock to be analyzed is irradiated with low-energetic gamma rays from an artificial radioisotope source (^{55}Fe, ^{238}Pu, ^{109}Cd, ^{3}H–Zr, and ^{147}Pu–Al are commonly used). The excited characteristic x-rays (see Chap. 9) are filtered and registered by a scintillation counter. Though only the outermost "skin" of the rock is analyzed satisfactory results have been obtained, especially in the case of tin ores.

3.3. Scaler and Ratemeter Circuits for Integral Counting, Pulse-Height Analysis

Detectors supply electric pulses with a frequency proportional to the incident radiation intensity. The number of pulses can be counted by a scaler or their time average can be indicated (or recorded) by a ratemeter. Usually the pulses are discriminated, amplified, and shaped (PS) before registration (Fig. 13). A threshold circuit (T) discriminates the pulses below a given pulse height; low pulses (e.g., from the dark current of the multiplier phototube) are eliminated. The high voltage supply (HV) must be checked and, if necessary, readjusted from time to time.

The scaler records the individual random pulses. Decimal scalers with five decade stages are most frequently used, since a large number of events

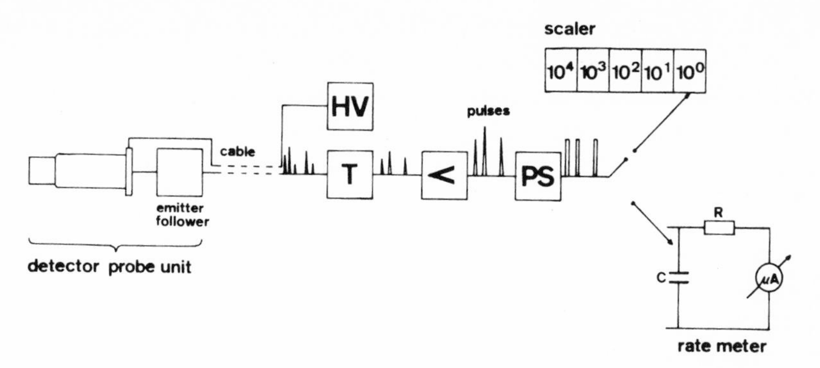

Fig. 13. Schematics of a scintillometer for integral gamma measurements.

must be recorded to attain adequate precision. The first stage records up to 10 pulses. The tenth pulse cleans out the record in the first stage and transmits one pulse to the second stage. When 10 pulses have been received by the second stage, it in turn transmits a pulse to the third stage, and so on. In general, with n decade stages 10^n pulses can be stored. In a counter with five decades the first stage counts the individual pulses, the second stage counts the tens, the third stage the hundreds, the fourth stage the thousands, the last stage the tens of thousands. If the numbers 6, 3, 0, 1, 0 show up, respectively, the total number of pulses counted is

$$6 \times 1 + 3 \times 10 + 0 \times 100 + 1 \times 1000 + 0 \times 10{,}000 = 1036 \text{ pulses} \quad (19)$$

The precision of this measurement (or of any measurement of N pulses) is

$$\sigma = \pm\sqrt{N} = \pm\sqrt{1036} = 32.2 \text{ pulses} \quad (20)$$

The probable error p is

$$p(\%) = \pm\frac{67}{\sqrt{N}} \quad (21)$$

which means a probability of 0.50 that the experimental value N differs from the true value by p. The count rate is given by

$$A = \left(\frac{N}{t} \pm \frac{\sqrt{N}}{t}\right) \text{ cpm or cps} \quad (22)$$

where t is the time (in minutes or seconds) during which the N pulses were accumulated.

Above a certain limit (>10 cps) direct indication of the count rate can be given by ratemeters. In an integrating RC circuit (Fig. 13) the current through the resistor R is a measure of the time average of the count rate A and is indicated on a meter or recorder. The time constant $R \cdot C$ is variable on most instruments (1–20 sec) and the precision of the reading is

$$\sigma(\%) = \pm \frac{100}{\sqrt{2ARC}} \tag{23}$$

This equation is valid if the reading has reached its "equilibrium" state (the capacitor C must be charged) which is established after a time of

$$t_n = RC(0.5 \ln 2ARC + 0.394) \tag{24}$$

t_n is short for short time constants; on the other hand, the precision of the reading will be relatively low. Usually a compromise must be made by selecting a medium time constant.

The time constant can be determined, if unknown, by following the decrease of the reading after removing a radioactive source: during the time constant RC the reading decreases to 37% of its original value ($1/e \simeq 0.37$).

More complicated electronics is necessary if pulse-height analysis (for α and γ spectrometry) should be performed. As described in Sections 3.1 and 3.2 NaI(Tl) crystals and Si or Ge(Li) solid-state detectors supply electric pulses with pulse heights proportional to the incident α or γ energy. The basic principle of pulse-height analysis is single-channel discrimination: since only pulses above a selectable limit can pass a discriminator, two disscriminators followed by an anticoincidence unit can select pulses with amplitudes between the upper (U_u) and lower (U_l) discriminator level (Fig. 14). Pulses above U_u (b, e, and f in Fig. 14) pass both discriminators and are eliminated by the anticoincidence circuit; pulses below U_l (a and d) are rejected by the discriminators; pulse c which lies between U_u and U

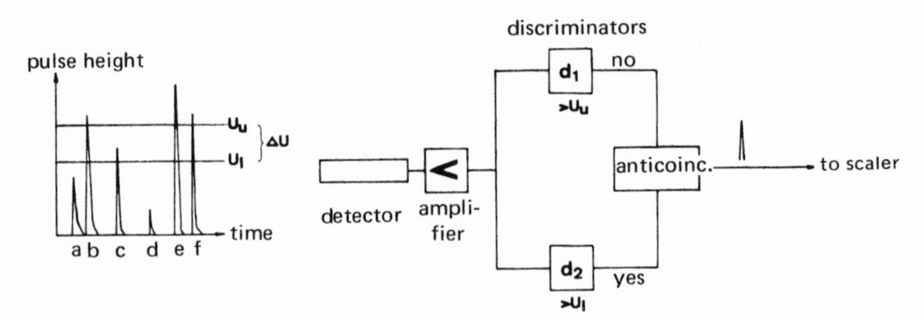

Fig. 14. Principle of pulse-height analysis (see text).

(within the "window" or "channel") appears only after d_2, and thus the anticoincidence unit will deliver a pulse to the counter. $U_u - U_l = \Delta U$ is called the channel width.

Operating several "windows" side by side the whole pulse-height spectrum can be analyzed at the same time. Multichannel analyzers with at least 100 channels are now routinely used in radiometric practice. During a preselectable time the pulses per channel are accumulated and the number is stored in a ferrite-core memory. The information is usually stored in BCD format and can be made available by various means: the decimal number may be recorded by using a typewriter or printer; BCD information may be recorded in computer format on punched cards, punched tape, or on magnetic tape. Digital information also may be converted to analog voltages which are used to display spectra on a cathode-ray tube and may also be used to drive an X–Y curve plotter.

3.4. Autoradiographic Methods

Autoradiography is certainly the oldest method for radioactivity determinations and is still the best way to study the distribution of radioactivity on the hand-specimen scale. In uranium ore-grade materials the radioactive constituents (e.g., pitchblende, brannerite grains) are often inhomogeneously distributed (Fig. 15) and very small (50–300 μ). Autoradiographs taken in direct contact with a polished section of a rock sample permit the exact localization of its radioactive mineral constituents.

The experimental procedure is very simple: the autoradiographs are taken in a light-proof box by placing the specimen under study on the photographic plate or film. In case of uncovered thin sections a good contact must be maintained by superimposing a weight or by clamping. Movement of the rock specimen relative to the emulsion during the exposure must be avoided. Exposure times depend upon the emulsion used, as well as on the specific activity of the radioactive minerals present, and range from a few minutes to several hours.

Using α-sensitive emulsions (Kodak NTB, Ilford K. 2) estimates of the U and Th contents can be made by counting the number of the individual α tracks (N_α) per cm^2 under a microscope:[16]

$$N_\alpha \propto T(c_{\mathrm{U}} + c_{\mathrm{Th}}) \qquad (25)$$

where T is the exposure time. The Th/U ratio (roughly 4 for common rocks, 0.03 for many types of uranium mineralization) must be assumed. For detailed discussion see Ref. 17.

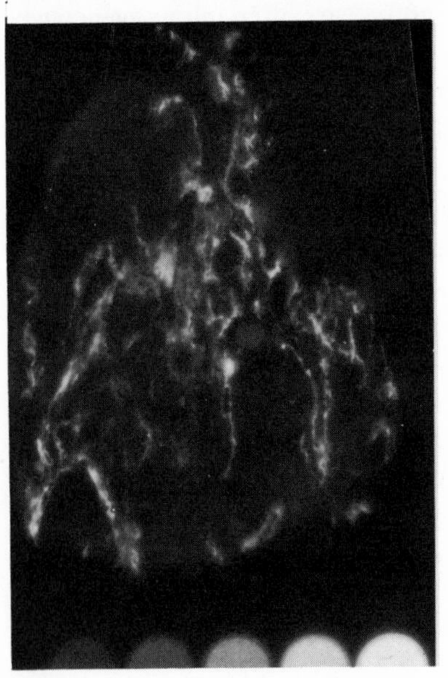

b

Fig. 15. (a) Contact autoradiograph of a quart conglomerate (permian Verrucano, Switzerland showing intergranular distribution of brannerit (white). Kodak single-coated medical x-ray film exposure 30 h. Natural size. (b) Contact auto radiograph (positive) of rock specimen and stan dards (bottom) for quantitative determinations Standard pellets have uranium concentration (from left to right) of 0.25, 0.5, 1.0, 2.0, an 4.0% U.

Comparing the image density produced by chemically analyzed standards to that of an unknown sample, quantitative assessments can be made. A photodensitometer is used for this purpose.[18]

3.5. Background Reduction for Laboratory Measurements

For all radiometric determinations the background count rate A_0 (cpm) must be subtracted from the measured, gross count rate (sample + background) A. The statistical standard deviation σ_A in the net-sample count rate is given by

$$\sigma_A = \pm \sqrt{\frac{A}{t} + \frac{A_0}{t_0}} \tag{26}$$

where t and t_0 are the counting times for the sample and background measurement, respectively. In normal laboratory practice A_0 is determined periodically by counting the background for much longer times than samples. To compare different counting setups the sensitivity A^2/A_0 is used; for assaying low-level samples improvements can be made either by increasing the count rate for a given sample or by reducing the background.

The simplest method of background reduction is the installation of a massive shield surrounding the detector. Different materials and shield thicknesses are used. The most common shielding material is lead. Steel and (triply distilled) mercury are also frequently utilized. Further background reduction can be obtained by installing the counting laboratory in the basement of a building.[19] Careful selection of the construction materials (concrete ingredients, cements) may also help.[20] A shield of 5 in. of lead or an equivalent thickness of other nonradioactive material provides about

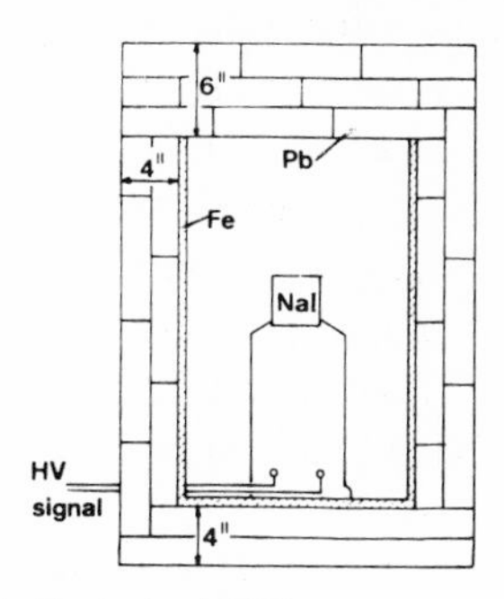

Fig. 16. Typical shielding construction for laboratory scintillation detectors.

the optimum shielding from the external natural radioelements. Figure 16 shows a shielding type which is used in many laboratories.

Cosmic radiation (see Section 2.2) can be drastically reduced by a Geiger anticoincidence ring,[21] but the intrinsic radioactivity of the detector itself (see Section 2.2) will still be present. Special care must be taken to select low-level detector components (Ref. 10, pp. 81–84).

4. RADIOMETRIC PROSPECTING

4.1. Airborne Radiometric Survey

The most typical geophysical field around a radioactive ore deposit (like the magnetic field caused by a magnetite ore body) is the *radioactive anomaly*. By definition the radiation intensity reaches at least *4× the "normal" background* at such locations. Various instruments, methods, and units are used, depending mainly upon the scale of the investigation.

Wide-range reconnaissance surveys are carried out with small airplanes or helicopters. Aeroradiometric prospecting provides a great amount of information in a short time. Several conditions must be fulfilled, however, to justify the high cost of such a survey: (a) moderate topographic relief to enable a nearly constant surveying altitude of 500 ft above ground; (b) the area to be surveyed should be free of glacial or alluvial deposits,* swamps, or extensive vegetation; (c) the airborne detection instruments must be sensitive enough to locate anomalies down to 0.1 μCi/m^2. One or more NaI(Tl) crystals of considerable size (4 × 2 in., 9 × 3 in.) with several photomultipliers connected in parallel are used.

Since the radiation intensity from the ground decreases rapidly with increasing flight altitude (Fig. 17) readings must be reduced to the standard 500-ft elevation: the accumulation time for the pulses coming from the detector (1 sec for 500 ft) is automatically changed according to the signals from the radar altimeter. The cosmic-radiation effect is also automatically subtracted. In aeroradiometric surveying the unit cps is used for integral counting as well as for gamma spectrometry discrimination.[22]

Various survey-line arrangements are employed depending on the problem encountered: short loops normal to the strike direction of length-extended rock units; grid flying over disjointed, patternlike outcrops; "con-

* The terrestrial component of gamma radiation found at 500 ft above terrain comes, due to self-absorption, from the radionuclides in the surficial 12 in. of earth materials within a circular area about 2000 ft in diameter.

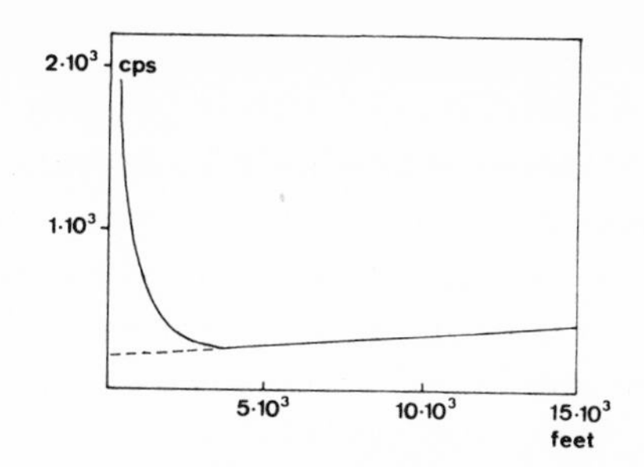

Fig. 17. Gamma-ray intensity *vs* flight altitude over rocks with no radiometric anomalies (after Hand[23]).

tour-line flying" over flatlying sedimentary formations. In order to couple radiometric data with ground locations a film is taken simultaneously with a vertical camera. Flight speed is commonly around 150 mph.

Calibration includes periodic testing of the crystal-detector-assembly response by a small ^{137}Cs source aboard the aircraft. Care must be taken to shield this source during actual measurements. Luminating dials on the flight control panel should make a minimum contribution to the measured radioactivity. The overall uncertainity associated with the measured radiation value is about $\pm 10\%$ in the aircraft position data of about $\pm 0.5\%$ of the distance flown.[23]

Aeroradiometric survey data (radioactivity profiles of the flight lines) are compiled either as a radioactivity unit map or as a contour map. The flight line spacing is usually 1 mile. Due to different radionuclide content of rock units their geologic mapping can be done in many cases by aeroradiometric measurements. Pitkin *et al.*[24] quote impressive examples. Under favorable conditions empiric calibration for concentration determinations can be made: Moxham[25] gives the equation

$$c_{eU}\ (\text{ppm}) = kI_{\gamma_a} \tag{27}$$

where c_{eU} is the mean uranium equivalent content (see Section 5.2) of the surface rocks in the 2000-ft circle below the aircraft, I_{γ_a} is the measured net aeroradioactivity (cps), and k a conversion factor.

More recently four-track registration has been employed:[26] three channels for spectral U, Th, and K discrimination and one track for the integral count rate. Aeroradiometric surveys are often performed simultaneously with aeromagnetic measurements.

4.2. Carborne Prospection

Interesting areas and anomalies found by aeroradiometric prospecting can be studied in more detail by carborne methods. Carborne radiometric traversing is especially successful in areas having extensive road networks. The equipment used is quite similar to the airborne devices. The detector is mounted over the roof of the vehicle to screen maximum area and to reduce the influence of the road pavement (a detector 20 in. above ground "sees" mainly a circular area 20 ft in diameter; from 6 ft this diameter is about 100 ft). Expandable masts can set the detector in a measuring position up to 20 ft above ground. Continuous ratemeter recording enables detecting of anomalies during and after survey. The vehicle speedometer is connected to the recorder chart-drive mechanism. Every mile a mark is automatically made on the record. The ratemeter time constant RC depends on the surveying speed v and must be selected to fulfill

$$RC \geqq \frac{d}{v} \tag{28}$$

where d is the minimum width of anomaly to be resolved. Surveying speeds up to 40 mph can be used. An acoustic alarm signal indicates anomalies above a certain level.

Interpretation of carborne measurements requires careful investigation of different factors influencing the radiometric reading at a given locality, of which the "geometry" (relative configuration of radioactive source-detector) is by far the most important one. The gamma-ray intensity I_γ in any given array is, according to Baranow,[27]

$$I_\gamma = k \frac{\Omega c}{a} \tag{29}$$

where Ω is the solid angle of the array, c is the content of gamma-emitting radioisotopes* (e.g., in U equivalent, see Section 5.2), $a = \mu/\varrho$, the mass absorption coefficient (see Section 3.1), and k a conversion factor relating I_γ (e.g., in cps) to c. For a small change in I_γ

$$\Delta I_\gamma = \frac{\partial I_\gamma}{\partial \Omega} \Delta\Omega + \frac{\partial I_\gamma}{\partial c} \Delta c + \frac{\partial I_\gamma}{\partial a} \Delta a = \frac{kc}{a} \Delta\Omega + k \frac{\Omega}{a} \Delta c - \frac{k\Omega c}{a^2} \Delta a. \tag{30}$$

* The gamma-emitting radioisotopes are assumed to be homogeneously distributed.

Since a varies only moderately over different gamma energies,[28] $\Delta a \approx 0$ and

$$\Delta I_\gamma = \frac{I_\gamma}{\Omega}\,\Delta\Omega + \frac{I_\gamma}{c}\,\Delta c = I_\gamma\left(\frac{\Delta\Omega}{\Omega} + \frac{\Delta c}{c}\right) \tag{31}$$

Thus, variation in the solid angle Ω can affect the reading seriously. If the detector is wholly surrounded by radiating-rock material (e.g., in a tunnel), $\Omega = 4\pi$. At a flat-rock surface $\Omega = 2\pi$. Under field conditions the geometry may vary rapidly which is in many cases difficult to compensate for exactly. Therefore field measurements have only a limited quantitative meaning.

4.3. Prospecting on Foot

Prospecting on foot is widely applied for small-scale surveys (to study radioactive outcrops in detail, to verify the anomalies found by airborne or carborne prospecting). Here also the influence of the geometry is important.

Measurements are made in profile points (e.g., for the investigation of a length-extended ore body) or in grid points (in case of irregularly shaped ore bodies). The detector is held at a distance of about 3 ft from the rock surface. For a detailed "spot examination" a closer position and the use of a lead collimator surrounding the detector head is convenient. The measured values are, after subtraction of the background, plotted and an *isoradiation map* is drawn on which the surface outlines of radioactive ore bodies will show up.

The unit on such surface maps is usually μR/h.* Since most portable instruments register in counts per second (cps) the converting calibration factor must be determined by measuring the radiation intensity of a source with known dose rate at a given distance from the detector. Commonly, small Ra needles (encapsulated in 0.5 mm Pt) are used for this purpose. The conversion factor k can be calculated according to

$$k = \frac{82.5\,q}{I_\gamma r^2}\ (\mu\text{R/h per cps}) \tag{32}$$

where q is the Ra source strength (μCi) and I_γ is the net gamma-ray intensity of the source (cps) measured at the distance r (cm).

Due to the influence of several factors which cannot be exactly enumerated (geometry changes, background variation) it is usually difficult to perform quantitative determinations *in situ*. Therefore, representative samples are taken in the field for laboratory investigations under standardized conditions. One exception is the interpretation of borehole measurements: here the constant geometry (4π) enables quantitative determinations.[29,30]

* μR/h = microroentgens per hour.

It must be emphasized at this point that because of the absorption of gamma rays in rock (90% of the radiation observed above an outcrop comes from the outermost 7 in. of the rock material) the methods mentioned above are in general unable to detect deep-seated ore bodies. Special techniques may help: radon sampling in soil and measuring its concentration;[22] infrared photographs taken from an airplane, which may show surface temperature anomalies caused by the heat produced in the radioactive body.

5. QUANTITATIVE DETERMINATION OF NATURAL RADIOELEMENTS

5.1. Sample Preparation

For quantitative radiometric determinations the measurement of gamma rays by scintillation detectors is most frequently used. Due to the relatively high penetrating power of gamma rays self-absorption within a relatively small sample can be kept fairly constant for materials of quite different compositions.[28]

Sample preparation commonly includes crushing, pulverizing, and homogenizing the material to be studied; by filling a convenient sample container with the rock or mineral powder the geometry can be standardized. There are no limitations on the grain size (which is in most cases <1 mm) if it is roughly constant for the series of samples under study. Sometimes solid specimens (rock cores with constant diameter, tektites with nearly spherical shape) can be measured directly.[31,32] Figure 18 shows different sample-detector arrangements.

5.2. Integral Counting for U and K Determinations

Integral γ counting has its limitations, but as a reconnaissance tool, its speed and simplicity are particularly valuable. In many cases, for example, if the main sample activity is due to its uranium content (ore-grade materials), a simple "total-γ" measurement (see Section 3.3) gives excellent results. With about 350 g of crushed and/or pulverized rock material filled into an 8-oz, aluminum-sample-container canister (Fig. 18g) the measurement can be performed under field conditions with a portable counter. The specific activity α of the sample, which is proportional to the concentration of the corresponding radioisotope in the sample, is given by

$$\alpha = \frac{I_\gamma - I_0}{G} \text{ (cps/g or } \mu\text{R/h/g)} \qquad (33)$$

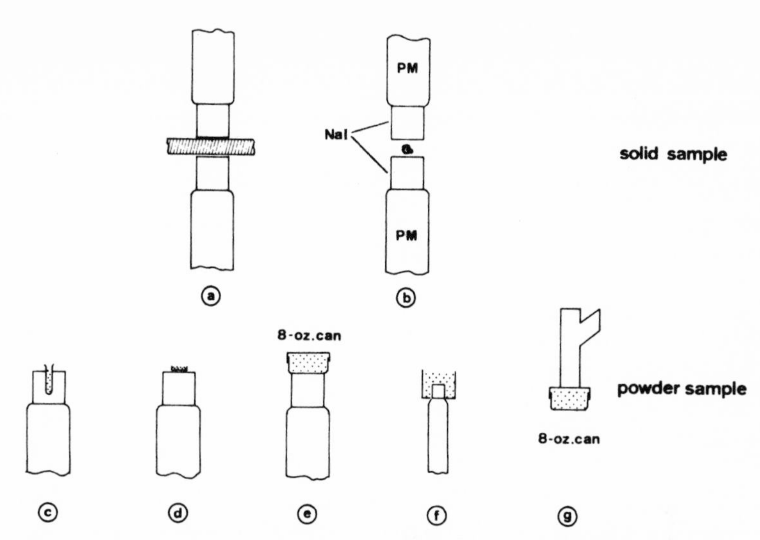

Fig. 18. Various sample-detector arrangements for (a) rock core, (b) tektite, (c) small sample (up to 10 g) in well-type detector, (d) small sample (few g), (e, f) large sample (200–1000 g), (g) measurement with portable counter (see Section 5.2). To reduce background shielding is used for the arrangements (a)–(f).

where I_γ is the measured "sample + background" activity, I_0 is the background (both in cps or μR/h), and G the sample weight (g). For rapid determinations the use of a calibration curve (α vs concentration) (Fig. 19a) is convenient; this can be drawn by measuring and plotting the α values for a series of samples (in the same type of container) with known concentrations* of the radioisotope in question.

For rapid determinations in U ores in radioactive equilibrium (see Section 2.1) this method is best suited. U concentrations >200 ppm can be determined without shielding. The precision depends mainly on the fluctuation of the ratemeter indicator (see Section 3.3) and is usually better than $\pm 10\%$. A measurement takes about 10 sec; an average reading is obtained by watching the fluctuations of the meter needle (the indication should be in the upper one-third of the meter scale; field ratemeter instruments have a range-selector switch for this purpose). Results are expressed in ppm eU, i.e., in "uranium equivalent" which is equal to the true uranium value only if there is no other significant radiation from Th and K present

* Powder standards containing known amounts of U and Th (in radioactive equilibrium) can be purchased from the USAEC New Brunswick Laboratory. Dry, pure KCl or KCO_3 is used as K standard.

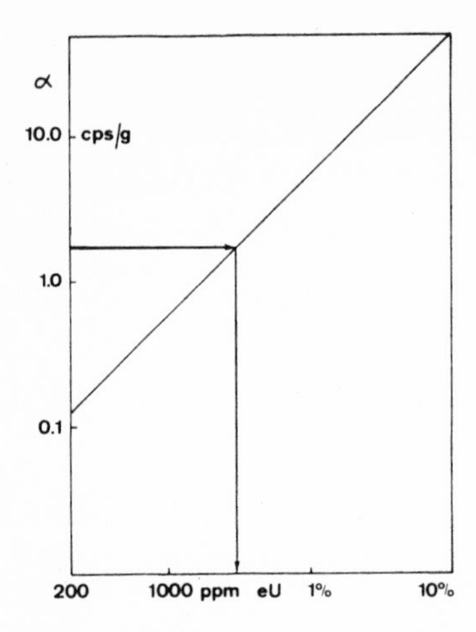

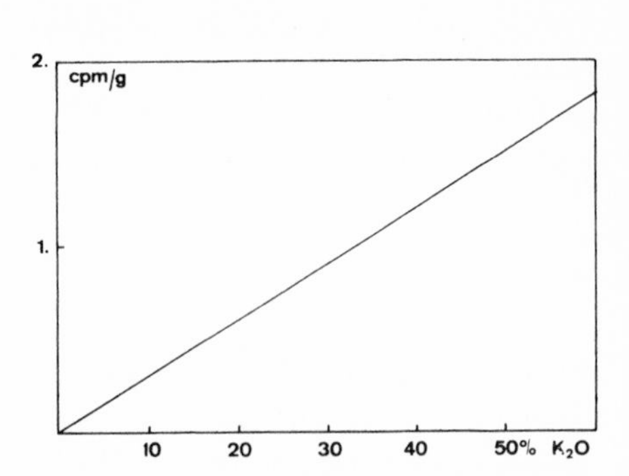

Fig. 19. Working curve for uranium determinations with a portable counter (a), potassium calibration curve for determinations with a laboratory scaler (b).

in the sample. This, of course, affects the accuracy. In Table IV a few examples are given.

With a better geometry (Fig. 18f, sample weight ca. 350 g) and a laboratory scaler (Fig. 13) instead of a portable ratemeter and a shielding of about 2 in. of lead surrounding the sample and detector the sensitivity can be improved and extended down to 0.1 ppm *eU*. In this low concentration

Table IV

Comparison of Equivalent Uranium (*eU*) with Effective Uranium Concentrations

Rock type	Portable field ratemeter, ppm *eU*	Laboratory scaler, ppm *eU*	ppm U	Reference
Sericite gneiss with U mineralization (Verbier, Switzerland)	750 1500	700 1450	730 1600	55
Triassic dolomite with U mineralization (Ferrera, Switzerland)	1100 500	1030 460	1010 450	56
Rotondo granite (Switzerland)	–	27	10.2	57
Mont Blanc granite (Switzerland)	–	28	12.5	54

region, however, the influence of K and Th in the sample is serious and can be resolved only by gamma-ray spectrometry (see Section 5.3).

Materials which are mined for potassium (usually a mixture of KCl and NaCl) contain only minute amounts of U and Th. K contents in such materials can be determined by integral counting (counting time in the order of 10–40 min) with a precision of 2–10%. The statistical error σ_α of a determination with a laboratory scaler is

$$\sigma_\alpha \simeq \pm \frac{\sqrt{N_T + tA_0}}{Gt} \text{ (cpm/g)} \tag{34}$$

where N_T is the number of counts of the sample (plus background) registered during the time t (min) and A_0 is the background count rate (cpm). α is here

$$\alpha = \frac{1}{G}\left(\frac{N_T}{t} - A_0\right) \text{ (cpm/g)} \tag{35}$$

5.3. Gamma-Ray Spectrometry for Simultaneous U, Th, and K Determinations

If the sample contains more than one radioelement gamma-ray spectrometry can resolve the individual radiations. For U and Th the most characteristic γ rays are emitted by the daughter isotopes ^{214}Bi and ^{208}Tl, respectively (see Table II). The K content can be determined by measuring the γ rays from ^{40}K (for isotopic composition and γ-ray abundance see Table I and Fig. 1). The measurements are most frequently carried out with the arrangement shown in Fig. 18f, using a 3 × 3-in. NaI(Tl) crystal. The γ-ray spectrum obtained by these means (upper solid line in Fig. 20) displays characteristic photopeaks at 1.46 MeV (^{40}K), 1.76 MeV (^{214}Bi of the ^{238}U series), and 2.62 MeV (^{208}Tl of the ^{232}Th series).

A "window" is opened at each of these peaks. The window is centered at the middle of the peak, and has a width $2\Delta E$, where ΔE is about $0.05E$: e.g., for potassium $\Delta E = 70$ KeV, and the window extends from 1.39 to 1.53 MeV. In multichannel pulse-height analysis these energies correspond to channel numbers; for a given gain setting the energy-calibration curve can be obtained by identifying the photopeaks in the different channels. For peak identification see the U, Th, and K spectra in Crouthamel[10] and Heath.[13]

The unknown U, Th, and K concentrations in the sample can be calculated after evaluating the calibration constants. This involves the measurement of standards with known U, Th, and K contents under the same

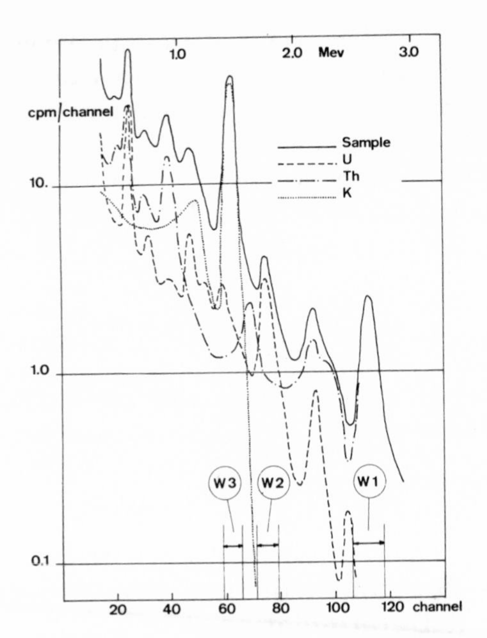

Fig. 20. Gamma-ray spectrum of a rock sample (solid line), contributions from K (. . .), U (—), and Th (-··-). Windows 1, 2, and 3 (see text).

experimental conditions as for the unknown samples. If one assumes that the effect of K is negligible (this is true in the case of accessory minerals like zircon or sphene which contain almost no K). The following symbols may be used

c_U = U content of the sample (ppm)
c_{Th} = Th content of the sample (ppm)
$c_U{}'$ = U content of the U standard (ppm)
c'_{Th} = Th content of the U standard (= 0 in common U standards)
c''_U = U content of the Th standard (ppm)
c''_{Th} = Th content of the Th standard (ppm)
α_1 = specific activity (cpm/g) of the sample in window 1 (at 2.62 MeV)
α_2 = specific activity of the sample in window 2 (at 1.76 MeV)
$\alpha_1{}'$ = specific activity of the U standard in window 1
$\alpha_2{}'$ = specific activity of the U standard in window 2
$\alpha_1{}''$ = specific activity of the Th standard in window 1
$\alpha_2{}''$ = specific activity of the Th standard in window 2

In this list the first two items are unknown, and the rest are known or can be measured. In the determination of the α's the background count rate in the corresponding window must be taken in account (see Table V).

For the sample one can write two equations with two unknowns

$$\begin{aligned} \alpha_1 &= f_1 c_U + f_2 c_{Th} \\ \alpha_2 &= f_3 c_U + f_4 c_{Th} \end{aligned} \tag{36}$$

which means that both the U and Th contents contribute to the activities measured in windows 1 and 2 (the f's are the calibration factors to be determined). Similarly for the U standard we have

$$\begin{aligned} \alpha_1' &= f_1 c_U' + f_2 c_{Th}' \simeq f_1 c_U' \\ \alpha_2' &= f_3 c_U' + f_4 c_{Th}' \simeq f_3 c_U' \end{aligned} \tag{37}$$

Table V

Digital Gamma-Ray-Spectrum Read Out (Counts in Channels 65-128), Calculation of α_1 and α_2[a]

1751	1728	1670	1610	1521	1462	1369	1392
1434	1683	2031	2128	2027	1790	1501	1232
1012	862	750	688	614	617	612	577
638	671	723	769	785	841	810	760
696	592	539	499	463	422	391	419
397	409	374	393	449	559	594	712
709	660	524	483	383	323	240	233
229	200	198	227	214	197	199	168

$$\alpha_1 = \frac{1}{\text{sample weight}}\left(\frac{\text{sum of counts in window 1}}{\text{time}} - \text{background in window 1}\right) =$$

$$= \frac{1}{1.9}\left(\frac{5840.0}{1183.0} - 1.50\right) = 1.808 \text{ cpm/g}$$

$$\alpha_2 = \frac{1}{1.9}\left(\frac{15{,}218.0}{1183.0} - 3.10\right) = 5.138 \text{ cpm/g}$$

[a] Spectrum of a mineral concentrate (1.9 g sphene) recorded by a Nuclear Data multichannel analyzer; accumulation time 1183 min. Window 1 for Th, channels 107–117; window 2 for U, channels 72–80. Background in window 1, 1.50 cpm; in window 2, 3.10 cpm.

from which the calibration factors f_1 and f_3 can be calculated:

$$f_1 = \frac{\alpha_1'}{c_U'}$$
$$f_3 = \frac{\alpha_2'}{c_U'} \tag{38}$$

in cpm/g/ppm. For the Th standard one can write

$$\alpha_1'' = f_1 c_U'' + f_2 c_{Th}''$$
$$\alpha_2'' = f_3 c_U'' + f_4 c_{Th}'' \tag{39}$$

leading to

$$f_2 = \frac{\alpha_1'' - f_1 c_U''}{c_{Th}''}$$
$$f_4 = \frac{\alpha_2'' - f_3 c_U''}{c_{Th}''} \tag{40}$$

From Eq. (36) it follows that

$$c_U = \frac{1}{f_3}(\alpha_2 - f_4 c_{Th}) = k_3(\alpha_2 - k_4 c_{Th}) \tag{41}$$

and

$$\alpha_1 = \frac{f_1}{f_3}(\alpha_2 - f_4 c_{Th}) + f_2 c_{Th} = \frac{f_1}{f_3}\alpha_2 + c_{Th}\left(f_2 - \frac{f_1}{f_3} f_4\right) \tag{42}$$

from which

$$c_{Th} = \frac{\alpha_1 - \dfrac{f_1}{f_3}\alpha_2}{f_2 - \dfrac{f_1}{f_3} f_4} = k_1(\alpha_1 - k_2\alpha_2) \tag{43}$$

where

$$k_1 = \frac{1}{f_2 - \dfrac{f_1}{f_3} f_4}$$
$$k_2 = \frac{f_1}{f_3} \tag{44}$$

k_2 and k_4 are also called the "stripping constants." In practical work c_{Th} is determined first; the result is then used to evaluate c_U.

The application of this method is demonstrated by U and Th determinations in accessory minerals separated from the Silvretta gneiss and from the Giuv syenite (Switzerland). The mineral concentrates as well as the standards were measured in small aluminum sample containers with a thin Mylar-foil bottom. Sample and standard weights were about 1 g. The sample containers (diameter, 1 in.) were centered on the top of a 3 × 3-in. NaI crystal (Fig. 18d). Counting times ranged from 5 to 24 h. The following calibration equations and factors have been found:

$$\begin{aligned} c_{\mathrm{Th}} &= 517(\alpha_1 - 0.0291\alpha_2) \\ c_{\mathrm{U}} &= 115(\alpha_2 - 0.00015c_{\mathrm{Th}}) \end{aligned} \tag{45}$$

The results are given in Table VI, together with mass-spectrometric U determinations. The precision of the results is in the order of ±5% as evaluated from repeated measurements. Taking the mass-spectrometric results as correct the accuracy is better than ±5%.

Table VI

Nondestructive U and Th Determination in Accessory Mineral Concentrates by Gamma-Ray Spectrometry[a]

Mineral	Rock	U, ppm	Th, ppm	U, ppm[b]
Zircon, 53–75 μ fraction	Silvretta gneiss (Switzerland)	550	160	525
Zircon, 42–53 μ fraction	Silvretta gneiss (Switzerland)	590	175	571
Zircon, 30–42 μ fraction	Silvretta gneiss (Switzerland)	600	190	615
Sphene	Giuv syenite (Switzerland)	575	850	–
Epidote–allanite	Giuv syenite (Switzerland)	315	1150	–
Apatite	Giuv syenite (Switzerland)	65	130	–
Zircon	Giuv syenite (Switzerland)	2800	2370	–

[a] Sample weight approximately 1 g.
[b] Mass-spectrometric determinations.[59]

For simultaneous potassium determinations a very similar procedure (solving three equations) gives the constants in the potassium equation:

$$c_K = k_5[\alpha_3 - (k_6 c_U + k_7 c_{Th})] \tag{46}$$

where c_K is the K content of the sample (%) and α_3 is the specific activity in window 3 (centered at 1.46 MeV).

For powdered rock samples (*ca.* 350 g) measured in the arrangement e of Fig. 18 the following factors have been found:

$$\begin{aligned} c_{Th} &= 1120(\alpha_1 - 0.021\alpha_2) \\ c_U &= 370(\alpha_2 - 0.00048 c_{Th}) \\ c_K &= 31.8[\alpha_3 - (2.2 \times 10^{-3} c_U + 5.6 \times 10^{-4} c_{Th})] \end{aligned} \tag{47}$$

It should be mentioned that the sample containers should always be filled to the top; a 2-mm deviation (incomplete filling) causes a 4% error in concentration. Density variations have a much slighter effect.[33]

Table VII indicates the general accuracy and precision of this method. The precisions indicated were calculated from repeated measurements. Counting times range from 40 min for an average granite sample to 24 h for basic rocks.

The method described above does not require too much calculation (a desk computer will do in most cases). Its major disadvantage is that it utilizes only a fraction (<25%) of the information held by the recorded

Table VII

Accuracy and Precision of Gamma-Spectrometric Results[a]

Rock	K, %	U, ppm	Th, ppm	Method
Standard diabase W-1	0.53	0.52	2.2–2.4	Recommended values[61]
	0.52	0.53	2.16–2.24	γ spectrometry
Granite G-2	3.71±0.05	2.1±0.3	25.7±0.5	γ spectrometry
	3.74±0.01			Flame photometry
		2.16±0.07	24.1±0.1	Neutron activation
Basalt BCR-1	1.39±0.03	1.6±0.1	6.1±0.4	γ spectrometry
	1.41±0.01			Flame photometry
		1.81±0.14	6.00±0.19	Neutron activation

[a] From Refs. 33 and 60.

"multichannel" spectrum which therefore must be measured with a relatively high statistical accuracy for long counting times.

In recent years the availability of computers facilitates the analysis of gamma-ray spectra and this analysis can be extended to the whole spectrum. The data obtained by multichannel analyzers offer an excellent opportunity for the application of least-squares methods. Salmon[34] used the unweighted least-squares method for the unscrambling of relatively simple gamma-ray spectra of monoenergetic, artificial radioisotopes. In our case, however, there is in addition to the complex components another complication present: the spectra extend over several orders of magnitude due to scattering effects (higher count rates in the low-energy channels). This fact suggests the introduction of weights.

Let us define the following symbols:

n = number of channels
a_i = count rate of the U standard in channel i (cpm)
b_i = count rate of the Th standard in channel i (cpm)
c_i = count rate of the K standard in channel i (cpm)
d_i = count rate of the sample in channel i (cpm)
v_i = a random error
w_i = weight of d_i
m_U = U in the sample (g)
m_{Th} = Th in the sample (g)
m_K = K in the sample (g)
M_U = U in the U standard (g)
M_{Th} = Th in the Th standard (g)
M_K = K in the K standard (g)
x = the ratio m_U/M_U
y = the ratio m_{Th}/M_{Th}
z = the ratio m_K/M_K
$\sigma_x, \sigma_y, \sigma_z$ = the corresponding standard deviations
Q_x, Q_y, Q_z = "weight factors"
G = sample weight (g)

Due to the linear superposition of the U, Th, and K contributions one can write for each channel (error equations)

$$d_i = a_i x + b_i y + c_i z + v_i, \quad \text{weight } w_i \tag{48}$$

Having n such equations for only three unknowns the application of the weighted least-squares method means the determination of the "best values" for x, y, and z in the sense of

$$\sum_{i=1}^{n} w_i v_i^2 = \sum_{i=1}^{n} w_i [d_i - (a_i x + b_i y + c_i z)]^2 = \text{minimum} \tag{49}$$

It is easy to show that for multichannel operation (constant counting time for each channel)

$$w_i = \frac{1}{d_i} \tag{50}$$

where with the normal equations (Ref. 35, p. 107)

$$\begin{aligned} x \sum_{i=1}^{n} \frac{a_i^2}{d_i} + y \sum_{i=1}^{n} \frac{a_i b_i}{d_i} + z \sum_{i=1}^{n} \frac{a_i c_i}{d_i} &= \sum_{i=1}^{n} a_i \\ x \sum_{i=1}^{n} \frac{a_i b_i}{d_i} + y \sum_{i=1}^{n} \frac{b_i^2}{d_i} + z \sum_{i=1}^{n} \frac{b_i c_i}{d_i} &= \sum_{i=1}^{n} b_i \\ x \sum_{i=1}^{n} \frac{a_i c_i}{d_i} + y \sum_{i=1}^{n} \frac{b_i c_i}{d_i} + z \sum \frac{c_i^2}{d_i} &= \sum_{i=1}^{n} c_i \end{aligned} \tag{51}$$

The matrix solution which is especially suitable for computer data processing gives the values x, y, and z with which the "best fit" can be computed:

$$d_i' = a_i x + b_i y + c_i z \tag{52}$$

With

$$v = \pm \left[\frac{1}{n-3} \sum_{i=1}^{n} \frac{(d_i' - d_i)^2}{d_i} \right] \tag{53}$$

the standard deviations of the unknowns are

$$\begin{aligned} \sigma_x &= \pm v Q_x^{1/2} \\ \sigma_y &= \pm v Q_y^{1/2} \\ \sigma_z &= \pm v Q_z^{1/2} \end{aligned} \tag{54}$$

The Q's can be calculated (Ref. 35, p. 95) from the matrix elements (normal equations).

Finally, the U, Th, and K concentrations in the sample are given by

$$\begin{aligned} \text{ppm U} &= (x \pm \sigma_x) \frac{M_{\text{U}}}{G} \times 10^6 \\ \text{ppm Th} &= (y \pm \sigma_y) \frac{M_{\text{Th}}}{G} \times 10^6 \\ \%\,\text{K} &= (z \pm \sigma_z) \frac{M_{\text{K}}}{G} \times 10^2 \end{aligned} \tag{55}$$

Experimental precautions must be taken to fulfill the assumptions necessary for this method: (a) the standard spectra should be measured to high statistical accuracy; (b) the background is constant; (c) there is no drift. The latter can be attained by utilizing a gain stabilizer.[36]

The measured spectra can be obtained in digital form on punched-tape readout which serves as the input for the computer. With a computer program the best fit and the U, Th, and K concentrations are calculated from the spectra using the channels above 0.35 MeV. Figure 21 gives an example of the best fit calculated by the least-squares method. The major advantage of this method is that counting times can be reduced by a factor of five to seven.

For the analysis of solid-rock core samples an automatic sample changer has been constructed.[19] Rapid analysis can be obtained by combining this device with computer-data interpretation.[31]

Several successful attempts have been made to develop portable γ-ray spectrometers for *in situ* U, Th, and K determinations.[37–39] The weight of

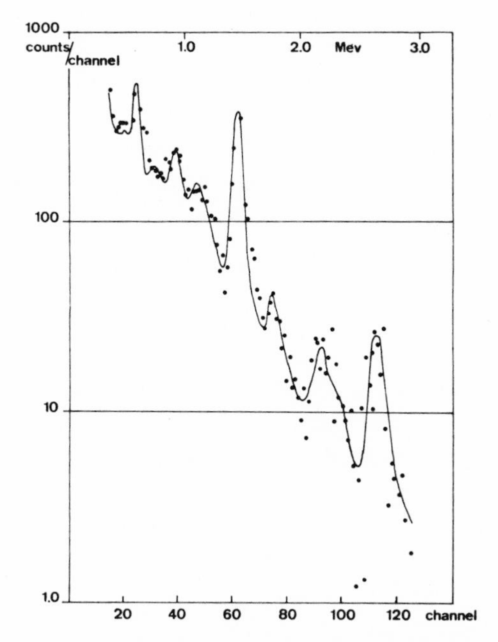

Fig. 21. Measured gamma-ray spectrum and computed best fit for a Mont Blanc granite sample (3.5 ppm U, 13 ppm Th, 2.5% K). Dots: counts in 1000 sec. Solid line: best fit. Sample weight: 595 g (from Rybach *et al.*[54]).

the equipment is in the order of 20 kg. The major benefit of determining U, Th, and K in the field is the greater volume sampled (approximately 300 kg, which is far more representative than a 300-g laboratory sample). Measurements have to be performed in 2π geometry (flat-rock outcrop). For an average granite a 5-min counting operation yields a precision of about $\pm 10\%$ for K and about $\pm 15\%$ for U and Th.

Measuring the natural radiation from U, Th, and K, radiometric methods give satisfactory results down to about 0.2 ppm U and Th and to about 0.1 ppm K (for 0.09-ppm Th and 0.05-ppm U precision is in the order of $\pm 50\%$.[40] For lower concentrations, as are common only in ultrabasic rocks, fission track counting (see Section 5.6) or neutron activation analysis are better suited.[41,42]

5.4. Beta-Gamma Counting for Equilibrium Studies

The foregoing methods assume radioactive equilibrium (see Section 2.1) between the parent isotopes ^{238}U and ^{232}Th (which are to be determined) and their γ-emitting daughter isotopes ^{214}Bi and ^{208}Tl (which can be detected). In most cases this assumption has been found to be valid (for unaltered, fresh rocks). In some cases, however (in altered rocks, leached U ores, miocene coals containing uranium, deep-sea sediments), serious errors can arise. To avoid tedious chemical determinations a simple radiometric method[43] can be used even in the case of completely disturbed radioactive equilibrium.

The method is based on the fact that disequilibrium affects the β and γ activity of a given sample differently. Thus, for example, the U content found by β and γ measurements will be different, depending on the state of equilibrium. For the γ measurement the arrangement f in Fig. 18 (open annular sample container) is used; the β's are counted with an end-window GM detector placed a few mm above the sample-powder surface. An extra calibration curve has to be drawn for the beta measurements by plotting the net β cpm against U concentration of the standards. If $c_{U\beta}$ is the U content as determined by β measurements and $c_{U\gamma}$ is the value found by γ counting, the true U content c_U can be calculated according to

$$c_U = 2.30\, c_{U\beta} - 1.30\, c_{U\gamma} \qquad (56)$$

The ratio $q = c_{U\gamma}/c_U$ is a measure of the state of equilibrium: $q = 1$ for equilibrium. Table VIII gives examples.

A special gamma-spectrometric method[44] enables determination of ^{235}U by measuring its 0.184-MeV photopeak. Since the ratio $^{235}U/^{238}U = 1/137$

Table VIII

Radioactive Equilibrium and Disequilibrium

Sample	$c_{U\gamma}$, ppm	$c_{U\beta}$	q	c_U	c_{UX}[a]
Triassic dolomite, with U mineralization (Ferrera, Switzerland)	445 895	435 875	0.95 0.95	460 920	440 940
Miocene coal (Sellenbüren, Switzerland)	310	76	0.12	615	–

[a] X-ray-fluorescence determinations from Ref. 56.

is constant in nature ^{238}U abundances can be determined even in case of completely disturbed equilibrium. Absorption effects cause difficulties, however, in this low-energy region.

5.5. Alpha Counting, Alpha Spectrometry

Most naturally occurring radioisotopes are α emitters. Despite the short range of the α particles (see Section 2.3) their detection possesses certain advantages as compared to β and γ measurements, the major advantage being the greater sensitivity due to the low background in α counting.

In total-α counting the sample is measured as a "thick source" (the thickness of the source is much greater than the maximum range of α particles, which is about 25–50 μ in solid samples). The measured α activity is expressed in units of $\alpha/cm^2/h$, taking the surface area into consideration. The number of α particles per unit area and time (N) is proportional to the uranium and/or thorium content. For a series of zircons Holland[45] found the relationship

$$c_U\ (\text{ppm}) = 0.463N\ (\alpha/\text{cm}^2/\text{h}) \tag{57}$$

In general two detector types have been applied for total α counting: ZnS screens optically coupled to a photomultiplier tube and proportional gas-flow counters with 2π geometry. Solid samples can be measured directly by placing a polished section on a thin ZnS foil (e.g., ATP-3 alpha strip with a 4.5-mg/cm² ZnS emulsion on Mylar) overlaying a photomultiplier. For an irregular sample-surface shape a matching piece of foil is cut; its weight is a measure of the sample surface area. For gas-flow GM counters,

Table IX

Total Alpha Counts vs Radioelement Content in Various Sedimentary Rocks[a]

Rock type	Net α counts per cm^2, h	Contents, ppm (by γ-ray spectrometry)	
		U	Th
Phosphate	108.0	240.0	4.0
Shale	28.3	80.0	5.0
Limestone	1.3	1.7	1.6
Bauxite	36.7	6.4	96.0

[a] From Ref. 62.

powdered samples (grain size <100 mesh) are filled into sample-holder planchets (typical dimensions, 2 in. in diameter, 1/8 in. deep, taking up about 1.5 g of powder). Background is very low for both detector types (about 0.15 α/cm^2/h); sensitivity is mainly a function of the ratio of the sample-to-detector area.

The counting time per sample ranges from 8 to 48 h in order to attain a reasonable statistical error. The major advantages of the method are the simple sample preparation and the small sample size required. Though less quantitative than γ-ray methods it gives informative results by simple means (see Table IX).

Since the introduction of multichannel analyzers, total-α counting has been frequently replaced by spectrometric techniques. After chemically separating uranium from thorium, single isotopes in the ^{238}U and ^{232}Th decay series can be detected (Figs. 22 and 12). Sample preparation is a rather

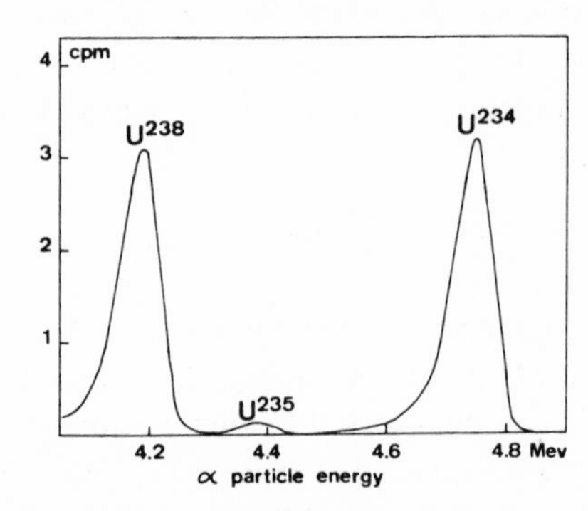

Fig. 22. Alpha spectrum of uranium isotopes. Resolution is 1.0% FWHM for the ^{234}U peak. Counting time 24 h (after Rosholt *et al.*[46]).

Table X

Uranium and Thorium Abundances, Isotopic Ratios Determined by Thin Source Alpha Spectrometry[a]

Sample	U, ppm	Th, ppm	$^{234}U/^{238}U$	$^{230}Th/^{232}Th$
Roanoke River sediment, 2–20 μ fraction	4.12 ± 0.15	13.37 ± 0.74	0.98 ± 0.03	1.01 ± 0.05
Mississippi River sediment, 2–20 μ fraction	3.29 ± 0.12	7.93 ± 0.57	0.94 ± 0.13	1.19 ± 0.06
Red River sediment, 2–20 μ fraction	2.73 ± 0.08	7.70 ± 0.39	0.94 ± 0.03	0.96 ± 0.02

[a] From Ref. 48.

elaborate chemical procedure (U and Th extraction) followed by electrodeposition to obtain a "thin source." Chemical extraction includes decomposing of 5–20 g of sample material in concentrated acids, sodium carbonate fusion of residues, anion exchange, extraction, purification, and determination of the chemical yield (for details see Ref. 46). For electroplating 1-in.-diameter discs (made of stainless steel, titanium, wolfram, or platinum) are used. Electrodeposition is carried out at 6–8 V and 1–2 A for 20–40 min (for details see Ref. 47).

During measurement the sample-holder discs are placed in a vacuum chamber (approximately 20 μ pressure) at a distance of 1–2 mm from a silicon-barrier semiconductor detector with a sensitive area of about 3 cm^2 (see Section 3.2). The spectra are recorded in digital form with a multichannel analyzer. For isotopic-abundance determination the numbers of counts/channel within the corresponding peak are summed up. For further details see Scott,[48] from which the figures in Table X are taken.

5.6. Fission Track Counting

Besides its application to age dating this method is especially powerful for the determination of the abundance and distribution of uranium in single mineral phases[63] or in whole rocks.[64] Polished thin sections or pellets pressed from sample powder and cellulose binder are covered with a piece of Lexan* (polycarbonate plastic). Good mechanical contact is essential between

* Available in sheets (10 μ to 100 μ thick) from Chemical Dept., General Electric Co., Pittsfield, Massachusetts, USA. The uranium content is very low ($<10^{-5}$ ppm).

sample and Lexan foil. Standards (glasses with known uranium content or pellets prepared from powder standards), covered the same way, and samples are mounted together into sandwiches which are irradiated in the thermal column of a nuclear reactor.

Thermal neutron doses of the order of 10^{15} n/cm^2 are applied for samples in the ppm range, lower uranium concentrations requiring correspondingly higher doses. The homogeneity of the neutron flux must be monitored. Irradiation induces fission of some ^{235}U nuclei in both sample and standard, and the massive, charged fission fragments strike the attached Lexan foils leaving tracks approximately 10 μ long.

After etching of the foils, these tracks can be seen with an ordinary optical microscope (Fig. 23). By means of location marks, the occurrence of fission tracks in the Lexan can be correlated with the mineral phases in the underlying thin section. The density (tracks per mm^2) is a measure of the uranium concentration. Track counting is performed visually, under the microscope,[65] or with a discharge counter[66]. The uranium content can be calculated from

$$c_U = c_S \frac{\varrho i}{\varrho iS} \tag{58}$$

where c_S is the uranium content of the standard and ϱ_i and ϱ_{iS} are the induced track densities in the Lexan foils covering sample and standard, respectively. Equation (58) assumes the same fission fragment range in sample and standard.

The major advantage of this method is, besides its simplicity, its great sensitivity; determinations in the ppb range can easily be performed.[67]

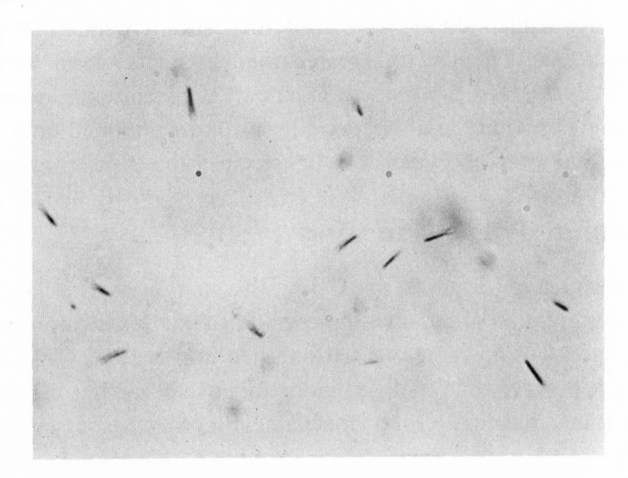

Fig. 23. Fission tracks in a Lexan foil (× 522). Etching conditions: 8 min in 6 *N* NaOH at 70°C.

6. RADIOMETRIC AGE DETERMINATIONS

In the early days of geochronology age dating was carried out on radioactive accessory minerals by the lead-alpha method[49] according to

$$t = \frac{kc_{\mathrm{Pb}}}{\alpha} \tag{59}$$

where t is the age of the accessoric mineral in 10^6 yr, c_{Pb} is the lead content determined by optical spectroscopy, α is the alpha radioactivity, and k is a constant depending on the Th–U ratio in the mineral to be dated. Thus, a *radiometric measurement* was necessary to evaluate the α value of a given mineral sample and, in turn, its age. Because of several serious limitations (one of which is the influence of nonradiogenic lead) this simple and rapid method was successively replaced by more sophisticated methods such as $^{238}U/^{206}Pb$, $^{235}U/^{207}Pb$, $^{206}Pb/^{204}Pb$, $^{87}Rb/^{87}Sr$, and $^{40}K/^{40}A$ dating.[50] These methods require, however, mass-spectrometric determinations. Though the term "radiometric age" appears everywhere in the literature one should be aware of the fact that most probably no radiometric technique was applied.

Radiometric methods are used only in very special cases. For instance gamma-ray spectrometry makes the dating of very young volcanic rocks possible (e.g., historic eruptions of Vesuvius;[51] from $^{234}U/^{238}U$ ratios determined by alpha spectrometry (see Section 5.5) ages in carbonates (e.g., corals) younger than 10^6 yr can be calculated.[52]

REFERENCES

1. Miesch, A. T., Methods of computation for estimating geochemical abundance, U.S. Geol. Survey Prof. Paper 574-B, 1967.
2. Adams, J. A. S., Osmond, J. K., and Rogers, J. J. W., The geochemistry of thorium and uranium, in *Physics and Chemistry of the Earth*, Pergamon Press, New York, 1959, Vol. 3, pp. 298–348.
3. Yokoyama, Y., Tobailem, J., Grjebine, T., and Labeyrie, J., Détermination de la vitesse de sédimentation océanique par une méthode non destructive de spéctrometrie gamma, *Geochim. Cosmochim. Acta* **32**, 347 (1968).
4. Birch, F., Heat from radioactivity, in *Nuclear Geology*, H. Faul, ed., John Wiley & Sons, New York, 1954, pp. 148–174.
5. Wollenberg, H. A., and Smith, A. R., Radiogeologic studies in the central part of the Sierra Nevada batholith, California, *J. Geophys. Res.* **73**, 1481 (1968).
6. May, H., and Marinelli, L. D., Cosmic-ray contribution to the background of low-level scintillation spectrometers, in *The Natural Radiation Environment*, Univ. of Chicago Press, Chicago, 1964, pp. 463–480.

7. Kraner, H. W., Schroeder, G. L., and Evans, R. D., Measurements of the effects of atmospheric variables on Radon-222 flux and soil-gas concentrations, in *The Natural Radiation Environment*, Univ. of Chicago Press, Chicago, 1964, pp. 191–215.
8. Pensko, E., Wardaszko, T., and Wochna, P., Natural atmospheric radioactivity and its dependence on some geophysical factors, *Atompraxis* **14**, 255 (1968).
9. Gustafson, P. F., and Brar, S. S., Measurement of γ-emitting radionuclides in soil and calculation of the dose arising therefrom, in *The Natural Radiation Environment*, Univ. of Chicago Press, Chicago, 1964, pp. 499–512.
10. Crouthamel, C. E., *Applied Gamma-Ray Spectrometry*, Pergamon Press, Oxford, 1960, 443 pp.
11. Taylor, J. J., Application of gamma-ray buildup data to shield design, WAPD-RM-217, 1954.
12. Taylor, D., *The Measurement of Radio Isotopes*, Methuen & Co., London, 1957, 132 pp.
13. Heath, R. L., *Scintillation Spectrometry Gamma-Ray Spectrum Catalogue*, Phillips Petroleum Co., 1964.
14. Hill, C. R., Osborne, R. V., and Mayneord, W. V., Studies of α radioactivity in relation to man, in *The Natural Radiation Environment*, Univ. of Chicago Press, Chicago, 1964, pp. 395–405.
15. Bowie, S. H. U., Portable X-ray fluorescence analysers in the mining industry, *Mining Magazine* **106**, 1 (1968).
16. Yagoda, H., *Radioactive Measurements with Nuclear Emulsions*, John Wiley & Sons, New York, 1949, 356 pp.
17. Ragland, P. C., Autoradiographic investigations of naturally occurring materials, in *The Natural Radiation Environment*, Univ. of Chicago Press, Chicago, 1964, pp. 129–151.
18. Bowie, S. H. U., Autoradiography, in *Physical Methods in Determinative Mineralogy*, J. Zussman, ed., Academic Press, London, 1967, pp. 467–473.
19. Adams, J. A. S., Laboratory γ-ray spectrometer for geochemical studies, in *The Natural Radiation Environment*, Univ. of Chicago Press, Chicago, 1964, pp. 485–497.
20. Wollenberg, H. A., and Smith, A. R., Radioactivity of cement raw materials, Symposium on Geology of Cement Raw Materials, Indiana University, 1966, pp. 129–147.
21. Sonntag, C., Extremely low-level scintillation spectrometer. *In*: Radioactive Dating and Methods of low-level Counting, IAEA Vienna, 1967, pp. 675–686.
22. Bowie, S. H. U., and Bisby, H., Methods of detecting and assessing low grade uranium deposits, *J. Brit. Nucl. Ener. Soc.* **14**, 169 (1967).
23. Hand, J. E., Instrumentation for aerial surveys of terrestrial γ radiation, in *The Natural Radiation Environment*, University of Chicago Press, Chicago, 1964, pp. 687–704.
24. Pitkin, J. A., Neuschel, S. K., and Bates, R. G., Aeroradioactivity surveys and geologic mapping, in *The Natural Radiation Environment*, Univ. of Chicago Press, Chicago, 1964, pp. 723–736.
25. Moxham, R. M., Some aerial observations on the terrestrial component of environmental γ radiation, in *The Natural Radiation Environment*, Univ. of Chicago Press, Chicago, 1964, pp. 737–746.
26. Morris, D. B., personal communication, 1969.
27. Baranow, V. I., Aeroradiometric prospecting for uranium and thorium deposits and the interpretation of gamma anomalies, *Proc. Int. Conf. on Peaceful Uses of Atomic Energy*, Geneva, 1956, Vol. 6, pp. 740–743.

28. Davisson, C. M., and Evans, R. D., Gamma-ray absorption coefficients, *Rev. Mod. Phys.* **24**, 79 (1952).
29. Scott, J. H., Dodd, P. H., Droullard, R. F., and Mudra, P. J., Quantitative interpretation of gamma-ray logs, *Geophysics* **26**, 182 (1961).
30. Scott, J. H., Computer analysis of gamma-ray logs, *Geophysics* **28**, 457 (1963).
31. Rybach, L., and Adams, J. A. S., Automatic analysis of the elements U, Th and K in solid rock samples by nondestructive gamma spectrometry, *Proc. Int. Anal. Chem. Conf. Budapest*, 1966, Vol. 2, pp. 323–330.
32. Rybach, L., and Adams, J. A. S., The radioactivity of the Ivory Coast tektites and the formation of the Bosumtwi Crater (Ghana), *Geochim. Cosmochim. Acta* **33**, 1101 (1969).
33. Heier, K. S., and Rogers, J. J. W., Radiometric determination of thorium, uranium and potassium in basalts and in two magmatic differentiation series, *Geochim. Cosmochim. Acta* **27**, 137 (1963).
34. Salmon, L., Analysis of γ-ray scintillation spectra by the method of least squares, AERE-Rept. 3640, 1961.
35. Grossman, W., Grundzüge der Ausgleichsrechnung, Springer Verlag, Göttingen, 1961, 406 p.
36. Comunetti, A. M., A new gain stabilizing system for scintillation spectrometry, *Nucl. Instr. Meth.* **37**, 125 (1965).
37. Adams, J. A. S., and Fryer, G. E., Portable γ-ray spectrometer for field determination of thorium, uranium and potassium, in *The Natural Radiation Environment*, Univ. of Chicago Press, Chicago, 1964, pp. 577–596.
38. Lovborg, L., A portable γ-spectrometer for field use, Danish Atomic Energy Commission, Risö Rept. No. 168, 1967.
39. Doig, R., The natural gamma-ray flux: *in-situ* analysis, *Geophysics* **33**, 311 (1968).
40. Heier, K. S., Uranium, thorium and potassium in eclogitic rocks, *Geochim. Cosmochim. Acta* **27**, 849 (1963).
41. Lovering, J. F., and Morgan, J. W., Uranium and thorium abundances in possible upper mantle materials, *Nature* **197**, 138 (1963).
42. Nagasawa, H., and Wakita, H., Neutron activation analysis of potassium in ultrabasic rocks, *Geochem. J.* **1**, 149 (1967).
43. Eichholz, G., Hilborn, I., and McMahon, C., The determination of uranium and thorium in ores, *Can. J. Phys.* **31**, 613 (1953).
44. Bunker, C. M., and Bush, C. A., Uranium, thorium and radium analyses by gamma-ray spectrometry (0.184-0.352 million electron volts), U.S. Geol. Survey Prof. Paper 550-B, 1966, pp. 176–181.
45. Holland, H. D., Radiation damage and its use in age determination, in *Nuclear Geology*, H. Faul, ed., John Wiley & Sons, New York, 1954, pp. 175–180.
46. Rosholt, J. N., Doe, B. R., and Tatsumoto, M., Evolution of the isotopic composition of uranium and thorium in soil profiles, *Bull. Geol. Soc. Amer.* **77**, 987 (1966).
47. Goldberg, E. D., and Koide, M., Geochronological studies of deep sea sediments by the ionium/thorium method, *Geochim. Cosmochim. Acta* **26**, 417 (1962).
48. Scott, M. R., Thorium and uranium concentrations and isotope ratios in river sediments, *Earth Plan. Sci. Letters* **4**, 245 (1968).
49. Larsen, E. S., Keevil, N. B., and Harrison, H. C., Method for determining the age of igneous rocks, using the accessory minerals, *Bull. Geol. Soc. Amer.* **63**, 1045 (1952).
50. Hamilton, E. I., *Applied Geochronology* Academic Press, London, 1965, 267 pp.

51. Vitozzi, P., and Rapolla, A., Radioactivity of vesuvian effusive products former to 1631, *Bull. Volc.* **32**, 136 (1969).
52. Thurber, D. L., Broecker, W. S., Blanchard, R. L., and Potratz, H. A., Uranium series ages of Pacific atoll coral, *Science* **149**, 55 (1965).
53. Cherry, R. D., Richardson, K. A., and Adams, J. A. S., Unidentified excess alpha-activity in the 4.4-MeV region in natural thorium samples, *Nature* **202**, 639 (1964).
54. Rybach, L., von Raumer, J., and Adams, J. A. S., A gamma spectrometric study of Mont Blanc granite samples, *Pure and Applied Geophysics* **63**, 153 (1966).
55. Hügi, Th., Köppel, V., De Quervain, F., and Rickenbach, E., Die Uranvererzungen bei Isérables (Wallis). Beitr. Geol. Schweiz, Geotechn. Serie, 1967, Lfg. 42.
56. Dietrich, V., Huonder, N., and Rybach, L., Uranvererzungen im Druckstollen Ferrera-Val Niemet. Beitr. Geol. Schweiz, Geotechn. Serie, 1967, Lfg. 44.
57. Rybach, L., Hafner, S., and Weibel, M., Die Verteilung von U-Th, Na, K und Ca im Rotondogranit, *Schweiz. Min. Petr. Mitt.* **42**, 307 (1962).
58. Adams, J. A. S., and Lowder, W. M., eds., *The Natural Radiation Environment*, Univ. Chicago Press, Chicago, 1964, p. 1069.
59. Grauert, B., and Arnold, A., Deutung diskordanter Zirkonalter der Silvrettadecke und des Gotthardmassivs (Schweizer Alpen), *Contr. Mineral. and Petrol.* **20**, 34 (1968).
60. Morgan, J. W., and Heier, K. S., Uranium, thorium and potassium in six U.S.G.S. standard rocks, *Earth Plan. Sci. Letters* **1**, 158 (1966).
61. Stevens, R. E., and others, Second report on a co-operative investigation of two silicate rocks, *U.S. Geol. Survey Bull.* **1113**, 1 (1960).
62. Adams, J. A. S., Richardson, J. E., and Templeton, C. C., Determinations of thorium and uranium in sedimentary rocks by two independent methods, *Geochim. Cosmochim. Acta* **13**, 270 (1958).
63. Price, P. B. and Walker, R. M., A simple method of measurering low uranium concentrations in natural crystals, *App. Phys. Let.* **2**, 23 (1963).
64. Fisher, D. E., Homogenized fission track analysis of uranium: A modification for whole rock geological samples, *Anal. Chem.* **42**, 414 (1970).
65. Kleeman, J. D. and Lovering, J. F., Uranium distribution in rocks by fission-track registration in Lexan plastic, *Science* **156**, 512 (1967).
66. Bertine, K. K., Chan, L. H., and Turekian, K. K., Uranium determinations in deep-sea sediments and natural waters using fission tracks, *Geochim. Cosmochim. Acta* **34**, 641 (1970).
67. Fisher, D. E., Homogenized fission track analysis of uranium in some ultramafic rocks of known potassium content, *Geochim. Cosmochim. Acta* **34**, 630 (1970).

Chapter 11

NUCLEAR ACTIVATION ANALYSIS

L. E. Fite, E. A. Schweikert, and R. E. Wainerdi

Texas A & M University
College Station, Texas

and

E. A. Uken

Scientific Advisory Council
Pretoria, South Africa

1. INTRODUCTION

The importance of the minor and trace elements in mineral chemistry is becoming more widely recognized with advances in technology and interpretation. The characteristics of many materials can be influenced by the trace-element impurities found in various systems. For example, the quality of steel is altered by its trace oxygen content[1]; the presence of trace vanadium in crude-oil poisons the platinum catalysts used in the cracking process by the oil industry[2]; sodium and chloride concentrations in sweat are used for the detection of cystic fibrosis in children today[3]; and the concentration of trace pesticides in foodstuffs have importance.[4] Numerous other practical applications of trace-element determinations have been investigated and reported on in the literature[5] in virtually all of the sciences.

The determinations of ultratrace, trace, and minor constituents in natural systems may also provide a useful framework for the classification of geochemical samples and may provide the basis of a method for mineral exploration. To obtain useful information pertaining to possible mineral deposits, or to the geochemical status of a given region, would require the analysis of many properly selected samples. One method by which this may be accomplished is atomic absorption spectroscopy (see Chap. 8) as described by West.[6] A generally more sensitive method is nuclear activation analysis. It can be automated under certain conditions; and, with the development of solid-state detectors and the use of computers, the technique can be made applicable to large-scale geochemical analyses.

Nuclear activation analysis is a technique for determining qualitative and quantitative elementary composition by means of nuclear transmutation and the subsequent measurement of emitted radiation from an unknown substance. Basically, when a material is irradiated by nuclear particles produced in a nuclear reactor, particle accelerator, cyclotron, or other suitable source, some of the atoms present in the material interact with the bombarding particles and, in many cases, are converted into radioactive isotopes with known nuclear characteristics.

The separation and identification of each source of induced radioactivity and a cataloging of the intensity of each becomes the basis for the measurement of the quantity of the original element present in the irradiated material. The radiation may be emitted either during the bombardment or in the course of radioactive decay, wherein elements (actually isotopes) present in a sample are identified by the characteristic energy and half-life of the emitted radiation. The interactions of major importance in nuclear activation analysis are obtained from nuclear-particle-induced decay gamma rays. The three basic steps in nuclear decay activation analysis are:

(a) Irradiation of the sample to induce the desired radioactivity.
(b) Measurement of the emitted radiations from the sample.
(c) Interpretation of the data to obtain the qualitative and quantitative analysis of the samples.

1.1. Irradiation

The activating irradiation can be neutrons, protons, deuterons, tritons, helions, or even high-energy gamma-ray photons.[7] The most common methods of irradiation are summarized in Table I. The technique of activation analysis involving neutrons can be further separated into three types of neutron-interaction analysis that produce gamma rays: inelastic scattering, capture, and decay studies. Each process emits a gamma ray that is analytically characteristic for a given interaction and radioisotope.[8] Analyses based on decay gamma rays are, by far, the most commonly used.

Table I

Methods of Irradiation

1. Nuclear reactor
    - Advantages
        - High thermal neutron flux density
        - Simultaneous irradiation of many samples
        - Can provide long irradiations
    - Disadvantages
        - High initial cost
        - Suitable reactors are not widely available for service irradiations
        - Flux variability from one sample location to another
2. 14-MeV neutron generators (Cockcroft–Walton type)
    - Advantages
        - Relatively low initial cost
        - Monoenergetic 14-MeV neutron output
        - Portability
    - Disadvantages
        - Low flux compared to reactors
        - Difficult to achieve uniform flux dosage in sample
3. Isotopic neutron sources
    - Advantages
        - Low cost
        - Extremely reliable, with no maintenance
        - Reasonably constant neutron output
    - Disadvantages
        - Neutron production cannot be turned off
        - Flux levels generally lower than either generators or reactors
4. Cyclotron (charged particles)
    - Advantages
        - Suitable for activating lighter elements
        - By proper selection of particle type and energy interferences can be avoided—surface analyses possible
        - Penetration depths can be controlled
    - Disadvantages
        - Comparatively expensive
        - Not presently suitable for large numbers of samples
        - Bulk analyses very difficult

1.2. Measurement of Induced Radioactivity

The two principal approaches to evaluating induced radioactivities are: (a) radiochemical separations followed by counting of the separated activities and (b) instrumental gamma-ray spectrometry. The instrumental approach to activation analysis, although generally simpler and more rapid than the radiochemical approach, usually requires that great care be given to the selection of bombarding particles, to the duration of bombardment, and to the counting procedures employed. It also tends to be more limited in application, particularly when the sample contains a number of trace elements in a matrix that activates very easily. The analyses of geochemical samples are usually performed instrumentally.

The various radioisotopes formed in a sample can be identified by a device called a "gamma-ray spectrometer," which usually consists of a gamma-ray detector, a pulse-height analyzer, and some form of read-out device, and which measures the energies and intensities of the various gamma rays that emanate from an activated sample. The two gamma-ray detectors most commonly used are thallium-activated sodium iodide crystals [NaI(Tl)][9] and high-resolution lithium-drifted germanium [Ge(Li)], a solid-state device.

1.3. Gamma-Ray Interactions

Gamma rays interact in matter by three principal processes: photoelectric capture, Compton scattering, and pair production.

A photoelectric interaction occurs when the complete energy of an incident gamma ray is deposited within the detector. The signal output of the detector can thus be proportional to the total energy of the interacting gamma ray.

The Compton-scattering process involves the transfer of some of the incident gamma-ray energy to an atomic electron, leaving the remainder of the scattered gamma ray with reduced energy. The scattered gamma ray may then interact by one of the three processes, or may escape the detector. The effect is, therefore, a count of less energy than results from a photoelectric interaction.

Pair production results in the formation of a positron and electron pair during the capture of a gamma ray. This requires that a minimum of 1.02 MeV of energy be possessed by the incident gamma ray, with any excess energy of the gamma ray appearing as kinetic energy shared between electron and positron that are produced as a pair. The positron will eventually recombine with some free electron, annihilating itself and the electron, and producing two gamma rays of 0.51-MeV energy, moving 180° apart. The annihilation radiation may interact in, or may escape from, the detector.

The Compton continuum, caused by Compton interactions, makes identification and recognition of low-energy gamma-ray photopeaks extremely difficult. Anticoincidence detectors have been designed for use with not only NaI(Tl) crystals[10] but also with Ge(Li) solid-state detectors.[11] These detectors usually consist of a second scintillation detector surrounding the primary detector and serve the purpose of reducing the number of Compton interactions that are counted.

When gamma rays interact with the atoms of a NaI(Tl) crystal, the crystal emits visible light, proportional in intensity to the gamma-ray energy deposited in the crystal. A photomultiplier tube, optically coupled to the detecting crystal, produces a voltage pulse proportional in magnitude to the intensity of the light striking the photocathode, and thus proportional to the energy of the interacting gamma ray.

The solid-state detector [Ge(Li)] is a reverse-biased diode, and when gamma rays interact with the atoms of this type of detector, free electrons are produced. The magnitude of the current flow from the solid-state detector is proportional to the energy deposited within the detector by the gamma ray that interacts with it.

Although the gamma-ray resolution of the Ge(Li) solid-state detector is superior by a factor of approximately 30–35 to that observed with a NaI(Tl) crystal, the efficiency of the Ge(Li) detector is significantly less than that of a NaI(Tl) crystal. The magnitude of this reduction in counting efficiency is, of course, dependent upon the respective sizes of the two detectors, but, typically, the efficiency of a 30-cc Ge(Li) detector is 30× lower than that of a 3 × 3 in. NaI(Tl) crystal.

The current pulse produced by the Ge(Li) solid-state detector or the voltage produced from a NaI(Tl) crystal, is then amplified and stored according to its height in the memory of a pulse-height analyzer. The channels of the analyzer correspond to energy intervals of the gamma-ray spectrum. The number of pulses counted and stored in the various channels are, therefore, a function of the characteristic gamma-ray spectrum of the radioactive material being examined. A plot of this information will show peaks in the channels corresponding to the particular gamma-ray energies that were detected and stored in the pulse-height analyzer.

The development of the high-resolution Ge(Li) solid-state detector has been very valuable for geochemical applications of neutron activation analysis. A gamma-ray spectrum from a geochemical sample can contain 50–70 distinct gamma-ray energies from 15–20 different elements.

The gamma-ray spectrum shown in Fig. 1 was obtained from a sulfide standard of the Canadian Association of Applied Spectroscopists (CAAS) using a 30-cc Ge(Li) solid-state detector. The sample had previously been bombarded in a thermal-neutron-reactor flux density of 2×10^{12} n cm^{-2} sec^{-1} for 4 h. Following a waiting time of 6 days to permit the decay of the short half-life interfering radioactivities, the sample was counted for 1 h. From the 26 observed photopeaks in the spectrum, 14 elements have been identified.

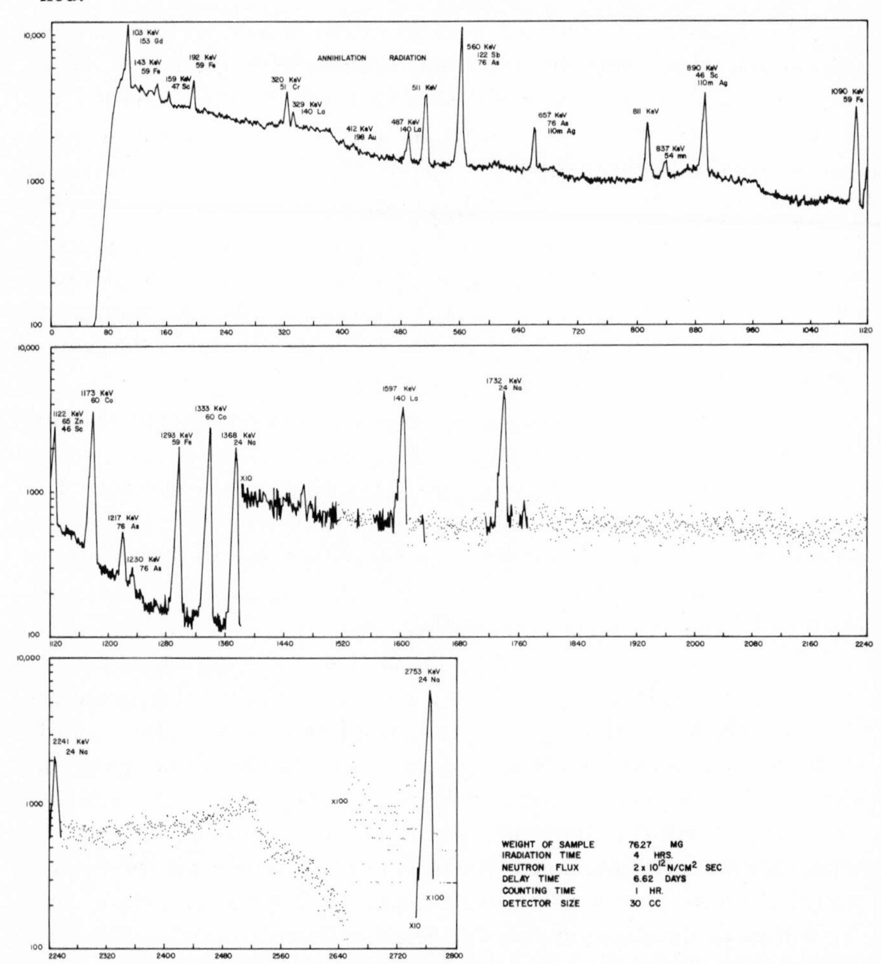

Fig. 1. Gamma-ray spectrum of CAAS sulfide ore-1 using germanium–lithium detector.

1.4. Data Interpretation

Computer techniques are highly suitable for the interpretation of data obtained through nuclear activation analysis and may even be considered essential for large numbers of geochemical samples in view of the speed with which data can be accumulated from automatic activation-analysis systems.

Mathematical techniques are used to resolve the complex spectrum into components that correspond to individual radioactivities. This procedure is based on the principle that the combined gamma-ray spectrum, measured under carefully controlled conditions, can be considered as a function of the sum of the individual constituent radioactivities.

The activation equation, which describes the absolute method of quantitative analysis, is

$$A = N\sigma\psi(1 - e^{-0.693T_1/t_{1/2}})\,(e^{-0.693T_2/t_{1/2}})$$

where A is the intensity of the activity or the disintegration rate of the radioisotope (dps), N is the total number of atoms of the target nuclide present in the sample, σ is the activation cross section (cm^2), ψ is the neutron flux density ($n\ cm^{-2}\ sec^{-1}$), $t_{1/2}$ is the half-life of the radioisotope, T_1 is the irradiation time (same units as $t_{1/2}$), and T_2 is the delay time from the end of irradiation to the beginning of the measurement of the induced activity (same units as $t_{1/2}$).

The activation cross section is the probability that a neutron will interact with the nucleus of a target atom and produce an unstable nuclide. The quantity $1 - e^{-0.693T_1/t_{1/2}}$ in the activation equation is known as the saturation factor and $e^{-0.693T_2/t_{1/2}}$ is designated as the decay factor.

Usually, instead of applying the above basic equation, the quantitative analysis of an unknown sample is accomplished by comparing the intensity of the radioactivity A_u induced in the unknown sample with A_s produced in a known standard of the same element. Thus,

$$\frac{N_u}{N_s} = \frac{A_u}{A_s}$$

where N_u is the number of atoms in the unknown and N_s is the number of atoms of the same element in the comparison standard.

Both the unknown and the standard must be subjected to identical irradiation and counting conditions in order to make this simple relation applicable.

Various computer programs have been developed to process analytical information on unknown samples. The principal mathematical methods of data reduction utilize linear superposition to resolve a complex spectrum into its component parts, thereby providing a qualitative and quantitative analysis. This procedure requires that the conditions under which the spectra are accumulated be such that the component spectra add linearly. Therefore, in any energy interval ΔE the total number of counts in the complex spectrum is due to the summation of the contributions of all the isotopes present. The most commonly used methods for separating a complex spectrum into its components are spectrum stripping,[12] matrix methods,[13] restricted and unrestricted least-squares curve fitting,[14,15] and peak-area calculations.[16] Each of these methods of data reduction requires that computers be supplied with "library" information concerning each radioisotope that may be observed in the complex spectrum. The library information contains the known characteristics of the radioisotope, such as its half-life and emitted gamma-ray energies, and a spectrum of the pure radioisotope. This spectrum is obtained by irradiating and counting the pure radioisotope and the unknown sample under identical conditions. A description of each of the four methods mentioned above follows.

1.4.1. Spectrum-Stripping Method

The spectrum-stripping method of data reduction is accomplished by first identifying the highest energy, statistically significant photopeak in a spectrum. The photopeak can be identified by the gamma-ray energy or energies and half-lives of the library radioisotopes. Once identified the library spectrum of the pure radioisotope is adjusted to yield the same peak area as observed in the complex spectrum; this component is stripped, or subtracted from, the complex spectrum. Since the library spectrum was obtained from a known amount of the pure element, the quantitative analysis is determined by the amount of the library radioisotope removed from the complex spectrum. This process is repeated until all the peaks have been examined and eliminated. The principal limitation of spectrum stripping is the inaccuracy resulting from the inherent compounding of statistical errors by successive subtractions.

1.4.2. Matrix-Solution Method

The matrix solution for a complex spectrum is based on the fact that the total activity in any given energy interval ΔE is the sum of all the constituent activities within that interval. Thus, for any energy interval

$$U_j = \sum_{i=1}^{i=p} \alpha_i S_{ji}$$

where U_j is the activity of the complex unknown spectrum for the energy interval j, S_{ji} is the activity in the energy interval j for the standard library i and α_i is the ratio of the activity of the unknown in the energy interval j to the standard library spectrum i.

The value of α_i is determined by solving a set of simultaneous equations that are constructed by measuring the height of the complex spectrum at different energy intervals. The number of equations is determined by the number of components in the spectrum.

1.4.3. Least-Squares Curve Fitting

The least-squares technique forms a synthetic spectrum from the library spectra in such a manner as to minimize the square roots of the variations between the observed spectrum and the synthetic spectrum. The factors by which the library spectra are adjusted to yield the best fit of the synthetic spectra to the observed spectra provide the quantitative analysis of the unknown sample. Certain restrictions should be employed when using the least-squares method to eliminate negative results. This can be accomplished by omitting the isotope or isotopes that yield negative results from the solution of the least-squares equations.

1.4.4. Peak-Area Calculations

The quantitative analysis of an unknown sample may also be determined by peak-area calculations. This is accomplished by comparing the peak areas observed in the complex spectrum to the photopeak areas obtained in the library spectra under the same irradiation and counting conditions. Various methods of determining the peak boundaries and the peak areas have been investigated and reported.[17] The peak-area method generally works very well unless there are interfering peaks, and, by using Covell's method,[18] the boundaries can be chosen to minimize the effect of the interference. The Covell and classical methods are widely used in activation-analysis calculations and generally give accurate results.

1.5. Automated Nuclear Activation-Analysis Systems

The automation of both radiochemical and instrumental separation methods has expanded the applicability of activation analysis to routine geochemical problems involving large numbers of samples. Automation of purely instrumental systems can be multipurpose[19] or of special design for a particular analysis, such as the determination of oxygen, aluminum, silicon magnesium, and iron in rocks.[20] Automated analysis systems capable of handling large numbers of similar samples using both reactors and accelerators as the activation source have been designed and are routinely used for the trace-element analysis of various materials.[21–23]

The sampling and preparation methods as well as the parameters for the analysis are determined by the trace elements of interest in the unknown sample, and by the interferences present in the sample matrix.

1.5.1. Sampling and Sample Preparation and Manipulation

Considerable attention must be given to the collection of the samples to obtain representative geochemically valid data. This is of particular importance when considering geochemical samples since the weathering process may extend several hundred feet into the earth's crust.[24] It may be desirable to crush the specimens to improve the homogeneity of the sample, but grinding the sample could also introduce contamination. As a general rule, preparation of the samples should be kept at a minimum, and it requires meticulous care to minimize contamination. Because the method of neutron activation analysis is a bulk method, it is usually desirable and sufficient to obtain samples in the mg range.

Since instrumental activation analysis is a nondestructive technique, it may be desirable to reanalyze a given sample to verify the previous results or to recount the sample at specified intervals of time to follow the decay of the radioisotopes produced by the bombardment of the sample. The ability to reactivate and/or recount samples as desired is one of the principal advantages of activation analysis in comparison to destructive methods.

Samples for activation analysis are usually placed in small, clean polyethylene vials for irradiation and counting. Very small and usually known amounts of sodium, chlorine, aluminum manganese, and vanadium are among the common contaminants observed in the polyethylene vials. Such vials become brittle and crack owing to irradiation damage during high reactor-power irradiations; therefore, quartz or aluminum containers are sometimes used where very high neutron fluxes or high temperatures are required. It is occasionally necessary to remove the sample from its irradiation container and place it in a nonirradiated vial for counting because of the impurities common to the different types of vials. The transfer step is difficult to automate and is, for the most part, done manually. Great care must be exercised to avoid partial or total loss of the irradiated sample; but postirradiation contamination is impossible, which is another principal advantage of activation analysis.

1.5.2. Analysis Parameters

Prior to the routine analysis of large numbers of samples it is desirable to optimize the parameters of analysis, which include irradiation, waiting and counting times, the type of irradiation source, and counting-system geometry. These parameters are dependent upon the gamma-ray energies of the emitted radioactivities, the activation cross section of the elements present in the sample matrix, and the half-life of the induced radioisotopes from not only the trace or minor elements of interest, but also the other elements within the sample that become radioactive. Large numbers of similar samples can then be processed automatically and routinely at rapid rates by unskilled personnel, provided no significant interferences are present.

2. REACTOR-THERMAL-NEUTRON ACTIVATION ANALYSIS

The basic components required for the analysis of short-half-life isotopes, i.e., radioisotopes with a half-life in seconds or minutes, are a pneumatic system, an irradiation and delay timer, a pulse-height analyzer with a suitable gamma-ray detector, and a computer-compatible read-out system. The automatic activation-analysis system shown in Fig. 2 is capable of irradiating, counting, and analyzing large numbers of similar samples. The system is primarily used for the determination of short-half-life radioisotopes, which are induced in samples by activation with thermal neutrons. The sample is pneumatically transferred to the reactor core for irradiation and returned for immediate counting and analysis. After the previously mentioned experimental parameters have been established, a technician following simple instructions can irradiate and count the activated samples on a routine basis. This system has been used to determine vanadium and titanium in geochemical samples on a routine basis.[25]

Standards of the radioisotopes of interest should be analyzed periodically to serve as flux monitors and also as checks on the activation-analysis-system performance. If the same samples are periodically reanalyzed, the technician operating the system can recognize a system malfunction by the altered location of the gamma-ray photopeaks and their relative peak heights after irradiation and counting, since the same composition should result for each analysis of the same sample.

The analysis of long-life radioisotopes can be performed on an automatic activation-analysis system, such as shown in Fig. 3, or by using the

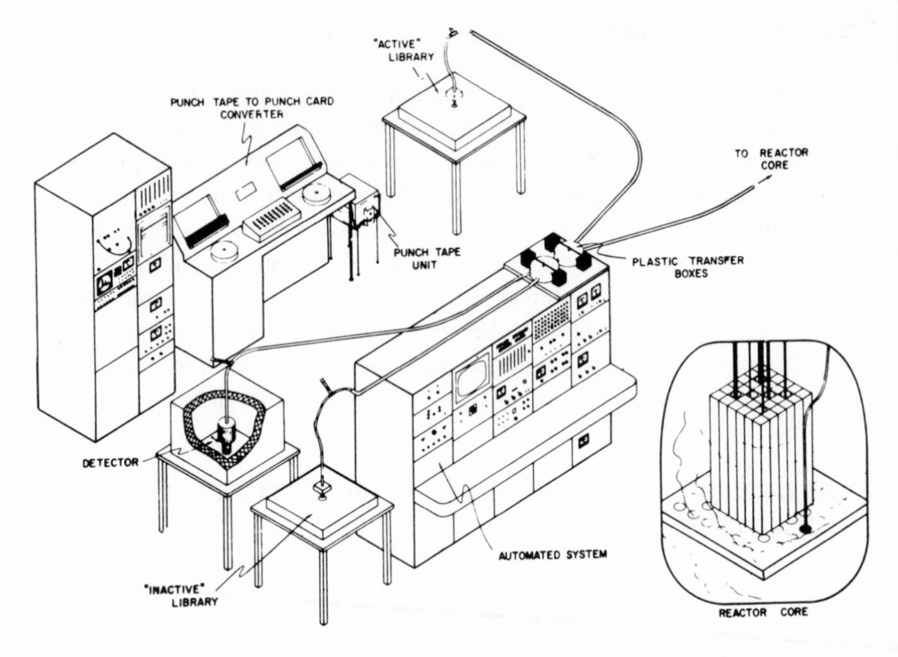

Fig. 2. Mark I–Ia automatic activation-analysis system.

high-resolution spectrometer, shown in Fig. 4. The Texas A & M Mark II Automatic Activation Analysis System shown in Fig. 3 is a very-high-capacity automated spectrometer. The system consists of three gamma-ray spectrometers, i.e., three 3×3 in. NaI(Tl) crystals and three 400-channel pulse-height analyzers with digital gain and baseline stabilization, two sample storage libraries, a pneumatic transfer system, a real-time clock, a computer-compatible magnetic tape read out, and a control system.

The samples located in the storage library are selected automatically by the control system and transferred to one of the detectors via the pneumatic tube. When the count starts, the time is recorded to the nearest 0.01 min on magnetic tape along with the sample identification number and the analyzer code number. The analyzer code number is used by the computer to determine which samples were counted by each pulse-height analyzer. At the end of the counting period the time and the data stored in the memory of the pulse-height analyzer are recorded on the magnetic tape. This system is capable of analyzing samples at the rate of 100/h and is routinely used in biomedical applications of activation analysis.[26] The use of the system for geochemical applications would be limited by the resolution of the NaI(Tl) crystals; alternatively, Ge(Li) detectors could be used.

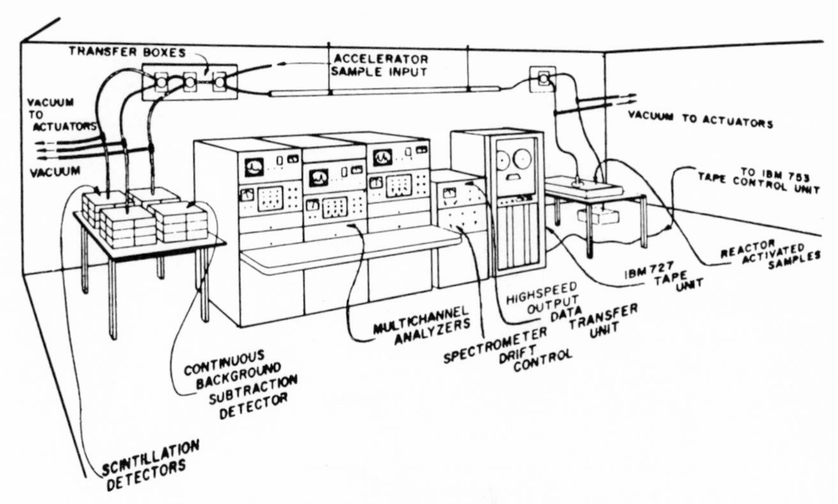

Fig. 3. Mark II automatic activation-analysis system.

Another automated system is shown in Fig. 4 and is designated as the high-resolution spectrometer. This device consists of a shielded 30-cc Ge(Li) detector, a 3200-channel pulse-height analyzer, a computer-compatible magnetic tape read out, a storage "library" for the samples after activation, a pneumatic system, a real-time clock, a sample identification, and a control system. The sample identification number, the time of day at the beginning and end of data accumulation, and the live time of counting are punched onto paper tape, and the spectral data is read onto magnetic tape. After converting the information on the paper tape to punch cards, the cards and the magnetic tape are processed by the computer to determine the analysis of the elements in the sample.

The routine analysis of 350 geochemical samples for gold content was recently completed using this automatic activation-analysis system. In addition, zirconium, lanthanum, and samarium have been routinely determined in geochemical samples with this device. The system was also used in a research program to study the distribution of iron, scandium, cobalt, lanthanum, europium, hafnium, and tantalum in a specific geologic area.[27]

Samples to be counted on the two automated systems described above are usually irradiated in a reactor-mounted rotisserie that is capable of holding 100 samples. Standards are also distributed throughout the rotisserie positions to serve as comparison mass standards for the analysis and as neutron flux monitors. It is assumed that all of the samples in the rotisserie are exposed to known levels of neutron flux, and this assumption is verified by analyzing known standards.

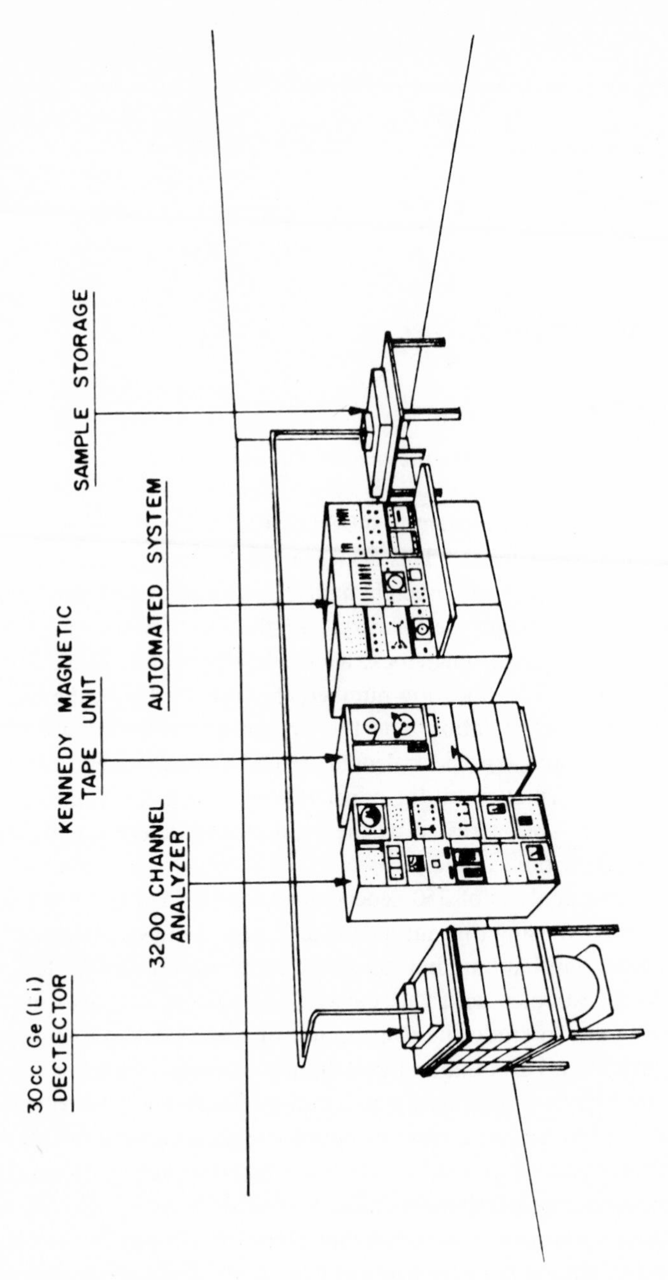

Fig. 4. High-resolution spectrometer.

Table II

Summary of Experimental Detection Limits Determined with Reactor Thermal Neutrons[a]

Atomic No. range	Minimum detection ranges, μg									
	10^{-5}	10^{-4}	10^{-3}	10^{-2}	10^{-1}	10^{0}	10^{1}	10^{2}	10^{3}	10^{4}
8–9					F				O	
11–18		Ar	Na, Al	Cl	Mg			Si, S		
19–35	Mn	Sc, V, Br	Co, Cu, Ga, As, Se	K, Ti, Zn, Ge	Cr, Ni	Ca		Fe		
37–53	In		Sr, Rh, Ag, Cd, Sn, Sb, I	Rb, Zr, Mo, Ru, Pd, Te	Y, Nb					
55–56		Cs, Hf	Ba, W	Ta, Pt, Os	Pb					
57–71	Eu, Dy	Sm, Ho, Er Lu	La, Nd, Gd, Yb	Ce, Pr,	Tm Tb					
72–82	Au	Re, Ir		Hg						

[a] The thermal neutron flux is 4×10^{12} n cm^{-2} sec^{-1}.

The detection limits for various elements that have been experimentally determined are given in Table II. These detection limits were obtained using a NaI(Tl) crystal as the gamma-ray detector and would vary somewhat for Ge(Li) solid-state detectors because of their lower counting efficiency.

Since most of the elements within a sample become radioactive when bombarded with thermal neutrons, often radioisotopes are produced that emit radiation that will interfere with the measurement of the desired radioactivities; therefore, it may be desirable to employ radiochemical separation techniques to remove the induced activity of interest from the other induced activities in a given sample. Generally, radiochemical techniques are based on postactivation radiochemical separations, and the automation of these techniques may expand the applicability of activation analysis. Automated radiochemistry systems are still in the developmental stage but tend to be specialized and designed to separate one or more elements from a given matrix. Radiochemical separation techniques have been applied to geochemical analysis of rock to reduce the matrix interference from the radioactivities of interest,[28] but for large numbers of analyses at modest cost instrumental methods are recommended at the present time.

3. 14-MeV NEUTRON ACTIVATION ANALYSIS

For the determination of certain minor elements in geochemical samples it may be desirable to use a 14-MeV neutron generator as the irradiation source. A typical automated activation-analysis system using a 14-MeV neutron generator is shown in Fig. 5. The sample is transferred to the 14-MeV neutron generator by the pneumatic tube and irradiated for a preset time as determined by the control system. Various neutron monitors are utilized to monitor the output of the neutron generator. At the end of the irradiation the sample is transferred to a position between two NaI(Tl) detectors, which are connected to a pulse-height analyzer, and, after a predetermined delay time, the data is accumulated. The spectral data is usually placed on paper tape while the next sample is being irradiated. This system is routinely used in the analysis of geochemical samples for their aluminum, iron, magnesium, silicon, and oxygen content,[29,30] including the deep-sea core samples recovered by drilling operations on the ocean floor.[31] Table III indicates the experimental detection limits for numerous elements[32] for a 14-MeV neutron flux of 2×10^8 n cm^{-2} sec^{-1}. This system, as well as the

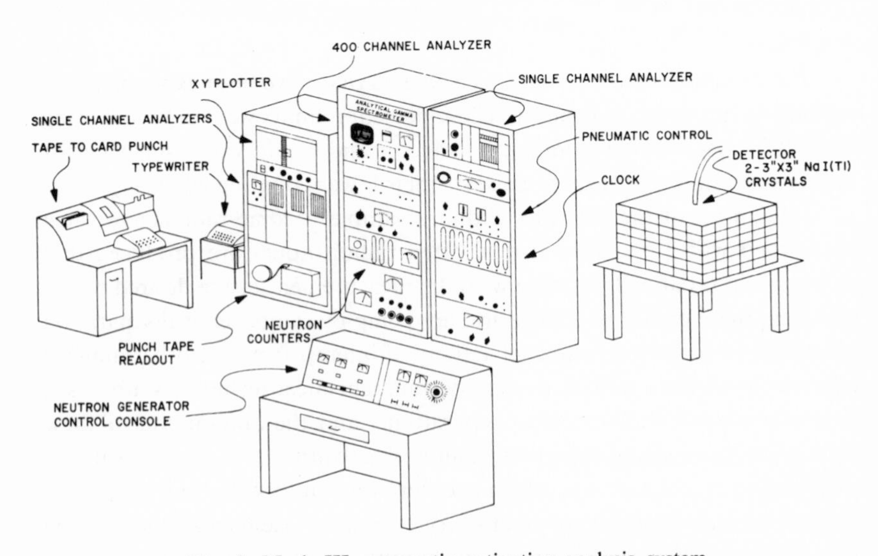

Fig. 5. Mark III automatic activation-analysis system.

Table III

Summary of Experimental Detection Limits Determined with 14-MeV Neutrons[a]

Atomic No. range	Minimum detection ranges, mg							
	0.001	0.01	0.05	0.1	0.5	1.0	5.0	10.0
8–9			O			F		
11–18			Si	P Al	Na		Mg, Cl	
19–36		Cr	Mn		Ni	Fe	K, Co	
			Cu	Se	Br			As
37–54				Sr Rb	Pd	Mo, Ru		Nb
				Sb	Cd	Sn	Ag	
							In	
55–71			Pr	Ba	Dy	Tb	Sm Eu	Cs Gd
			Nd	Ce			Er	Lu
72–86					Ta	Hf		Ir
					Hg			Au Pb

[a] The neutron flux is 2×10^8 n cm^{-2} sec^{-1}.

others previously described, uses polyethylene pneumatic tubes, and all transfer boxes are built from plastic without lubricants to further reduce the possibility of sample contamination.

A portable system similar to the one shown in Fig. 5 has been successfully used for the routine analysis of samples in the field. A sealed-tube neutron generator was placed in a 6-ft-deep, 3-ft-sq. hole while the detector and the counting equipment were located in a trailer adjacent to the hole. By placing the neutron generator in the hole, adequate shielding was provided for the system operator and the counting equipment from the neutrons produced by the generator.

An automatic activation-analysis system that is capable of performing *in situ* compositional analysis of a geologic area of interest is shown in Fig. 6 and is called the "turn-around" system. This system consists essentially of a small 14-MeV neutron generator and a NaI(Tl) crystal. The system is rotated about the center axis of the device to allow either the detector or the neutron source to be placed directly above and close to the sample. This configuration permits the maximum practical distance between the neutron generator and the detector, to prevent activation of the detector crystal, while providing the optimum geometry for irradiation and counting. The gamma rays detected in the scintillation crystal are stored in a pulse-

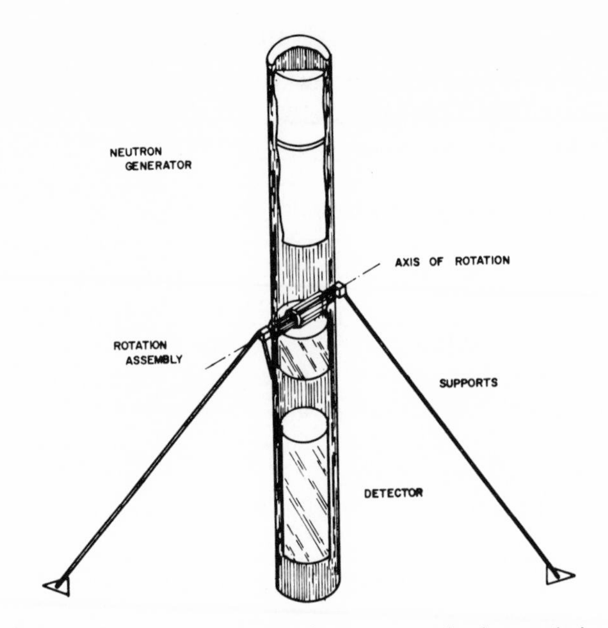

Fig. 6. "Turn-around" system for remote activation analysis.

height analyzer (not shown). This system has been used in the analysis of numerous rock samples[33,34] and was developed as part of a proposed moon probe.

A similar system, which has no moving parts, is shown in Fig. 7. In addition to decay activation analysis, this system can be used for inelastic

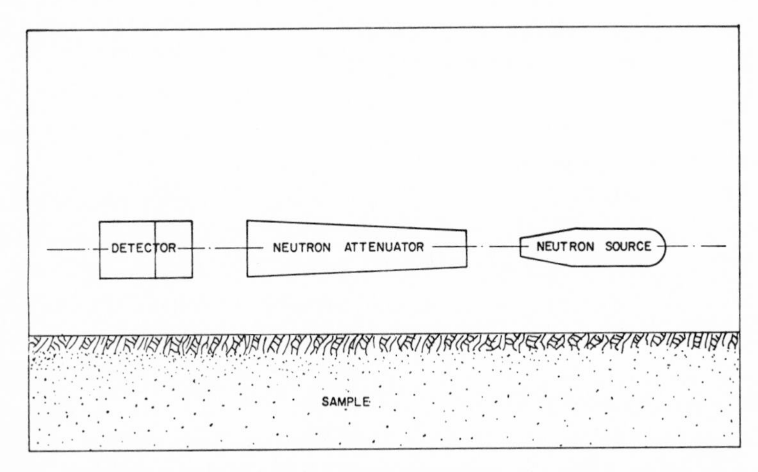

Fig. 7. Physical arrangement of major components for a combined activation analysis system.

scattering, capture gamma ray, and cyclic activation analysis of large geochemical samples. When considering decay activation analysis, this system has the disadvantage of poor irradiation and counting geometry when compared with the previously mentioned system.[35]

3.1. Cyclic Activation Analysis

In this method of analysis the sample is bombarded with a short burst of neutrons and the induced activity is measured between bursts. A suitable delay between the neutron burst and the counting must be made to allow for the depletion of the neutron-capture gamma rays. This technique has been demonstrated in the analysis of rock samples for aluminum, silicon, and oxygen content.[36] A serious limitation of the method results from the poor irradiation and counting geometry arrangement; however, the method is well suited for the determination of radioisotopes with a relatively short half-life.

4. CHARGED-PARTICLE AND PHOTON ACTIVATION ANALYSES

4.1. Introduction

In addition to thermal and fast neutrons, photons and charged particles can also be used for activation analysis. Almost all stable elements can be determined with excellent sensitivities (10^{-6}–10^{-10} g) using photons, protons, deuterons, ^{3}He, or ^{4}He ions of energies up to 40 MeV. More specifically the interest for these activating particles lies in that they offer high sensitivity for most of the elements which cannot be determined with neutron activation analysis or can only be determined with limited sensitivity. This is the case for the elements with atomic numbers 3 to 23 and a few heavier elements such as iron and lead.

For a comprehensive survey on the field of photon and charged-particle activation analysis the reader is referred to the detailed discussions given by Albert,[37,38] Engelmann,[39,40], Baker,[41–43] and Tilbury.[45]

4.2. Photon Activation Analysis

4.2.1. Nuclear Reactions and Sources of Photons

The type of nuclear reaction induced by photons depends upon their energy. At low photon energies (3–6 MeV) generally only γ, γ' reactions can occur. At higher photon energies ($E_\gamma > 7$ MeV), the most prominent reaction is γ, n, which for photon energies 7–10 MeV has cross sections typically 10^3 higher than γ, p and 10^5 higher than γ, α reactions, for example. However, even for γ, n reactions the absolute values of the cross sections are usually small (10–100 mb), thus for determinations featuring high sensitivity, high-intensity sources of photons are needed. The threshold energies for γ, n reactions on light elements range from 10 to 15 MeV and from 7 to 10 MeV for the heavier elements, the resonance energies giving maximum cross sections are usually 10 MeV above the threshold values.

Photons of the required energy and high intensity can be obtained as bremsstrahlung photons produced by bombarding a high-Z-number target with electrons. The accelerators used for producing high-energy electrons are Van de Graaff accelerators, cyclic machines, such as the betatron, or linear accelerators, "linacs." Linacs are used mostly for photon activation analysis since they feature two important advantages, production of high-intensity (up to 250-μA average electron currents) and high-energy (up to 45-MeV electron energies) electrons and thus bremsstrahlung fluxes.

4.2.2. Analytical Applications

This discussion is divided into two parts, arbitrarily termed low-energy photon activation using photons up to 6 MeV, and high-energy photon activation with photons of energies higher than 7 MeV, typically photons of 15–40 MeV and as high as 70 MeV.

Low-Energy Photon Activation Analysis. This type of activation has been used for the determination of heavier elements (Se to Mg) where γ, γ' reactions yield short-lived isomers as, for example $^{77}Se(\gamma, \gamma')^{77m}Se$ or $^{179}Hf(\gamma, \gamma')^{179m}Hf$. The cross sections for γ, γ' reactions are low (a few mb), thus limiting the sensitivity which is generally in the range of a few milligrams.[45,46]

Two γ, n reactions can also occur in the low-energy range: $^9Be(\gamma, n)^8Be$ and $^2H(\gamma, n)^1H$. In both cases the reaction products are not radioactive, and the analytical applications developed are based on the detection of the neutrons emitted during irradiation.[47] The thresholds for these reactions are sufficiently low, 1.67 MeV and 2.23 MeV, respectively, that photons from radioactive sources can be used for activation. Beryllium can be determined at concentration levels as low as 1 ppm[48,49] using relatively low intensity sources of up to 1 Ci of ^{124}Sb (E_γ, 2.1 MeV) and ^{88}Y (E_γ, 1.84 MeV) Deuterium down to 0.02 vol. % of D_2O has been measured[50,51] using ^{24}Na (E_γ, 2.75 MeV).

High-Energy Photon Activation Analysis. A very large range of elements can be determined with excellent sensitivities using γ, n reactions with bremsstrahlung photons of 20–35 MeV. To illustrate this, some detection limits reported[52] are given in Table IV. These detection limits are calculated for an interference-free irradiation equal to one half-life or 1-h maximum using a 27-MeV electron beam of 50 μA with the following counting conditions: minimum of 100 counts/min for radioisotopes with half-lives of less than 6 days and 500 counts/600 min for radioisotopes of half-lives longer than 6 days, counting on a 3×3 in. NaI(Tl) detector. Additional information on calculated detection limits for a large number of elements at different irradiation energies can be found in the papers published by McGregor,[54] Engelmann,[39,40] Wilkniss and Linnenboom,[55] Baker,[42,43] Anderson *et al.*,[56] and Owlya *et al.*[57]

The analytical procedures for destructive or nondestructive determinations are similar to those used in neutron activation analysis. Two particular points must be mentioned, however, one concerning the correct and reproducible positioning of the sample in the bremsstrahlung beam and the other concerning the selection of the proper irradiation energy. These questions have been discussed by Engelmann,[39,40] Baker *et al.*,[58] Wilkniss and Linnenboom,[55] and Lutz.[44]

Most applications reported have dealt with light-element analyses such as the determination of oxygen and especially carbon and nitrogen in high-purity materials.[59–61] Rocco *et al.*[62] have used this technique for determining nitrogen in diamonds. The sensitivity limit for nondestructive analyses was estimated at 0.2 ppm of nitrogen for 40-mg samples.

Several investigators[41–43,52,53,55,56] have studied interesting analytical applications for determining heavier elements ($Z > 10$), where photon activation has definite advantages over neutron activation for the following reasons:

(a) Photon activation gives higher sensitivity or the reaction product has a more readily identifiable gamma spectrum or half-life (de-

Table IV

Examples of Detection Limits in Photon Activation Analysis

0.1–1 μg	Cr, Mn, Te, Co, Ni, Zn, Mo, Hf, Pb
0.01–0.1 μg	C, N, O, F, Ti, As, Sr, Zr, Nb, Ag, Sb, Ta, Pt
0.001–0.01 μg	Cu

terminations of titanium, iron, magnesium, and calcium, for example).

(b) An interference is eliminated (analysis of nickel in copper).

(c) Self-shielding problems as occurring in thermal-neutron activation with matrices composed of boron, tungsten, manganese, cadmium, and hafnium are eliminated.

Specific applications in geochemical analysis are illustrated in the references cited below. Berzin *et al.*[63] have reported on the analysis of rocks and ores with selective irradiations using successively increasing photon energies. With this procedure interfering contributions could be accounted for. Copper, zinc, barium, silicon, and iron were determined in ores using successively a 10.8-MeV irradiation giving barium, an 11.6-MeV irradiation giving barium and copper, a 12.8-MeV irradiation giving barium, copper, and zinc, and, finally, a 15.6-MeV irradiation giving copper, zinc, barium, silicon, and iron. Schmitt *et al.*[64] have described elegant nondestructive methods combining photon activation and high-resolution Ge(Li) gamma-ray spectrometry. Several meteorites and USGS Standards were analyzed for their major- and minor-element content. Magnesium, calcium, titanium, and nickel were determined nondestructively. In addition, this method offered similar sensitivities to those attained in instrumental thermal-neutron activation for sodium, scandium, chromium, manganese, iron, cobalt, rubidium, yttrium, and cerium.

4.3. Charged-Particle Activation Analysis

4.3.1. Nuclear Reactions and Sources of Charged Particles

As for photon activation, this technique provides a wide analytical potential complementing the capabilities of neutron activation. A number of analytically useful nuclear reactions can be induced by charged particles of low (a few hundred keV to a few MeV) and medium energies (up to 50 MeV).

Each reaction has its characteristic threshold and resonance energies with cross sections which can be quite large (up to the barn range). Examples of the different types of nuclear reactions which can be used for the determination of oxygen are given in Table V.

The minimum energy required for a nuclear reaction to occur (except for low-energy elastic scattering or Rutherford scattering) is set by the Coulomb barrier of the target nucleus with respect to the incident and outgoing particles and, in addition, for endothermic reactions, by the Q value (energy balance) of the reaction.[44] A variety of sources of charged particles

Table V

Low-energy reactions	Medium-energy reactions
$^{16}O(p, p)^{16}O$	$^{18}O(p, n)^{18}F$
$^{16}O(p, \gamma)^{17}F$	$^{16}O(p, \alpha)^{13}N$
$^{18}O(p, \alpha)^{15}N$	$^{18}O(d, n)^{18}F$
$^{18}O(p, \gamma)^{19}F$	$^{16}O(^{3}He, p)^{18}F$
$^{16}O(d, p)^{17}O$	$^{18}O(^{3}He, pn)^{19}F$
$^{18}O(d, p)^{19}O$	$^{16}O(^{4}He, pn)^{18}F$
$^{16}O(d, \alpha)^{14}N$	
$^{18}O(d, \alpha)^{16}N$	

are used for charged-particle activation analysis depending upon the type and energy of particle required: isotopic sources (α-emitting heavy radioisotopes), Cockcroft-Walton and Van de Graaff accelerators, heavy-ion linear accelerators ("hilacs"), and cyclotrons.

Activation techniques based on the use of charged particles must take into account the fact that their penetration is very limited and dependent upon the kind and energy of the bombarding particle and the chemical composition of the target. From the standpoint of analytical applications, two important consequences result from the rapid loss of energy of charged particles in matter:

(a) The limited penetration depth and the corresponding small sample volumes analyzed are of disadvantage in analytical applications directed toward obtaining meaningful average values of an impurity content. On the other side the energy-dependent penetration opens interesting possibilities for investigating impurity distributions in successive sample thicknesses. The only applications in this direction have, however, been limited so far to utilizing low-energy charged particles for surface-impurity determinations. It must also be noted here that the energy lost through the stopping of the bombarding particles is given up in the form of heat (for example, a 10-MeV proton beam of 10 μA intensity dissipates 100 W). Thus, the target must be heat resistant or cooled in an appropriate way.

(b) A nuclear-reaction cross section varies with the energy of the bombarding particle; thus, the specific radioactivity induced will vary with the thickness of the sample. This nonuniform activation implies that particular attention be given to ensure correct quantization. This question has received thorough attention and different investigators have proposed various quantization methods.[65–67] A discussion on quantization and flux monitoring in geochemical analyses has been given by Sippel *et al.*[68]

4.4. Analytical Applications

4.4.1. Prompt Detection Methods

These methods, developed for surface analysis, use low-energy charged particles and are based on the measurement of the prompt γ radiation emitted or of the elastically or inelastically scattered particles.

Nuclear interactions emitting prompt γ radiation such as p, γ reactions have strong resonances for protons up to 4 MeV. Excellent sensitivities for p, γ activation have been reported for a number of light elements by Pierce *et al.*[69] A specific application studied by these authors dealt with the magnesium, aluminum, silicon, and iron distribution in copper pyrites and the aluminum distribution in quartz–tourmaline samples.[70]

Turkevich[71] has demonstrated the usefulness of elastic and inelastic scattering of charged particles in geochemical analysis. He has, in particular, applied this technique for the first determination of the chemical composition of the lunar surface. A Cm-242 source emitting α particles of approximately 6 MeV was used for this purpose. On the heavier elements these α particles undergo Rutherford scattering, whereas on the light elements nuclear interactions take place; in particular, α, p reactions were used for their detection. The energy of the scattered particles is dependent upon the mass number of the scattering nucleus; thus, by measuring the energies and intensities of the particles scattered at a given optimum angle with respect to the incident beam, qualitative and quantitative information on the chemical composition is obtained. Three different sites were analyzed on three different Surveyor missions[72–75] and eight elements or groups of elements were determined: carbon, oxygen, sodium, magnesium, aluminum, silicon, "calcium" or elements with mass numbers between 28 and approximately 45, and "iron" or elements with mass numbers ranging from approximately 45 to 65.

4.4.2. Techniques Involving Delayed Counting

Most analytical work utilizing high-energy charged particles has dealt so far with the determination of trace amounts of the light elements (boron, carbon, nitrogen, and oxygen, particularly) in high-purity metals and semiconductors.[76–81]

The main roadblock preventing, up to now, the application of this technique to geochemical problems has most probably been the lack of access to sources of high-energy charged particles such as Van de Graaff accelerators and cyclotrons. It can be expected that with the multiplication of these machines and the advent of small cyclotrons,[82] this technique will certainly be used more often. Most likely future applications should be forthcoming on such problems as the determination of the isotopic ratios of light elements as, for example, ^{12}C–^{13}C in meteorites and the quantitative analysis of trace elements, as, for example, lead, for which charged-particle activation analysis appears to offer outstanding sensitivity.[83]

5. ACTIVATION ANALYSIS IN THE GEOSCIENCES

Geoscientists are making ever-increasing use of the various methods of activation analysis. A summary[84] of results obtained for the complete analysis of standards G-1 and W-1 indicates that over 50 elements were determined by neutron activation analysis. In most cases the reported results compare favorably with those obtained by other analytical procedures, especially for those elements that are present in trace amounts only. The applications are, however, no longer confined to laboratory assaying, but a number of field tests and *in situ* analyses have already been and are still being developed. Some examples of these trends are briefly outlined in the following sections.

5.1. Mineral Prospecting

A feasibility study[85] has revealed that some 30 elements may be detected by irradiating the ground surface for periods of a few minutes with fast neutrons. It has been found that 3-MeV neutrons in penetrating the ground are thermalized to produce a relatively uniform flux of neutrons of energy below the 1-keV range, at an approximate depth of 20 in. There is a strong possibility that elements that are not amenable to neutron activation analysis may still be detected on grounds of their association with certain other elements in nature. The distribution of lead, for example, may be mapped by determining silver. Fluorine and chlorine may be indicative of mineralization, thus leading to the detection of an ore deposit.

The formal study of the deposition and distribution of minerals in the earth's crust, called metallogeny, may be carried out by investigating the areal distribution of the trace elements. According to proponents of metallogenesis,[86–90] the concentrations of groups of elements in ores are basically determined by the presence of superimposed structures in so-called "metallogenic provinces." A study of the distribution of some trace elements, as determined by purely instrumental, reactor-thermal-neutron activation analysis, has revealed the potential of this approach.[91,92] The simultaneous determination of the elements Fe, Sc, Co, La, Eu, Hf, and Ta in uncrushed, whole-rock specimens, not only revealed the compositional trends of these elements across the different zones of a batholith, but a later intrusive was also identified.[91] An extension of this mapping procedure has indicated the existence of a particular gold deposit in the Philippines.[92] With some refinements at least 15 elements may be determined simultaneously in rock samples, by instrumental activation analysis.[93] Large-scale mineral exploration based on activation analysis therefore appears feasible. It has been suggested that a fast-neutron source may be used for the remote analysis of major constituents of relatively inaccessible areas, including lunar and planetary surfaces.[94] The availability of the high-thermal-neutron output, ^{252}Cf spontaneous-fission source, will favor the determination of minor rock constituents in the field. The continental-drift theory and other related studies that are aimed at establishing the similarity of rock types may also be studied. Activation-analysis techniques are particularly useful where only a small amount of sample is available, as in meteorites. After the analysis the sample material may still be studied mineralogically, since this instrumental technique is nondestructive.

5.2. Logging Techniques

The major advantage of nuclear logging techniques over other methods is that they may be applied to encased holes. Both neutrons and gamma rays will penetrate steel or cement casings.

In the natural-gamma-ray-logging technique naturally radioactive elements such as uranium, thorium, and potassium may be detected by passing a sodium iodide probe down the borehole. A full description of radiometric techniques is given in Chap. 10.

In the gamma–gamma or bulk-density-logging technique a gamma-ray source is used such as Co-60, for example. The gamma rays emitted are reflected from the surrounding material to reach a sodium iodide detector, which is shielded from the source. The number of reflected gamma rays reaching the detector is statistically related to the density of the surrounding material. For a fixed-source strength, distance between source and detector, and borehole geometry, the instrument may be calibrated in units such as g/cm^3.

In neutron–neutron or moisture logging a neutron source emitting fast neutrons is used in conjunction with a thermal-neutron detector. The principle underlying the slowing down of neutrons may be illustrated as follows: if a tennis ball (i.e., a fast neutron) is thrown against a solid or heavy wall it bounces back, and little energy is lost. If the ball is thrown against a curtain which is relatively light, the ball does not recoil, but drops straight down. The fast neutron loses its energy in a similar fashion if it collides with a hydrogen nucleus. If a large number of hydrogen atoms are present, a large number of incident neutrons will be slowed down. The number of fast neutrons surviving may therefore be said to be inversely proportional to the hydrogen content. On an atom-per-volume basis water contains much more hydrogen than any other naturally occurring inorganic material. Neutron-flux variations as determined by the neutron detector are therefore a measure of the variation in the water or moisture content of the material surrounding the borehole. The instrument may be calibrated to determine the weight of water per unit volume. This technique has been successfully applied by the Kennecott Copper Corp., Utah, to study dump leaching processes.[95] If a gamma-ray detector is used in conjunction with a neutron source, the activation products may also be determined, of course.

5.3. Ore Sorting

The major advantage of sorting by nuclear techniques rather than by optical methods based on some physical property such as color, for example, is that activation analysis is not restricted to the shape, appearance, or surface composition of a particular sample. Shielding requirements do create a problem, but these are surmountable.

The principles involved in the continuous on-stream monitoring of a particular isotope have already been layed down.[96,98] A good example of an automated neutron-activation-analysis system in industry may be found in the routine determination of oxygen in steel,[97,1] as is currently carried out by the South African Iron and Steel Corp., Pretoria. The gamma-ray activity resulting from the $^{16}O(n, p)^{16}N$ reaction with 14-MeV neutrons is measured by integrating the gamma-ray activity above 4.5 MeV.

Signals from such a monitoring device may be fed into a sorter, in order to pick out or divert a particular fraction of the material passing along a belt. A promising application is the separation of rock from coal, by monitoring the aluminum content.[99] The determination of carbon, oxygen, aluminum, and silicon in coal has also been investigated.[100] Taconite ore may be sorted on the basis of the continuous analysis of iron and silicon by 14-MeV-neutron activation analysis. Copper[101] and fluorine[102] in their respective ores may also be measured. A method for monitoring the addition of fluorine to drinking water has been described.[103] An isotopic neutron source such as americium–beryllium may be used for the determination of fluorine, since the isotopic neutron source is not only cheaper and more plant compatible than a neutron generator, but the reaction cross section for fluorine with 3–5-MeV neutrons is also larger.

Other elements with short-lived radioactive products such as Au-97*m*[104] are currently receiving attention. More sophisticated techniques such as neutron time-of-flight methods may also be employed in order to extend the application of nuclear techniques in industry.

REFERENCES

1. Van Wyk, J. M., *et al.*, A study of the macroscopic distribution of oxygen in a steel rod by neutron activation and vacuum fusion techniques, *Analyst* **91** (1082), 316 (1966).
2. Lenihan, J. M. A., Activation analysis in the contemporary world: questions and answers, *Modern Trends in Activation Analysis*, James R. De Voe, ed., National Bureau of Standards, Washington, 1969 (special publication 312, Vol. 1, pp. 10–12).
3. Fite, L. E., *et al.*, Sampling methodology development on a large scale cystic fibrosis screening program based on automated neutron activation analysis, *Modern Trends in Activation Analysis*, James R. De Voe, ed., National Bureau of Standards, Washington, 1969 (special publication 312, Vol. 1, pp. 147–155).
4. Wainerdi, R. E., and Menon, M. P., Comparison of the various instrumental methods for the analysis of traces of bromine pesticides in plant materials, by activation, in the limit of sensitivity range, *Nuclear Activation in the Life Sciences*, International Atomic Anergy Agency, Vienna, 1967, pp. 33–50.
5. Guinn, V. P., The current status of neutron activation analysis applications, *Modern Trends in Activation Analysis*, James R. De Voe, ed., National Bureau of Standards, Washington, 1969 (special publication 312, Vol. 2, pp. 679–697).
6. West, T. S., Atomic flame spectroscopy in trace analysis, *Minerals Sci. Engng.* **1** (2), 3 (1969); **2** (1), 31 (1970).
7. Wainerdi, R. E., and Dubeau, N. P., Nuclear activation analysis, *Science* **139** (3559), 1027 (1963).
8. Taylor, D., *Neutron Irradiation and Activation Analysis*, Van Nostrand, Princeton, N. J., 1964.
9. Siegbahn, K., *Alpha-, Beta-, and Gamma-Ray Spectroscopy*, North Holland, Amsterdam, 1965.
10. Lam, C. F., and Wainerdi, R. E., A CsI-NaI dual crystal Compton reduction detector, *1965 International Conference on Modern Trends in Activation Analysis*, College Station, Texas, 1965, pp. 189–194.
11. Kantele, J., *et al.*, Gamma spectrometer systems employing an anti-Compton annulus, *Nucl. Instrum. Meth.* **39** (2), 194 (1966).

12. Kuykendall, W. E., and Wainerdi, R. E., The analysis of a complex gamma spectrum utilizing the IBM 704 computer, *Trans. Am. Nucl. Soc.*, 95 (1960).
13. Fite, L. E., *et al.*, Computer coupled activation analysis, Washington Office of Technical Services, TEES 2671-1, May 1969.
14. Drew, D. D., *et al.*, Computer coupled activation analysis, *Application of Computers to Nuclear and Radiochemistry*, National Academy of Sciences - National Research Council, Washington, 1962, pp. 237–242.
15. Salmon, L., Analysis of gamma rays scintillation spectra by the method of least squares, Harwell, Atomic Energy Research Establishment, R-3640, April 1961.
16. Yule, H. P., HEVESY, a computer program for analysis of activation analysis gamma-ray spectra, *Modern Trends in Activation Analysis*, James R. De Voe, ed., National Bureau of Standards, Washington, 1969 (special publication 312, Vol. 2, pp. 1108–1110).
17. Yule, H. P., Computation of experimental results in activation analysis, *Modern Trends in Activation Analysis*, James R. De Voe, ed., National Bureau of Standards, Washington, 1969 (special publication 312, Vol. 2, pp. 1155–1204).
18. Covell, D. F., Determination gamma-ray abundance directly from the total absorption peak, *Analyt. Chem.* **31** (4), 1785 (1959).
19. Wainerdi, R. E., and Fite, L. E., Automated systems for nuclear activation analysis. *Automation and Instrumentation.* L. Dadda, and U. Pellegrin, eds., Pergamon Press, New York, 1967, pp. 557–569.
20. Fite, L. E., *et al.*, A study of the feasibility of using nuclear activation analysis for lunar compositional analysis, Office of Technical Services, Washington, TID-199999, May 1963 (NASA).
21. Menon, M. P., *et al.*, Activation analysis for nondestructive localization of impurities in foils, *Nucl. Applic.* **2**, 335 (1966).
22. Wainerdi, R. E., *et al.*, The use of high speed, computer coupled automatic systems for non-destructive medical and biological nuclear activation analysis, *Proceedings of the Third International Conference on Biology*, Saclay, France, University of France Press, pp. 171–198.
23. Dawson, E. B., *et al.*, Activation analysis of placental trace metals, *J. Nucl. Med.* **9** (4), 160 (1968).
24. Bowen, H. J. M., and Gibbons, D., *Radioactivation Analysis*, Claredon Press, Oxford, 1963.
25. Yule, H., private communication.
26. Fite, L. E., *et al.*, Special considerations in the large scale analysis of biomedical samples by automated nuclear activation methods, *Proceedings of the Ninth Japan Radio-Isotopes Conference*, Tokyo (in press).
27. Wainerdi, R. E., *et al.*, Neutron activation analysis and high resolution gamma-ray spectrometry applied to area-elemental distribution studies, *Nuclear Techniques and Mineral Resources*, Vienna, International Atomic Energy Agency, 1969, pp. 507–532.
28. Higuchi, H., *et al.*, Use of a Ge(Li) detector after simple chemical group separation in the activation analysis of rocks samples, *Modern Trends in Activation Analysis*, James R. De Voe, ed., National Bureau of Standards, Washington, 1969 (special publication 312, Vol. 1, pp. 334–338).
29. Santos, G. G., *et al.*, Preliminary study on the use of fast-neutron activation analysis for sea-floor compositional mapping, *Nuclear Techniques and Mineral Resources*, Vienna, International Atomic Energy Agency, 1969, pp. 463–487.

30. Santos, G. G., and Wainerdi, R. E., Determination of silicon in rocks by fast-neutron activation analysis using internal standards, *J. Radioanalyt. Chem.* **1**, 509 (1968).
31. Hoffman, B., *et al.*, The 14 MeV neutron activation analysis of deep sea core samples, *Trans. Am. Geophys. Un.* **20** (4), 201 (1969).
32. Cuypers, M. Y., and Cuypers, J., Gamma ray spectra and sensitivities for 14 MeV neutron activation analysis, *J. Radioanalyst. Chem.* **1**, 243 (1968).
33. Cuypers, M. Y., *et al.*, The measurement of the lunar surface composition using nuclear activation analysis, *Proceedings of the XIIIth Congresso Scientifica Internazionale per l'Elettronica*, Rome, 1966, pp. 17–28.
34. Hislop, J. S., and Wainerdi, R. E., Extraterrestrial nuclear activation analysis, *Analyt. Chem.* **39** (2), 29A (1967).
35. Cuypers, M. Y., *et al.*, Lunar and planetary surface analysis using neutron activation, *Radioisotopes for Aerospace*, J. C. Dempsey and P. Polishuk, eds., Part 2, Systems and Applications, Plenum Press, New York, 1966, pp. 292–308.
36. Givens, W. W., *et al.*, Cyclic activation analysis, *Modern Trends in Activation Analysis*, James R. De Voe, ed., National Bureau of Standards, Washington, 1969 (special Publication 312, Vol. 2, pp. 929–937).
37. Albert, P., *Chimia (Aarau)* **21**, 32 (1967).
38. Albert, P., Proc. of 2nd Int. Conf. on Practical Aspects of Activation Analysis with Charged Particles, Euratom Report, Eur 3896 d-f-e, 1, 1967.
39. Engelmann, C., Commissariat a l'Energie Atomique Report, CEA 2559, 1964.
40. Engelmann, C., Commissariat a l'Energie Atomique Report, CEA R-3307, 1967.
41. Baker, C. A., U.K. Atomic Energy Authority Report AERE R-5265, 1967.
42. Baker, C. A., Hunter, G. J., and Wood, D. A., U.K. Atomic Energy Authority Report R-5547, 1967.
43. Baker, C. A., and Wood, D. A., U.K. Atomic Energy Authority Report AERE R-5818, 1968.
44. Lutz, G. J., *Anal. Chem.* **41**, 424 (1969).
45. Tilbury, R. S., Activation analysis with charged particles, NAS-NS 3110, USAEC.
46. Lukens, H. R., Jr., Otvos, J. W., and Wagner, C. D., *Int. J. Appl. Rad. and Isotopes* **11**, 30 (1961).
47. Kaminishi, K., *Japan J. Appl. Phys.* **2**, 399 (1963).
48. Goldstein, G., *Anal. Chem.* **35**, 1620 (1963).
49. Rook, H. L., *Anal. Chem.* **36**, 2211 (1964).
50. Guinn, V. P., and Lukens, H. R., *Trans. Am. Nucl. Soc.* **9**, 106 (1966).
51. Faries, R. A., Johnston, J. E., and Miller, R. J., *Nucleonics* **12** (10), 48 (1954).
52. Khristianov, V. K., *Isotopenpraxis* **3**, 235 (1967).
53. Schweikert, E. A., and Albert, Ph., *Radiochemical Methods of Analysis*, Int. Atomic Energy Agency, 1964, Vol. 1, 405 pp.
54. McGregor, M. H., *Proc. 2nd Int. Conf. on Peaceful Uses of Atomic Energy*, 1958, Vol. 20, p. 1771.
55. Wilkuiss, P. E., and Linnenboom, V. I., Proc. of 2nd Int. Conf. on Practical Aspects of Activation Analysis with Charged Particles, Euratom Report, Eur 3896 d-f-e, 1967, p. 119.
56. Andersen, G. H., Graber, F. M., Guinn, V. P., Lukens, H. R., and Settle, D. M., *Nuclear Activation Techniques in the Life Sciences*, Int. Atomic Energy Agency, 1967, p. 99.

57. Owlya, A., Abdeyazdan, R., and Albert, P., Proc. of 2nd Int. Conf. on Practical Aspects of Activation Analysis with Charged Particles, Euratom Report, Eur 3896 d-f-e, 1967, p. 161.
58. Baker, C. A., Pratchett, A. G., and Williams, D. R., U.K. Atomic Energy Authority Report, AERE R-5363, 1967.
59. Engelmann, C., Fritz, B., Gosset, J., Graeff, P., and Loeuillet, M., Proc. of 2nd Int. Conf. on Practical Aspects of Activation Analysis with Charged Particles, Euratom Report, Eur 3896 d-f-e, 1967, p. 319.
60. Engelmann, Ch., Gosset, J., Loeuillet, M., Marschal, A., Ossart, P., and Boissier, M., *Proc. of Int. Conf. on Modern Trends in Activation Analysis*, NBS Spec. Publ. 312, 1968, Vol. II, p. 819.
61. Revel, G., Chaudron, T., Debrun, J. L., and Albert, Ph., *Proc. of Int. Conf. on Modern Trends in Activation Analysis*, NBS Spec. Publ. 312, 1968, Vol. II, p. 838.
62. Rocco, G. G., Garzon, O. L., and Cali, J. P., *Int. J. Appl. Rad. and Isotopes* **17**, 433 (1966).
63. Berzin, A. K., Kunetsov, K. F., Sulin, V. V., Belov, V. I., Vitozhents, G. Ch., Martynov, Yu. T., Suslov, V. G., and Shoruikov, S. I., USAEC Report ORNL-tr-1097.
64. Schmitt, R. A., and Loveland, W. D., *Geochim. Cosmochim. Acta* **33**, 375 (1969).
65. Albert, P., Sue, P., and Chaudron, G., *Bull. Soc. Chim. France* **2016**, 97 (1953).
66. Engelmann, C., *Acad. Sci. Paris* **258**, C4279 (1964).
67. Ricci, E., and Hahn, R. L., *Anal. Chem.* **40**, 54 (1968).
68. Sippel, R. F., and Glover, E. D., *Nucl. Instr. and Method* **9**, 37 (1960).
69. Pierce, T. B., Peck, P. F., and Cuff, D. R. A., *Analyst* **92**, 143 (1967).
70. Pierce, T. B., Peck, P. F., Cuff, D. R. A., *Nucl. Instr. and Meth.* **67**, 1 (1969).
71. Turkevich, A. L., *Science* **134**, 672 (1961).
72. Turkevich, A. L., Franzgrote, E. J., and Patterson, J. H., *Science* **158**, 635 (1967).
73. *Ibid.*, *Science* **160**, 1108 (1968).
74. Franzgrote, E. J., Patterson, J. H., and Turkevich, A. L., *Science* **162**, 117 (1968).
75. Turkevich, A. L., Anderson, W. A., Economou, T. E., Franzgrote, E. J., Griffin, H. E., Grotch, S. L., Patterson, J. H., and Sowinsku, K. P., Jet Propulsion Laboratory, Technical Report 32-1265, 1968.
76. Lamb, J. F., Lee, D. M., and Markowitz, S. S., Proc. of 2nd Int. Conf. on Practical Aspects of Activation Analysis with Charged Particles, Euratom Report, Eur 3896 d-f-e, 1967, p. 225.
77. Revel, G., and Albert, P., Proc. of 2nd Int. Conf. on Practical Aspects of Activation Analysis with Charged Particles, Euratom Report, Eur 3896 d-f-e, 1967, p. 261.
78. Debrun, J. L., Barrandon, J. N., and Albert, P., *Proc. of Int. Conf. on Modern Trends in Activation Analysis*, NBS Spec. Publ. 312, 1968, Vol. 2, p. 774.
79. Engelmann, C., and Cabane, G., *Proc. of Conf. on Modern Trends in Activation Analysis*, Texas A & M University, College Station, Texas, 1965, p. 331.
80. Ricci, E., and Hahn, R. L., Proc. of 2nd Int. Conf. on Practical Aspects of Activation Analysis with Charged Particles, Euratom Report, Eur 3896 d-f-e, 1967, p. 15.
81. Rook, H. L., and Schweikert, E. A., *Anal. Chem.* **41**, 958 (1969).
82. Fliesher, A. A., Hendsy, C. O., Smith, C. G., Tom, J. L., and Wells, D. K., Proc. of 2nd Int. Conf. on Practical Aspects of Activation Analysis with Charged Particles, Euratom Report, Eur 3896 d-f-e, 1967, p. 435.
83. Schweikert, E. A., *Trans. Am. Nucl. Soc.* **13**, 58 (1970).

84. Fleisher, M., Summary of New Data on Rock Samples G-1 and W-1, 1962–1965, *Geochim. Cosmochim. Acta* **29**, 1263 (1965).
85. Senftle, F. E., and Hoyte, A. F., Using in situ Neutron Activation, Nucl. Instrum. Meth. 1, 42, 1966, pp. 93–103.
86. Bilibin, Y. A., Metallogenic Provinces and Epochs, *Gosgeolteknizdat*, 1955, Dept. of the Secretary of State of Canada, Foreign Languages Division, Bureau of Translations, No. 58972.
87. Bryner, L., Metallogeny in Russia's Drive for Ore Deposits, *Mining Engineering*, 59, June 1963.
88. Burnham, C. W., Metallogenic Provinces of the Southwestern United States and Northern Mexico, *N. Mex. Bur. Mines and Mineral Res. Bull.*, 65 (1959).
89. Rose, A. W., Trace Elements in Sulfide Minerals from the Central District, New Mexico, and the Bingham District, Utah, *Geochim. Cosmochim. Acta* **31**, 547 (1967).
90. Schuiling, R. D., Tin Belts on the Continents Around the Atlantic Ocean, *Econ. Geol.* **62**, 540 (1967).
91. Uken, E. A., Santos, G. G., and Wainerdi, R. E., Neutron Activation Analysis for the Study of Metallogenic Provinces, *Talanta* **15**, 1097 (1968).
92. Wainerdi, R. E., Uken, E. A., Santos, G. G., and Yule, H. P., Neutron Activation Analysis and High Resolution Gamma-Ray Spectrometry Applied to Areal Elemental Distribution Studies, Presented at IAEA Symposium on the Use of Nuclear Techniques in the Prospecting and Development of Minerals Resources, Buenos Aires, Argentina, November 4–8, 1968.
93. Gordon, G. E., Randle, K., Goles, G. G., Corliss, J. B., Beeson, M., and Okley, S. S., Instrumental Activation Analysis of Standard Rocks with High Resolution Gamma-Ray Detectors, *Geochim. Cosmochim. Acta* **32** (4), 369 (1968).
94. Caldwell, R. L., Combination Neutron Experiment for Remote Analysis, *Science* **152**, 457 (1966).
95. Howard, E. V., China Uses Radiation Logging for Studying Dump Leaching Processes, *Min. Eng.* **20** (4), 70 (1968).
96. Anders, O. U., Activation Analysis for Plant Stream Monitoring, *Nucleonics* **20** (2), 78 (1962).
97. Coleman, R. F., and Perkin, J. L., The Determination of the Oxygen Content of Beryllium Metal by Activation, *Analyst* **84**, 233 (1959).
98. Anonymous, Employing nuclear methods to automate the grading of ores from gold deposits, *Sov. J. Non-ferr. Metals* **5** (8), 20 (1964).
99. Rudanovskii, A. A., The Use of the Method for Induced Radioactivity for Automation of Rock Separation, *Nucl. Sc. Abstr.* **15**, 15775 (1961).
100. Martin, T. C., Mathur, S. C., and Morgan, I. L., The Application of Nuclear Techniques in Coal Analysis, USAEC, Report TED-20080, 1963, pp. 1-1–2.
101. Gaudin, A. M., and Ramdohr, H. F., Induced Radioactivity for Course Copper - Ore Concentration, *Can. Metall. Quarterly* **1**, 173 (1962).
102. Jeffery, P. G., and Bakes, J. M., Neutron Activation Analysis with Neutron from Isotopic Sources: Analytical Control of Fluorite Production, Warren Spring Laboratory Report, Stevenage, Herts, U.K.
103. Nargolwalla, S., and Jervis, R. E., Continuous analysis of trace fluorine by 14 MeV neutron activation, *Trans. Am. Nucl. Soc.* **8**, 86 (1965).
104. Uken, E. A., Watterson, J. I. W., Knight, A., and Steele, T. W., The Application of Neutron Activation Analysis to Sorting Witwatersrand Gold-Bearing Ores, *J.S.A. Inst. of Mining and Metallurgy* **67** (3), 99 (1966).

Chapter 12

MASS SPECTROMETRY

J. N. Weber and P. Deines

Pennsylvania State University
University Park, Pennsylvania

1. INTRODUCTION

The field of mass spectrometry originated with the work of J. J. Thomson in the early part of this century. He anticipated the use of this tool for chemical analysis and discussed some of its possible applications in his book *Rays of Positive Electricity and their Application to Chemical Analyses*. After Thomson's discovery that the element neon consisted of isotopes, Dempster and Aston initiated a thorough study of the phenomenon. Aston's research later resulted in the first precise determination of atomic masses by mass spectrometry. The field advanced rapidly when the potentialities of mass spectrometry were recognized by the petroleum and nuclear industries, and presently the instrumentation is used in a wide spectrum of industrial and scientific research. Geochemical applications range from the study of variations in isotopic composition produced by radioactivity, spallation, fission, or physicochemical processes to investigations of chemical composition by means of thermal-source isotope dilution or rf spark-source mass spectrometry. In organic geochemistry mass spectrometry is also of value in structural studies of organic compounds. The instruments used in such research have been described in detail by Inghram and Hayden,[1] Duckworth,[2] McDowell,[3] Herzog *et al.*,[4] and Roboz.[5] In this chapter determination of carbon and oxygen stable-isotope ratios and spark-source analysis of trace-element concentrations are discussed.

2. STABLE CARBON- AND OXYGEN-ISOTOPE MASS SPECTROMETRY

2.1. Principles Underlying the Construction of an Isotope-Ratio-Recording Mass Spectrometer

Over 20 years elapsed after Aston and Dempster developed techniques to measure the relative abundance of stable isotopes before mass spectrometers were produced commercially. Modern instruments have achieved a high degree of sophistication but the basic principles of design remain little changed. The mass spectrometer contains three essential parts in addition to a vacuum system: an ion source, an analyzer, and an ion detector. The ion source converts neutral molecules into ions, which can then be affected by electric and magnetic fields. If the sample is gaseous (CO_2, for example) ionization can be effected readily by electron impact, the electrons coming from a heated filament. The purpose of the analyzer is to separate these ions into a mass spectrum using a magnetic field or a combination of magnetic and electric fields. The ion beams comprising the mass spectrum impinge upon plates or "collectors" in the ion-detection system and the currents produced are amplified and measured.

For carbon- and oxygen-isotope-ratio analysis the gas most preferred is CO_2, although CO[6] and molecular oxygen[7] can also be fed directly into the mass spectrometer. The use of oxygen has some advantages in special applications, because to obtain $^{18}O/^{16}O$ measurements in natural samples using CO_2, an assumption is made about the concentration of ^{17}O. In tracer experiments the relative abundance of ^{17}O can be quite large and the formulae[7] derived for use with molecular oxygen are mathematically correct for all ^{16}O–^{17}O–^{18}O abundance ranges. Samples of geochemical interest contain carbon and/or oxygen in many different forms of chemical combination, such as carbonates, water, silicates, oil, etc., but techniques have been developed to prepare carbon dioxide from virtually any type of starting material.

Absolute measurements of isotope ratios can be made only with great difficulty,[8] whereas for certain elements, *differences* in isotopic composition can be determined easily and with high precision. By introducing two gases alternately into the mass spectrometer over a short period of time, it is possible to distinguish, for example, between a sample containing 2040.1 ppm ^{18}O and another containing 2040.0 ppm ^{18}O, that is, a difference of 0.1 ppm in the natural range of ^{18}O concentration. The pressure in the ionization region of the ion source is low ($\sim 10^{-4}$ mm Hg) and only a small quantity of CO_2 is actually admitted to the source. The gas-inlet system must therefore be designed so as to reduce the pressure between the source and the sample by several orders of magnitude with a minimum of isotope fractionation in the process. Furthermore, a device to permit the admission of one or the other of two gas samples in alternate sequence must be incorporated into the inlet system.

Pressure reduction is accomplished by inserting a constriction or leak between the sample reservoir and the source. The type most commonly used is a long length of capillary tubing which forms a viscous leak. Isotope fractionation is minimal if the gas moves by viscous flow, that is, where the mean free path of the molecules is much smaller than the dimensions of the leak. As the mean free path is inversely proportional to pressure, it is desirable to design the leak so that the gas flows through the greater part of it at a pressure high enough to maintain viscous flow. Kistemaker[9] emphasized the importance of careful design in the construction of gas-inlet systems to avoid fractionation, and demonstrated how fractionation effects can be calculated for various combinations of leak dimensions, pressure, and size of the sample container. Similar calculations for viscous-flow gas-inlet systems are given by Boerboom and Tasman.[10] A period of time (usually minutes) is required to reach steady-state conditions after the flow of gas through the capillary leak has begun. In order to compare two gas samples rapidly, each sample flows continually through its own leak and a solenoid-operated gas valve is employed at the low-pressure end of the leak to switch the inflowing gas either into the ion source or into a "waste" pump. As soon as steady-state flow is attained at the beginning of an analysis, one sample or the other can be directed into the mass spectrometer by the solenoid valve, and the change from one gas to the other is almost instantaneous. In this way, the effects of small and uncontrollable secular changes in instrumental conditions are essentially eliminated, as these variations should alter the isotope ratio of both gas samples to the same extent, leaving the difference unchanged.

Smaller gas samples can be accommodated by using a molecular leak such as a pinhole in the sample reservoir. Unlike the viscous leak arrangement, the pressure of the gas in the reservoir is low, but unfortunately, isotope fractionation occurs in molecular flow. This is not a random error but a regular change with time, and the fractionation effects can be largely compensated for by automatic inlet valves which permit both gases to flow through their respective leaks for equal periods of time. Shackleton[11] reports a reproducibility of $\pm 0.01\,^{0}/_{00}$ for such a system, using the CO_2 from $CaCO_3$ samples weighing as little as 0.4 mg.

The major disadvantage of the viscous leak is that samples larger than a micromole are necessary, but this can be overcome by adding an inert gas such as nitrogen or helium to very small CO_2 samples. In this way, the mean-free path of the gas molecules through the capillary leak is smaller because the total gas pressure is high. Han and Fritz[12] analyzed submicromole samples of CO_2 by this technique and reported no detectable isotope fractionation in the leaks with partial pressures of CO_2 as low as 0.1 μ.

Most modern CO_2 mass spectrometers are equipped with a two-channel (dual) gas-inlet system and capillary viscous leaks. The sample reservoirs are fitted with mercury pistons which permit accurate and rapid adjustment of the gas pressure, and automatic recycling timers can be used to activate the solenoid gas-switching valve at a preselected time interval. Different gas-inlet systems have been described,[13–15] including a three-channel design.[16]

A heated filament in the ion source emits electrons which are accelerated by an electric field (40–100 V) and then collimated by a weak magnetic field. These electrons, whose emission is carefully regulated, collide with the CO_2 molecules of the sample, forming positive ions. The latter are drawn out of the ionizing region by an electric field and accelerated to form a beam of ions traveling along the analyzer tube between the poles of a large magnet. The charged particles are forced into circular paths whose radius of curvature is a function of m/e (mass divided by charge) for a given magnetic field strength. Ion collectors are appropriately located to intercept the ion beams of interest. For a constant magnetic field strength the mass of the ions collected is inversely proportional to the accelerating potential.

For natural samples, CO_2 ions of mass 44 ($^{12}C^{16}O^{16}O$) to mass 49 ($^{13}C^{18}O^{18}O$) are formed in the mass spectrometer but it is not necessary to collect and measure the intensity of each of the ion beams in this mass range in order to obtain $^{13}C/^{12}C$ and $^{18}O/^{16}O$ ratios. The ion collector is constructed with two Faraday cups called the "single collector" and the "multiple collector." Their design is such that for an appropriate value of the accelerating potential, the output of the single collector can be attributed to the mass-45 ion beam while that of the multiple collector derives from the mass-44 ion beam. The potential drop across a high resistance in the single and multiple collector circuits is amplified by a vibrating-reed electrometer and a dc amplifier, respectively. A decade-resistor potentiometer is connected with the two collectors so that with proper selection of the decade resistors, no current flows through the circuit (Fig. 1). By this "null" method, the ratio of the number of mass-45 ions collected to the number of mass-44 ions collected is obtained from the value of the resis-

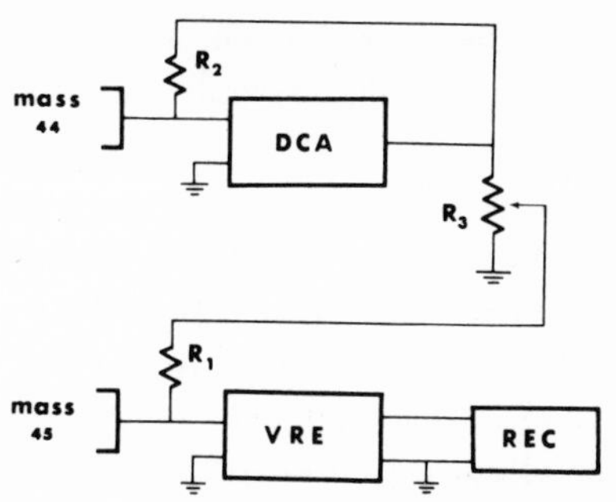

Fig. 1. Ion-collector circuitry for determining the ratio of mass-45 to mass-44 ion beams. DCA, dc amplifier; VRE, vibrating-reed electrometer; REC, recorder; $R_1 = 1.5 \times 10^{11}\ \Omega$; $R_2 = 1.5 \times 10^{10}\ \Omega$; R_3 = decade resistance potentiometer. (After McKinney *et al.*[13])

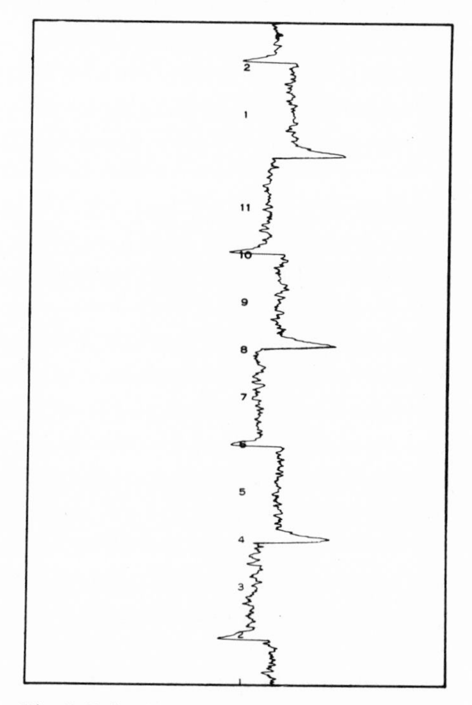

Fig. 2. Strip-chart record obtained from the recorder shown in Fig. 1. Traces for sample and standard gases alternate at about 1.5-min intervals. The displacement of the sample trace from the standard trace corresponds to a difference in isotopic composition of about 0.25 permil.

tance inserted in the circuit. At a different accelerating potential, the mass-46 ion beam enters the single collector while ions of masses 44 and 45 strike the multiple collector.

The small residual current which cannot be reduced to zero by the decade resistors is displayed by a strip-chart recorder (Fig. 2). From this information and the values of the decade resistors, the following ratios are determined: (a) number of mass-45 ions to number of mass-44 ions, and (b) number of mass-46 ions to number of mass-44 and mass-45 ions. A number of instrumental correction factors[17–19] are determined and applied to the analytical data. Isotope ratios in delta notation are then calculated from the two ion ratios using equations derived by Craig.[17] More detailed descriptions of isotope-ratio-recording mass spectrometers have been published.[13,14,20–24]

2.2. Reporting of Carbon- and Oxygen-Isotope Data

Determination of the absolute isotope ratio is an exceedingly difficult task, but it is possible to measure *differences* in isotope ratio between two samples with a high degree of accuracy and precision. For this reason, isotope data are reported in delta notation as deviations from an arbitrarily selected reference (standard):

$$\delta^{13}C = \left(\frac{^{13}C/^{12}C_{\text{sample}}}{^{13}C/^{12}C_{\text{standard}}} - 1 \right)$$

Usually the difference is expressed in parts per thousand ($^0/_{00}$) and the expression on the right-hand side of the equation is then multiplied by 1000.

The rapidly expanding application of stable isotopes to geochemical problems has created some difficulties and confusion in the area of isotope standards. Early investigators in different laboratories selected their own standards or reference samples. The form of the standard chosen, such as water, carbonate, silica, etc., depended upon the type of materials to be analyzed. While this situation was adequate for a number of years, increased precision and accuracy resulting from developments in analytical techniques and instrumentation could not be fully realized in comparing different types of samples or data obtained in different laboratories unless certain criteria for adequate standards were fulfilled. A great many standards, such as "Moscow tap water," "Stockholm $BaCO_3$," or even tank CO_2, have been used to report isotope data, but most of the carbon- and oxygen-isotope-ratio measurements published today are related to two important "reporting standards."

Isotope standards should serve two major purposes and permit investigators to report their measurements with respect to the same reference point so that isotope-ratio data obtained in different laboratories or for different types of materials can be compared, and calibrate their mass spectrometers and determine instrumental correction factors. Theoretically, only one standard would be necessary to serve the first purpose, but in practice it is desirable to have *at least* two so that a standard with an isotopic composition not too far removed from those of the samples can be chosen. For the second purpose, i.e., calibration of a mass spectrometer, it is *necessary* to have at least two standards.

The criteria for an adequate isotope standard are: (a) the isotopic composition must not change with time or be subject to alteration during any handling or sample-preparation procedures; (b) the isotopic composition should be within the range of compositions of the analyzed samples; and (c) the standard must be readily available to all investigators and each aliquot of standard must have exactly the same isotopic composition. Most of the standards used by earlier workers have not met the third criterion and these standards either no longer exist or are available in such limited quantity that they cannot be distributed.

The U.S. National Bureau of Standards established a program[25] to distribute a stock of isotope reference samples and to serve as a clearinghouse for data concerning these "standards." Included in the isotope reference samples are water, air, $CaCO_3$, C, and SiO_2 [26] The two widely used "reporting standards" are PDB-I and SMOW, neither of which is actually available. However, the differences in isotopic composition between these reporting standards and reference samples that can be obtained by an investigator are known. Thus it is possible to use one or more of the reference samples as a measuring or "working standard" (i.e., for determining the difference in isotope ratio between a sample and the measuring or working standard in a mass spectrometer) and then mathematically convert the result to a delta value expressing the difference in isotope ratio between the sample and the reporting standard. For example, NBS-20 can be used as a "working standard" in the laboratory, but for the purpose of publication, the measurements can be converted mathematically so that the isotopic data are referred to the PDB-I or SMOW standards.

The PDB-I standard is CO_2 prepared from PDB $CaCO_3$ by reaction with 100% H_3PO_4 at 25.2°C. The standard is not the carbonate itself, which is a particular sample of a belemnite rostrum (*Belemnitella americana*) collected from the Cretaceous Peedee formation of South Carolina. A large quantity of data, especially in the isotopic paleotemperature field, has been published using PDB CO_2 as the standard. For this reason, many investigators prefer to continue the use of PDB as a reporting standard even though the standard itself can no longer be distributed.

SMOW, or "standard mean ocean water," was proposed by Craig[27] as a standard for both $^{18}O/^{16}O$ and hydrogen–deuterium measurements. Although most of the isotope-ratio data for carbonates was referred to PDB at this time, most of the $^{18}O/^{16}O$ determinations for water were expressed with respect to a sea-water standard. Use of the latter was advantageous for two reasons: (a) it provided a reference standard whose isotopic composition was similar to most of the analyzed samples, and (b) the preferred analytical technique for water then involved the equilibration of CO_2 (the gas to be fed into the mass spectrometer) with water at a known temperature. If a standard water sample were treated in exactly the same manner, *differences* in the isotopic composition of the CO_2 equilibrated with standard and sample waters would be identical to *differences* in the isotopic composition of the sample and standard waters; it is therefore not necessary to know the fractionation factor defining the distribution of oxygen isotopes between water and CO_2 in equilibrium at a given temperature.

For the isotopic analysis of water some investigators had been using an "average-ocean-water" standard.[28] Unfortunately, no single sample of "average ocean water" which could be circulated widely actually existed. An NBS water sample (No. 1) is, however, readily available for distribution, but it differs appreciably from sea water in isotopic composition. The problem was solved by creating a hypothetical standard suitable for reporting sea-water analyses but which is defined in terms of NBS-1. This hypothetical reporting standard, SMOW, is related to NBS-1 by the equation:[27] $^{18}O/^{16}O$ (SMOW) $= 1.008\,^{18}O/^{16}O$ (NBS-1), and is very close to the "average ocean water" of Epstein and Mayeda.[28] Using delta notation, the $\delta^{18}O$ value of NBS-1 with respect to SMOW is $-7.94‰$. Attention should be drawn to a minor oversight[27] where the statement is made that the $\delta^{18}O$ of PDB-I (i.e., CO_2 from PDB carbonate obtained by reaction with anhydrous H_3PO_4 at 25.2°C) is $+0.22‰$ on the SMOW scale. The difference ($+0.22‰$) referred to by Craig is between PDB CO_2 and CO_2 in isotopic equilibrium with SMOW at 25°C, not between PDB CO_2 and SMOW.

Many of the longer established laboratories use secondary standards for the actual mass-spectrometric measurements and have calibrated their working standards against PDB-I when PDB was still available for this purpose. A number of these standards have been compared.[17,29] NBS Isotope Reference Samples can be used as working standards by those who do not have adequately calibrated working standards. In many cases the working standards are for the laboratory's internal use only and are never referred to in publications. Establishing new working standards by calibration against some other working standard which in turn may have been calibrated by still other standards is not a recommended practice, nor are other complicated procedures[30] used by some to relate their measurements to the PDB standard.

Isotope geochemists are not completely satisfied with all of the isotope reference samples presently available. The problems involving isotope standards were discussed at both the Third International Conference on Nuclear Geology (Spoleto, Italy, 1965) and the Fifth International Conference on Stable Isotopes (Leipzig, DDR, 1967). There was unanimous agreement on the need for additional standards. A new calcite standard closer to PDB in isotopic composition than is NBS-20 was proposed to eventually supplant NBS-20, and a new ocean-water standard to complement NBS-1 has also been suggested. In addition, there is a need for two CO_2 samples about $40^0/_{00}$ different in isotopic composition to permit the calibration of mass spectrometers. The International Atomic Energy Agency (IAEA) has distributed three water samples to each of 12 laboratories to determine the extent of systematic analytical errors and unreliable working standards.[31] The IAEA is establishing two new water standards for the isotopic analysis of both hydrogen and oxygen.

Isotope-ratio data expressed relative to one standard can be converted so that the same sample data are expressed relative to another standard. δ values with respect to standard S, for example, can be recalculated to give δ values relative to PDB by using the equation

$$\delta_{(x-\mathrm{PDB})} = \delta_{(x-S)} + \delta_{(S-\mathrm{PDB})} + 10^{-3}\,\delta_{(x-S)}\,\delta_{(S-\mathrm{PDB})}$$

where $\delta_{(x-\mathrm{PDB})}$ is the isotopic composition of the sample x relative to the PDB standard, $\delta_{(x-S)}$ is the isotopic composition of the sample relative to a standard S, and $\delta_{(S-\mathrm{PDB})}$ is the isotopic composition of standard S relative to the PDB standard, all δ values being expressed in $^0/_{00}$. δ values for various standards relative to PDB have been reported by Craig.[17]

2.3. Sample-Preparation Procedures

2.3.1. Water

It is undesirable to introduce water directly into the mass spectrometer.[32] An oxygen-containing gas which is very convenient for mass-spectrometric isotopic analysis of oxygen is CO_2 The problem then is to relate the $^{18}O/^{16}O$

ratio of a water sample to the $^{18}O/^{16}O$ ratio of a CO_2 sample so that the analysis of the latter will provide a measurement of the ^{18}O content of the water. A variety of techniques to accomplish this have been developed and it is possible to obtain high-precision isotopic analyses of water samples ranging from many liters to a fraction of a μl in size. These techniques can be divided into two categories: (a) techniques that yield O_2, CO, or CO_2 by chemical reaction including the subsequent conversion of O_2 and CO to CO_2, and (b) techniques that involve isotopic equilibration of water and CO_2.

Procedures of the first type should produce yields as close to 100% as possible, as isotope fractionation generally occurs. Methods such as electrolysis are thus excluded. CO_2 is evolved directly by heating water with guanidine–HCl and good results have been obtained with water samples as small as 0.1 μl.[33,34] Also suitable for very small samples is a procedure by which H_2O is reduced to yield CO and H_2 on a graphite rod heated to 1100–1200°C.[35] Hydrogen is removed from the system by diffusion through a palladium tube and the CO remaining is converted to CO_2 on a nickel catalyst. Water vapor can also be reduced by a mixture of Fe and C at 500°C.[36] Iron oxide and H_2 are formed without producing a significant amount of CO. The iron oxide is later reduced by carbon at 1000°C and the resultant CO is catalytically converted to CO_2 over nickel. With this method, samples of water as small as 10 mg may be analyzed.

Molecular oxygen may be liberated from water by a variety of chemical reactions.[37] The use of bromine pentafluoride for this purpose provides many advantages,[38] among which are 100% yields, good reproducibility ($\pm 0.1‰$), lack of a measurable memory effect, and rapidity (15–20 min per sample). The reaction $BrF_5 + H_2O \rightarrow BrF_3 + 2HF + \frac{1}{2}O_2$ takes place explosively in a nickel vessel, but yields are erratic at room temperature. Between 80 and 100°C, however, the reproducibility is greatly improved. A cold trap cooled by liquid nitrogen removes all of the reaction products except O_2, which is then quantitatively converted to CO_2 over a heated carbon rod.

Equilibration methods are based on the principle that the ratio, $\alpha = (^{18}O/^{16}O)_A/(^{18}O/^{16}O)_B$ is a constant for two phases A and B in isotopic equilibrium at a given temperature. If the quantities of A and B and the fractionation factor α are known, the isotopic composition of one phase can be calculated from mass-balance considerations if the isotopic composition of the other phase has been measured before and after equilibration. CO_2 is most widely used for this purpose although for solid-source mass spectrometry, compounds such as K_2CO_3 and $KMnO_4$ have been utilized.[39,40] A considerable period of time, usually days, is required for isotopic equilibrium to be established between H_2O and CO_2 at room temperature. In the procedure reported by Epstein and Mayeda,[28] the water samples are frozen and the air is quickly pumped away. Approximately 175 cc (STP) of CO_2 are added to 25 g of water, the pH is adjusted to 6 or lower to facilitate equilibration, and the containers are placed in a water bath maintained at 25.3°C. Aliquots of gas are withdrawn at 2-day intervals and analyzed until a constant value of $^{18}O/^{16}O$ indicates that isotopic equilibrium has been obtained.

If a "standard" water is treated exactly as the water samples, the delta value calculated from the $^{18}O/^{16}O$ ratio of the CO_2 equilibrated with standard water and the $^{18}O/^{16}O$ ratio of the CO_2 equilibrated with a water sample will be the same as the delta value expressing the difference in the $^{18}O/^{16}O$ ratio of the two water samples themselves. It is thus unnecessary to determine the fractionation factor, $\alpha = (^{18}O/^{16}O)_{CO_2}/(^{18}O/^{16}O)_{H_2O}$. In order to compare the oxygen-isotope ratios of water with those of oxides, silicates, carbonates, etc., the value of α must, however, be known. It is not possible to determine the difference in the $^{18}O/^{16}O$ ratio of a sample of water and the $^{18}O/^{16}O$ ratio of a sample of quartz by a direct mass-spectrometric comparison of CO_2 containing all of the oxygen originally present in the sample of SiO_2 with the CO_2 which has been equilibrated with the sample of water, unless α is known. Experimentally determined values of α at 25°C are 1.0407,[36] 1.0424,[41] 1.0417,[35] and 1.04073.[38] The value of α also depends on the $^{13}C/^{12}C$ ratio of the CO_2 and the H/D ratio of the water, but variations in carbon- and hydrogen-isotope ratios in natural samples are too small to affect the value of α used for practical purposes. The oxygen isotope fractionation factor at 25°C has been calculated by statistical mechanics for the various isotopic end members in the CO_2–H_2O system,[42] and for the system $CaCO_3$–H_2O–CO_2.[43]

The major disadvantage of the CO_2–H_2O isotopic equilibration method is the length of time required to achieve isotopic equilibrium at room temperature. A number of attempts to increase the rate of equilibration by using higher temperatures have been reported,[44,45] but comparative studies have shown that greater reproducibility is obtained at lower temperatures.[46]

2.3.2. Carbonates

CO_2 can be prepared from carbonates in three different ways: (a) thermal decomposition, (b) acid decomposition, and (c) liberation of oxygen which is subsequently converted to CO_2. In thermal decomposition two of every three oxygen atoms form CO_2 while the third combines with the cation or cations to form an oxide. If the reaction is carried to completion the carbon originally present as carbonate is quantitatively converted to CO_2, hence any fractionation of carbon isotopes during decomposition is of no consequence. In the case of oxygen, however, fractionation effects are important as the $^{18}O/^{16}O$ ratio of the two products, CO_2 and oxide, is appreciably different. If the fractionation of oxygen isotopes between CO_2 and oxide were highly reproducible, thermal decomposition would serve as a rapid and simple method of obtaining CO_2 for both carbon- and oxygen-isotope analysis. McCrea[47] demonstrated, at least for calcite, that the kinetics of thermal decomposition are such that CO_2 with desired reproducibility of isotopic composition cannot be obtained. Although thermal decomposition of carbonates is used,[48,49] decomposition by phosphoric acid or BrF_5 yields better results.

As in the case of thermal decomposition the reaction of acids with carbonates liberates only two-thirds of the total oxygen of the carbonate, but for certain acid reactions the fractionation of oxygen isotopes is highly reproducible. McCrea[47] investigated a number of different acids and obtained the best results with anhydrous H_3PO_4. In this method powdered carbonate (generally between 80- and 200-mesh grain size) and H_3PO_4 are mixed under vacuum and permitted to react in a closed vessel until the evolution of CO_2 appears to cease. The configuration of the reaction vessel is such that acid and sample remain separate until the vessel is evacuated and closed. A convenient design is shown in Fig. 3. The acid is placed in the side arm of the reaction vessel and can be admixed with the carbonate by tipping the vessel on its side. After the reaction appears to be complete, the CO_2 and other volatile phases, which include water and acid vapor, are frozen in a trap cooled by liquid nitrogen and any vapors that are non-condensable at this temperature are rapidly pumped away. The trap is then brought to the temperature of a mixture of dry ice–chloroform–carbon tetrachloride in order to release the CO_2 while retaining water.

The fractionation factor for oxygen in the phosphoric acid reaction is $\alpha = (^{18}O/^{16}O)_{CO_2}/(^{18}O/^{16}O)_{carbonate}$, where $(^{18}O/^{16}O)_{CO_2}$ refers to the carbon dioxide liberated from the carbonate by H_3PO_4. The value of α is dependent upon the temperature of the reaction and the type of carbonate, but the time of reaction for a monomineralic carbonate sample and the ratio of sample to excess acid have no measurable effect.[50] Reactions for Ca, Sr, Ba, and Pb carbonates are essentially complete in 1 day, but 2–3 weeks are required for dolomite and for the carbonates of Zn, Cd, and Mn.[51] Complete reaction, however, is not necessary as no variations in $^{13}C/^{12}C$ or $^{18}O/^{16}O$ with time have been observed.[51] The relationship between α and temperature for calcite over the range 15–40°C is $\alpha = -4.1 \times 10^{-5}t + 1.0111$,

Fig. 3. Vacuum system for the preparation of CO_2 gas samples from carbonates by reaction with anhydrous phosphoric acid. Reaction vessels with side arms for H_3PO_4 are shown above a series of sample tubes used to transfer CO_2 from the reaction system to the mass spectrometer.

where t is the temperature in °C.[50] The fractionation factor is different for different carbonates, and this fact must be taken into account when the $^{18}O/^{16}O$ of two different carbonates, each prepared for analysis by the H_3PO_4-reaction method, are compared. Values of α at 25°C have been determined for the carbonates of Zn, Pb, Sr, Ba, Cd, and Mn, in addition to calcite, aragonite, and dolomite.[51]

The duration of the reaction time is important when the sample contains two or more carbonate minerals of different isotopic composition. Some carbonates react much more slowly than others, and both the carbon- and oxygen-isotope ratios of the liberated CO_2 will change during the reaction. In some cases, advantage may be taken of the differential reaction rates to determine the isotopic compositions of the individual carbonates comprising a mixture. Some carbonate rocks contain both dolomite and calcite so fine-grained and intimately admixed that physical separation is precluded. As dolomite reacts more slowly than calcite with H_3PO_4, the CO_2 released at the beginning of the reaction is almost entirely from the calcite, whereas that evolved near completion of the reaction is almost entirely derived from the dolomite. Good results have been obtained for calcite–dolomite mixtures, about -200 mesh in particle size, prepared from calcite and dolomite each of known isotopic composition.[52] The CO_2 extracted after a reaction time of 1 h represents a 100% yield from calcite but only a 5% yield from the dolomite.

The reproducibility of $\delta^{18}O$ measurements is greatly improved if biogenic carbonate samples are heated in flowing helium prior to reaction with phosphoric acid.[53] Numerous contaminants will contribute to ion beams between masses 43 and 47, and the heating process apparently removes such organic impurities. The measured ^{18}O content is usually lower in samples that have been heated. Twenty to 40 min at 420–430°C is generally sufficient.[54] A vacuum may be substituted for the helium atmosphere.[55]

Oxygen can be extracted quantitatively from carbonates by carbon reduction or by reaction with fluorine or halogen fluorides such as BrF_3 or BrF_5.[51,56] At 700°C the decomposition of calcite by BrF_5 is represented by $3BrF_5 + CaCO_3 \rightarrow CaF_2 + 3BrF_3 + CF_4 + \frac{3}{2}O_2$. This method is especially useful for carbonates that react very slowly with H_3PO_4. Carbon reduction and halogen fluoride extraction are described in more detail in the following section.

2.3.3. *Silicates, Oxides, Phosphates, Sulfates, etc.*

Three types of methods can be used to extract oxygen from a variety of inorganic compounds: (a) reaction with agents such as mercuric cyanide or guanidine–HCl to yield CO_2; (b) reduction by carbon at high temperatures to yield CO; and (c) oxidation by fluorine or halogen fluorides to yield O_2. For most minerals the latter method offers more advantages.

Anbar and Guttmann[57] described techniques by which CO_2 is formed when various inorganic compounds are heated at 400°C with a mixture of mercuric chloride and mercuric cyanide. They reported good results for oxides of Fe, Co, Cu, Zn, Pb, Hg, B, Ba, U, Ag, and Mn, and for some phosphates, sulfates, and arsenates, but the method is unsuitable for silicates and for the oxides of Al and Cr. AgCN can also be used in this way.[58]

With guanidine–HCl, oxygen can be extracted from orthophosphate samples as small as 1–2 mg.[33,34,45]

The carbon-reduction method[59–61] has been used to convert the oxygen of silicates, oxides, and some carbonates to CO, but recent studies have demonstrated the superiority of halogen fluorides, especially BrF_5, for this purpose. To produce CO the oxygen-bearing sample is intimately mixed with finely powdered graphite and reacted at a temperature in the range 1600–2000°C. The major disadvantage of carbon reduction is that complete reaction and 100% yields are sometimes difficult to obtain, resulting in considerable isotopic fractionation for some minerals. A modified version of carbon reduction has been successfully applied to the isotopic analysis of oxygen in sulfates.[62,63]

The reaction of fluorine or chlorine trifluoride with silicates is illustrated by the following examples:[64]

$$3Mg_2SiO_4 + 8ClF_3 \xrightarrow{430^\circ C} 6MgF_2 + 3SiF_4 + 4Cl_2 + 6O_2$$

$$Mg_2SiO_4 + 4F_2 \xrightarrow{420^\circ C} 2MgF_2 + SiF_4 + 2O_2$$

These oxidizing agents are hazardous, however, and the oxygen yields for some minerals are below 100%. Better results can be obtained with bromine pentafluoride, which at room temperature is a colorless liquid. As BrF_5 reacts readily with glass, a metal vacuum system of monel or nickel is required. In the procedure developed by Clayton and Mayeda,[65] commercially available BrF_5 is distilled from a steel cylinder and transferred to a nickel reaction vessel containing between 5 and 30 mg of the mineral to be analyzed. The quantity of BrF_5 used is approximately $5\times$ the stoichiometric requirement.

The reaction between BrF_5 and orthoclase is represented by

$$KAlSi_3O_8 + 8BrF_5 \rightarrow KF + AlF_3 + 3SiF_4 + 8BrF_3 + 4O_2$$

For a 12-h period of reaction, Clayton and Mayeda[65] recommend a temperature of 450°C for quartz, micas, and feldspars; 600°C for magnetite, hematite, and ilmenite; and 690°C for garnet and olivine. A trap cooled by liquid nitrogen retains all of the volatile reaction products except oxygen. The latter is transferred by means of a Toepler pump into a calibrated volume to determine the yield and is subsequently passed over a carbon rod heated to 550–600°C to obtain CO_2. In a method specifically developed for the oxygen-isotope analysis of orthophosphate the sample is first precipitated as $BiPO_4$ and then reacted with BrF_3.[66]

2.3.4. Carbon, Organic and Biologic Compounds, Atmospheric CO_2, and Volcanic Gases

The preparation of CO_2 from carbon, coal, petroleum, organic matter, and hydrocarbon natural gases for carbon-isotope analysis is a simple procedure in which all of the carbon of the sample is oxidized to CO_2.[67–69] Samples are oxidized at 800–900°C over CuO in the presence of oxygen, and if the gases are recycled through a trap cooled by liquid nitrogen, 100% yields in the form of CO_2 can be attained. Volatile impurities are separated from the CO_2 by one or more distillations using a trap cooled by a mixture of dry ice, carbon tetrachloride, and chloroform, and also by passing the gas over manganese dioxide and copper at 500°C. This method is useful for a wide variety of carbon-containing compounds such as organic matter in sediments,[70] natural gases,[71] hydrocarbons,[72] and biogenic organic compounds.[73]

Carbon dioxide can be prepared from organic matter by heating the sample at fairly low temperatures with a strong oxidizing agent. A mixture of chromium trioxide, potassium iodate, phosphoric acid, and fuming sulfuric acid is capable of oxidizing many carbon-containing substances in a few minutes at 150°C.[74]

Isotopic analysis of the oxygen of organic compounds is more difficult than for carbon. The $^{18}O/^{16}O$ ratio of some organic compounds can be measured by directly introducing the sample into the mass spectrometer without any chemical conversion,[75,76] but both sensitivity and precision are insufficient for the determination of natural variations in ^{18}O abundance. The method can be used, however, in biogeochemical tracer experiments provided tracer enrichment is 5% or more. Greater sensitivity is obtained by chemically converting the oxygen of organic compounds into CO_2, but yields close to 100% are required because of isotope fractionation during the conversion process. Rittenberg and Ponticorvo[77] oxidized oxygen-containing organic compounds by heating the sample with $HgCl_2$ at 360–530°C. Isotopic analysis of the CO_2 produced by this reaction yields good results for amino acids and a variety of other organic compounds. Gas–liquid chromatography can be used to purify the CO_2.[78]

Atmospheric carbon dioxide can be absorbed in basic solutions or extracted from the air by freezing with liquid nitrogen. As large isotopic fractionation occurs with the first method, condensation by cooling is more desirable. Keeling[79,80] used evacuated 5-liter flasks to collect air samples in the field. In the laboratory CO_2 was extracted from these samples by passing the air through a trap cooled with liquid nitrogen. Water is easily removed by fractional distillation at dry-ice temperature. This procedure, however, does not separate atmospheric N_2O from the CO_2. These two gases have almost identical physical properties, including similar vapor pressure; N_2O follows CO_2 through the purification stages and enters the mass spectrometer as a contaminant having molecular masses in the same range as CO_2. Although the abundance of N_2O in the atmosphere over the continents is only about 0.5 ppm by volume, i.e., about $1.6^0/_{00}$ of atmospheric CO_2, the effect of N_2O on the measured $\delta^{13}C$ value should be corrected for.[81] Evacuated gas-sample bulbs or flasks are also used to collect other natural gases such as those emanating from volcanos and fumaroles.[82]

2.4. Geochemical Applications

Isotopes differ in both chemical and physical properties. Although these differences are small, for carbon and oxygen they are sufficiently large to give rise to measurable "isotope effects" in many geologic systems. The behavior of two isotopes in a chemical reaction, which may be part of a geologic process, is not exactly the same, and by measuring natural variations in isotope-abundance ratios, information may be obtained to either elucidate the process or determine conditions under which the process operated. Equilibrium and kinetic isotope effects depend on the relationship between isotopic mass and chemical bonding. If, for example, a heavy isotope, ^{18}O, is substituted into a molecule of oxygen, $^{16}O_2$, at ordinary temperatures, the vibrational frequency between the two atoms is decreased and the bond is strengthened. More energy is therefore required to dissociate $^{16}O^{18}O$ than $^{16}O^{16}O$. In addition to chemical reactions isotopes can be fractionated by physical processes such as diffusion, evaporation, etc. $H_2{}^{16}O$ and $H_2{}^{18}O$, for example, have different physical properties, including specific gravity and vapor pressure. In Thode's[83] words, stable isotopes are a "key to our understanding of natural processes." A summary of isotope effects, including their quantitative treatment in terms of equilibrium constants and free-energy changes in isotope-exchange reactions, has been published by Bigeleisen.[84]

The types of problems to which carbon and oxygen stable isotopes have been applied cover almost every aspect of geochemistry. The literature of this field up to 1968 comprises at least 1800 papers and publications.[85] In this summary only a few papers can be cited to exemplify the different applications to geochemistry.

2.4.1. *Ore Deposits and High-Temperature Isotope Thermometry*

By means of stable-isotope ratios the mechanisms by which native sulfur deposits form in salt domes have been elucidated.[86] Isotopes can be used as a guide to the location of mineral deposits as well as to reveal their origin.[87–92] From the fractionation of oxygen isotopes between two or more coexisting minerals such as quartz, calcite, magnetite, etc., it is often possible to determine the temperature at which the mineral deposit was formed.[93–98] In certain cases the isotopic composition of the hydrothermal fluid can be estimated.

2.4.2. *Petroleum, Coal, and Natural Gas*

Carbon-isotope data have yielded much information on the fascinating problems concerned with the origin of petroleum and the mechanisms by which oil evolves,[70,99–101] and the technique may be used in the search for carbonate reservoir rocks.[102,103] $^{13}C/^{12}C$ ratios of natural gases, coal, other fossil fuels, and organic matter in sediments have been extensively studied.[104–113]

2.4.3. *Biogeochemistry*

Abelson and Hoering[114,115] showed how carbon-isotope-ratio measurements are useful in various biogeochemical investigations. One of the more interesting aspects involves the interpretation of the isotopic composition of carbon found in highly metamorphosed Precambrian rocks, implying the existence of very ancient life forms.[116–132] Assigning a biologic or nonbiologic origin to ancient graphites solely on the basis of carbon-isotope data has, however, been criticized.[133,134]

2.4.4. *Oceanography, Limnology, Glaciology, Hydrology, and Meteorology*

The hydrologic cycle is analogous to a large and complex fractional distillation column. Because of the effect of $^{18}O/^{16}O$ on the vapor pressure of water, considerable fractionation of oxygen isotopes occurs in meteorological and related processes. Isotope-ratio measurements can thus be used to help elucidate processes involving the oceans[28,135–141] and lakes.[142–144] As the ^{18}O content of meteoric precipitation depends on such variables as temperature, latitude, altitude, and origin of the air mass and its previous history, $^{18}O/^{16}O$ measurements on snow, firn, and glacial ice can often provide information about past climatic conditions and the history of glaciers.[145–160] Other hydrological applications involve ground and formational waters,[161–166] and hot springs and geysers.[167–174]

2.4.5. *Paleoenvironmental Analysis, Paleotemperatures, Sedimentation, and Diagenesis*

Under some circumstances, $^{13}C/^{12}C$ and/or $^{18}O/^{16}O$ ratios of dissolved bicarbonate are directly or indirectly related to the salinity of the water, or to the nature of the environment in which carbonate sediments are deposited. Isotope studies of modern carbonates, including the skeletons of sediment-contributing organisms, have made it possible to interpret the isotopic composition of ancient carbonate rocks in terms of the depositional environment.[175–189] Paleotemperature analysis is based on the fact that the fractionation of oxygen isotopes between two phases is temperature dependent. In the method established by Epstein *et al.*,[53,54] the two phases are biogenic calcite (fossil invertebrate skeletons) and sea water. Long-term trends in climates of the past have been revealed by paleotemperature determinations.[190–195] The interpretation of some of this work, however, has been questioned.[196–199] Longinelli[200,201] is investigating the possibility of using biogenic phosphate in sedimentary rocks for isotopic paleotemperature measurements. The abundance of dolomite in ancient sedimentary rocks has never been adequately accounted for, and numerous isotope studies, both of natural occurrences and by hydrothermal synthesis, have been directed toward this problem.[52,202–213] Evaporite deposition has also received considerable attention.[214–218] Carbon and oxygen isotopes can provide information concerning the diagenesis of sediments.[219–224]

2.4.6. *Magmatic Processes and Metamorphism*

Isotope-fractionation effects are smaller at higher temperatures but they can be measured and used to elucidate the nature of magmatic processes.[225–229] Of particular interest are the origin of carbonatites,[230–233] the sources of carbon in diamonds,[234] and the nature of metamorphic processes.[235–242]

2.4.7. *Meteorites and Tektites*

A number of studies are described in the literature.[243–252]

2.4.8. *Miscellaneous*

$^{13}C/^{12}C$ analysis can be used to "correct" radiocarbon age determinations for possible fractionation effects.[253,254]

2.5. Future Developments of Stable-Isotope Mass Spectrometry

Stable-isotope mass spectrometry has advanced to the stage where further improvement may be limited by sample-preparation techniques rather than instrument performance. Unless great care is taken it is possible to alter the isotopic composition of the sample, either during the preparation and purification procedures, or by isotope fractionation during gas handling and transfer, to the extent that the differences to be measured are smaller than the variations introduced. This is especially true of very small samples where contamination and isotope fractionation by adsorption on surfaces of the sample reservoir, etc., become more important.[255]

In the future the laborious task of manually recording the many measurements made during analysis will likely be taken over by an analog-to-digital converter system which will prepare punched cards, punched tape, or magnetic tape records for computer processing. A recorder chart-to-punched card converter has already been developed for this purpose.[256] McCullough and Krouse[257] incorporated a five-figure integrating voltmeter–ratiometer and a digital printer in the measuring circuit of a conventional, simultaneous-ion-collection mass spectrometer. They reported increased precision in addition to ease of measurement, as the device eliminates human measurement errors involved in drawing lines through null balance traces on a recorder chart, and averaging of the signals may be accomplished electronically. Other data-processing techniques have also been proposed.[258–260]

Improvements are expected in the use of isotope standards, especially in the availability of adequate standards for mass-spectrometer calibration.

3. TRACE-ELEMENT-CONCENTRATION DETERMINATIONS BY SPARK-SOURCE MASS SPECTROGRAPHY

The determination of trace-element concentrations by conventional thermal- and gas-source mass spectrometry, using stable-isotope dilution techniques, has become a widely used analytical method in geochemistry.

In spite of its superior accuracy the method has two serious limitations: (a) it may be applied only to gaseous elements and those which may be efficiently ionized in a thermal-ion source and (b) only one or at the most a few elements can be determined simultaneously (the rare earths are an exception).

These two limitations have been overcome by the recently developed technique of spark-source mass spectrometry.[5,261,262] In this method a very large number of elements, potentially all, may be determined simultaneously, and a wide concentration range from major elements down to trace constituents may be covered in a single analysis. The sensitivity is approximately, within a factor of two to three, the same for all elements, and the detection limits achieved today for most of the elements are in the ppb range. A decided advantage of this technique is the limited amount of sample preparation required; conductors and semiconductors may be analyzed directly, whereas nonconductors require mixing with a conductor before analysis. The amount of sample consumed during a trace analysis is very small, and 1-ppm detection limits have been achieved on 0.01-mg samples. Whereas this may be of considerable value in the case of small samples, it will pose a problem if the bulk composition of a large sample has to be determined, since it requires that the sample be homogeneous at the scale of volume which is vaporized and ionized in the analysis.

3.1. The Radiofrequency Spark-Source Mass Spectrograph

The mass spectrographs used for trace-element determinations by spark-source techniques (a schematic drawing of a typical instrument is shown in Fig. 4) incorporate as essential components: (a) an ion source, which produces an ion beam representative of the total sample; (b) a mass analyzer (electrostatic analyzer and magnetic analyzer), that resolves the ion beam into a spectrum of mass lines; (c) a detection system (photoplate), which permanently records the lines of the resolved mass spectrum and their intensities; and (d) a vacuum system, that encloses the whole apparatus.

A rf spark source is commonly used as the ion source for trace analysis by mass spectrography because it ionizes all elements with approximately the same efficiency and demonstrates, in addition, high sensitivity, both in terms of the amount of sample consumed during the analysis and the concentration of the element detected. Two disadvantageous features of this type of ion source largely determine the design of the rest of the instrumentation. Firstly, the ions leaving the source show a wide range in energy.

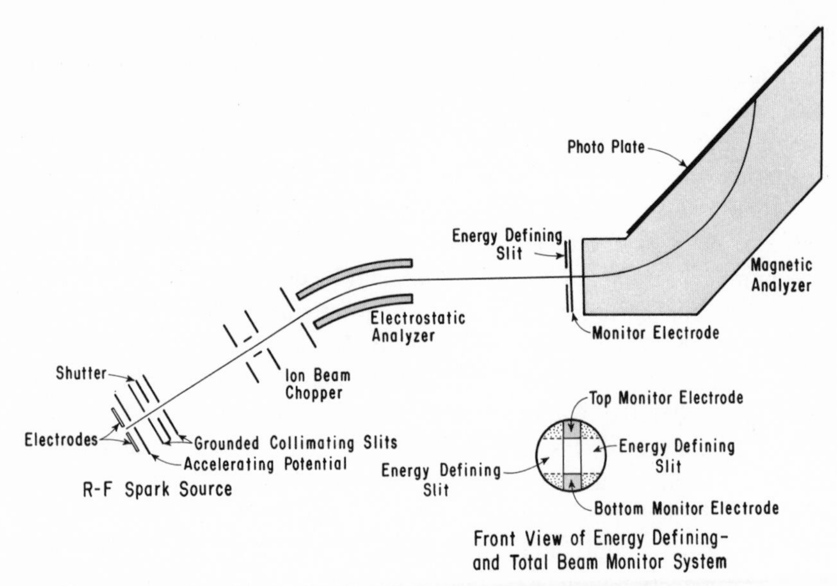

Fig. 4. Diagram of a Mattauch–Herzog double-focusing mass spectrograph.

A mass analyzing system is therefore required that separates and focuses ions irrespective of their initial direction and energy. A combination of cylindrical, or spherical, electric field and magnetic sector field has been found to have this property. The mass-analyzing system most frequently used is based on the design of a double-focusing mass spectrograph by J. Mattauch and R. Herzog. Secondly, the ion beam leaving the source also shows erratic intensity variations in time; hence, the abundance of the mass-resolved ion beams of all elements of interest should be recorded simultaneously and averaged over a certain length of time. This is done conveniently with a photographic plate, which is hence widely used as the ion detector, although electric recording systems have also been employed.

A wide variety of sample types may be analyzed by spark-source mass spectrography; materials studied include conductors, semiconductors, insulators (ceramics and rocks), liquids, and organic materials.

3.1.1. The rf Spark Source

In the rf spark source ionization of the sample is accomplished by the application of a rf potential (50–100 kV at about 10^6 cps) to two conducting electrode rods (1 mm diameter, 10 mm long) of the sample material. A discharge occurs resulting in localized heating of the solid, vaporization, and ionization. In order to keep the temperature of the electrodes low, since excessive heating would cause selective losses of the more volatile elements, the rf field is pulsed. The discharge is sustained only about 3% of the time at most, usually smaller duty ratios are used.

At the effective temperatures in the spark (50,000–60,000°K) differences in the ionization efficiencies between elements are substantially reduced (to within a factor of two to three) compared to other types of ion sources. Thermal-source ionization efficiencies, for example, vary over many orders of magnitude. The relative uniform ionization efficiency is in fact the most attractive feature of the rf spark source for trace analysis. The overall efficiency of the spark-source method is given by the ratio of atoms vaporized to the number of ions detected, a typical atom–ion ratio may be of the order of $5 \times 10^7/1$.

In the rf discharge singly charged positive and negative ions as well as multiply charged ions are formed. For a given element the singly charged species always produces the most intense lines, and the abundance of the multiply charged ions decreases by approximately a factor of 10 for each added charge. The relative-intensity distribution of the multiply charged ions differs from element to element and changes with the operating characteristics of the spark.

Polyatomic ions of the electrode material have been observed; charged carbon clusters as large as C_{34}^+ and C_{33}^- are formed from carbon electrodes, and different polyelemental combinations have also been reported. In general one can count on the formation of polyatomic clusters containing the major constituents of the sample electrodes, and the relative concentration of different species can be related to bond characteristics of the sample compounds. The occurrence of this type of ion will complicate the mass spectra obtained with a rf spark source and has to be taken into account in their interpretation.

The residual gas in the ion source, as well as the gas adsorbed on the electrode surfaces, is ionized in the rf discharge. Whenever the concentration of elements that are present in the background gas have to be determined, thorough cleaning of the electrode surfaces under vacuum and low-ion-source pressures (10^{-9} torr) are required.

The ions formed in the spark acquire their energy by acceleration in the rf field between the electrodes; accordingly, the ion energy will depend on where and when in the rf discharge ionization took place. Ions of different elements, charge states, or polyatomic clusters are formed at different locations between the electrodes and at different times in the rf cycle; hence, they experience different accelerations and accordingly show different energy distributions. Due to the high potential of the rf field ionization in the spark leads to a considerable energy spread (several thousand V) of the ions. Not only must a double-focusing mass spectrometer be therefore used in conjunction with the rf source, but an additional problem is also introduced. As, on the one hand, the ions of different elements formed in the rf spark show varying energy distributions, on the other hand the energy range which can be accepted by the mass-analyzing system is limited ($\pm$300 V), the composition of the total resolved ion beam is not necessarily identical with the ion composition of the spark plasma, and hence corrections may have to be applied.

The ion-acceleration system of the spark source is relatively simple. Since ionization takes place rather erratically, and the ions show a considerable energy spread, little would be gained by employing complex-ion optical-focusing arrangements. Hence, the electrodes are usually placed in front of a large plate with a small, round hole at the optical axis of the instrument. The main accelerating potential (about 20 kV dc) is applied to this plate, and one of the electrodes is electrically tied to it. At a suitable distance two narrow, grounded, collimating slits are provided to limit the width and angular spread of the ions drawn from the source. In addition a wider split plate may be located between the two grounded slits to be used, as an electronic shutter, by the application of an electric potential.

For maximum analytical reproducibility all controllable conditions under which the samples are analyzed have to be kept constant. This includes spark voltage, pulse length and repetition frequency of the rf field, and the width of the spark gap and its position with respect to the accelerating system, as well as the accelerating potential, the electrostatic analyzer potential, and their ratio.

Since the use of the rf spark source for the analysis of solids is associated with difficulties, other types of ion sources have been considered for this purpose; these include the vacuum-vibrator source, the triggered, dc arc source, the laser-discharge source with or without auxiliary electron beams, and ion-bombardment sources. However, presently only the rf spark source is widely used for trace analysis by mass spectrography.

3.1.2. The Mass-Analyzing System

Since the ion beam that leaves the source possesses an angular divergence and a certain spread in energy, the mass-analyzing system should be able to focus ions of the same mass at one point irrespective of their initial direction and energy. This can be accomplished with an electric cylindrical field and a magnetic field in series. Since such an arrangement focuses an ion beam regardless of variations in two of its properties, it is called a double-focusing mass-analyzing system.

The Magnetic Field. In many mass spectrometers mass separation is accomplished with a sector magnetic field. Ions entering such a field are deflected on circular paths of radius a_m, which is given by

$$a_m = \frac{c}{H} \sqrt{\frac{2mV}{ez}}$$

where c is the velocity of light, H is the magnetic field strength, m is the mass of the ion, V is the electric potential through which the ion has been accelerated, e is the elementary charge, and z is the number of unit charges carried by the ion (units in cgs–esu system). For constant H, V, e, and z the deflection of ions in the magnetic field will produce a mass separation. Since ion beams that are drawn from a rf discharge show energy variations (the potential V is not the same for all ions) a magnetic field is insufficient to perform a mass separation on them.

In addition to the deflecting properties the sector magnetic field possesses focusing properties, i.e., ions emerging with a small angular divergence from a point in front of the magnetic field will be recombined in a point behind the magnetic field. If H, e, z, and V are constant for the ion beam entering the magnetic field, ions of a given mass will be focused at this point. If H, e, z, and m are constant, and the ion beam contains ions of two distinct energies, eV_1 and eV_2, where $eV_2 - eV_1 = e\Delta V$, two separate images of ions of the same mass will result. The spacial separation of the two images will be proportional to the relative energy difference $\Delta V/V$. This action of the magnetic field is called energy dispersion.

The Electric Field. The electric cylindrical field is produced by a capacitor consisting of two cylindrical plate sections with a gap of about 1 cm between the two plates. The dimension of the capacitor in the direction of the axis of the cylinder is about 5–10 × larger than the gap width. A potential of 1000–2000 V is applied between the two plates producing a homogeneous, radially symmetrical field. Positive ions entering the field are deflected towards the center if the outer plate is kept at a positive and the inner plate at a negative potential. The amount of deflection will depend on the kinetic energy of the charged particle. The actual paths taken by individual ions are rather complex and cannot be determined as easily as in the case of the magnetic field. The electric cylindrical field also possesses focusing properties. A monoenergetic ion beam of small angular divergence

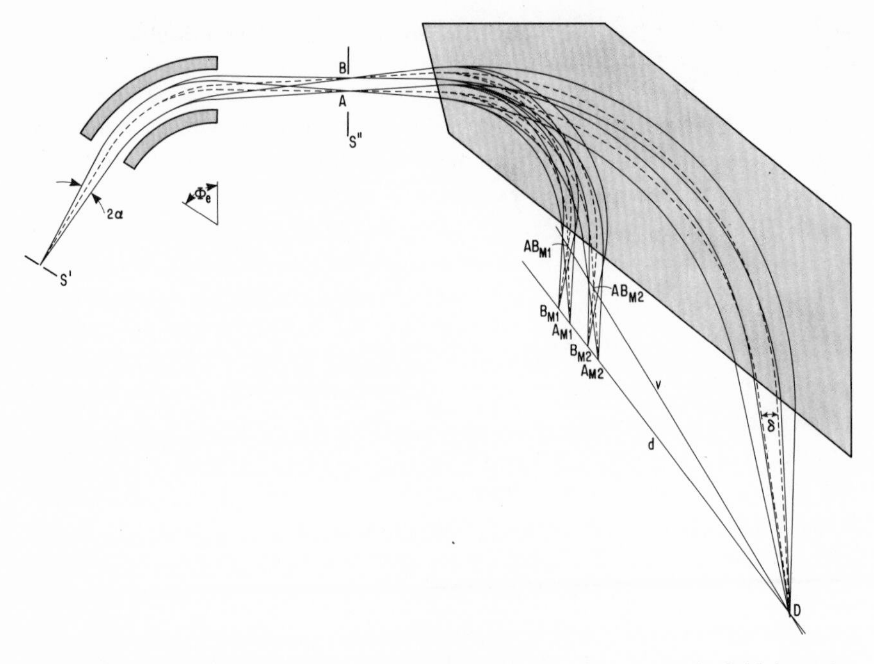

Fig. 5. Principle of double focusing with an electric and a magnetic field in series (schematic).

emerging from a point in front of the electric field will be focused at a certain distance behind the field, as shown in Fig. 5. For a given deflecting potential of the electric field ions of a certain kinetic energy will be focused on the central ion path, indicated by point A in Fig. 5; if ions of a different energy are present in the beam their image will still lie at the same distance behind the electric field, however, it will be displaced from the central path by a certain amount (point B in Fig. 5). The distance between the two points of focus is directly related to the relative energy difference between the ions that are recombined in them. This action of the electric field is called energy dispersion.

The Principle of Double Focusing. Consider an ion beam containing two masses M_1 and M_2 that shows a certain energy spread. If such a beam enters the mass-analyzing system shown in Fig. 5, ions of the smallest energy present in the beam may be focused at point A, and ions with the largest energy at point B. If a slit S″ is provided in the plane of focus of the electric field and adjusted so that only ions focused at point A will pass through it, these ions will be separated in the magnetic field according to their mass and will come to a direction focus behind the magnetic field, for example, at points A_{M_1} and A_{M_2}, corresponding to the masses M_1 and M_2, respectively. Mass separation in the magnetic field is possible because ions passing through A are monoenergetic. Likewise, if the slit is set such that only ions focused at point B will pass through it, the ions will be recombined at points B_{M_1} and B_{M_2} corresponding to the same two masses M_1 and M_2, representing, however, ions of a different energy. The line along which ions of a given energy and small initial angular divergence are focused according to their mass, is called the direction focusing curve *d*. If the slit is opened so that ions of an energy range between the limits given by A and B pass through it the resulting mass line for mass M_1 would be spread over the distance A_{M_1} to B_{M_1}. The paths of the central beams through A and B intersect at points AB_{M_1} and AB_{M_2}; likewise, all other central beams of ions of masses M_1 and M_2 and of energies between A and B will pass through these two points. At points AB_{M_1} and AB_{M_2} the central beams of ions of the same mass but different initial energies are recombined. The line defined by the focus of ions of the same mass but different initial energies and vanishingly small angular divergence (remember, central beams were considered only) is referred to as the velocity focusing curve, *v*. In general the velocity and direction focusing point of a mass do not coincide. Where the velocity focusing curve and the direction focusing curve intersect one another, ions of the same mass are, however, brought to focus, irrespective of their initial direction or energy, i.e., double focusing occurs (point D). By consideration of the paths of the ions which are recombined in point D it is found that double focusing of an ion beam is achieved whenever the energy dispersion introduced into the ion beam in the electric field equals that of the magnetic field, traversed in the opposite direction.

By judicial choice of instrument parameters J. Mattauch and R. Herzog were able to design a mass spectrograph in which the line of direction focus and velocity focus coincide for all masses, the line of double focusing is straight and located at the exit of the magnetic field. This basic design, with only slight modifications, is still being used in most of the mass spectrographs today.

3.1.3. Detection System

Although electrical detection of the resolved ion beams has been used in spark-source mass spectrographs and will become more important in the future, the mass spectra are presently still commonly recorded on photographic plates. Photoplates used in spark-source mass spectrographs consist generally of 2-in.-wide and 10, 12, or 15-in.-long pieces of thin glass covered with a thin layer of gelatin containing an emulsion of silver halide salts.

When ions strike the photographic emulsion they affect the silver halide grains and a latent image is formed. Development of the latent image reduces to metallic silver, the silver salts in those grains which have been affected by ion impact. The resulting cluster of silver grains darkens the emulsion locally. To make the darkening on the plate permanent and to clear the unexposed plate portion, the unreduced silver halide grains are dissolved in a suitable chemical solution; this process is known as fixing. The fixing solution contains, also, a hardening agent which toughens the gelatin. After developing and fixing the plate is washed and dried.

The intensity of darkening on the photoplate is determined with a microdensitometer by measuring the attenuation of a light beam when passing through the darkened area, relative to that when passing the same light beam through an unexposed, clear portion of the plate.

The photographic response measured with the densitometer is in some manner related to the number of ions which have struck the plate over the line area. The exact form of this relationship has to be established by plate calibration. Higher ion-beam intensities and longer exposures will cause more intensely blackened lines than weak ion beams or short exposures. If the blackening of the plate is a function of the product of signal intensity and exposure time, i.e., if the reduction of the exposure length by a given factor may be compensated for by increasing the intensity of the ion beam by the same factor, the law of reciprocity is valid for the plate emulsion.

Special types of photographic plates are required for successful ion-beam-intensity measurements. Desirable emulsion properties include fine, uniform grain; low gelatin content; and uniform, near-surface distribution of the silver halide grains. The Q2 plates (one of the Q emulsions developed for ion detection by the Ilford Company in cooperation with F. W. Aston) are commonly used today for analytical work with spark-source mass spectrographs. These more conventional gelatin-containing plates show, however, a number of undesirable characteristics: (a) emulsion deterioration at room temperatures; (b) sensitivity dependence on mass and energy of the ions; (c) unsharp images due to scattering of ions in the emulsion; (d) line broadening due to poor electrical conductivity of the emulsion; (e) excessive fogging around the lines of heavy exposure; (f) poor vacuum characteristics due to the absorption of water by the gelatin; and (g) susceptibility to abrasion. Since a number of these poor characteristics are related to the presence of gelatin, gelatin-free photoplates have been developed. Their photosensitive silver salts are deposited without a binder directly on the glass backing. The use of these plates has not become widespread, however.

The validity of the law of reciprocity for the Q emulsions has been investigated and verified a number of times. Studies of the ion specific properties of the plate emulsions have shown that their sensitivity is reduced with increasing ion mass and decreasing ion energy, i.e., for a given ion-beam intensity the blackening of a line will be more intense for ions of lower mass and higher energy. The saturation density decreases with increasing mass and decreasing ion energy. The contrast of the Q2 emulsion is also affected by the ion energy; with increasing ion energy the contrast of the plates is reduced.

In order to use the photographic plate darkening as a measure of the size of the ion current producing it, the relationship between ion exposure and photographic response should be determined over the total range of darkening of the emulsion from barely visible images to saturation blackening. A number of methods have been proposed to accomplish this calibration; although they differ considerably in the details of their execution, they are all based on the same principle, namely, that on the plate a number of lines of different densities are produced which possess known relative abundances. The relative intensities of the signals producing a set of calibration lines of varying density on a photoplate may be known from: (a) the length of time an ion beam of constant intensity was striking the plate; (b) the relative abundance of different molecular ion species in the beam; (c) the isotopic composition of elements; and (d) by electrical measurement of part of the total ion beam which is producing the calibration spectra. This can, for example, be done with the total beam monitor, as shown in Fig. 4.

Quite frequently the Churchill two-line method, utilizing the known relative abundance of two isotopes of an element, has been employed for emulsion calibration. The use of the Churchill two-line method has been criticized, however, since only a fraction of the total information usually available on an analytical plate is evaluated for plate calibration. The method also is quite sensitive to small deviations of the assumed isotopic composition from the isotopic abundance which is actually producing the lines on the photoplate (instrumental bias). This latter point is of special importance because of the additive manner in which the emulsion calibration is established by the Churchill method.

A calibration which avoids this problem and utilizes all the available information on the plate may be carried out as follows. In a series of mass spectra of increasing exposure length, recorded below one another on a photoplate, the density of the lines of an element including all isotopes in all exposures is measured. The relative-exposure values of a given mass line in different exposures are given by the exposure ratios determined with the total beam monitor. The relative-exposure values of different isotopes in one exposure are given by their isotopic abundances. The optical density of the lines plotted *versus* the logarithm of the known relative-exposure values yields the plate calibration. Background, line-width variations, and ion specific properties of the emulsion have to be considered.

The calibration methods can be made absolute by measuring simultaneously one of the isotopes of an element electrically while another is recorded photographically. Typically approximately 10^4–10^5 ions/mm^2 must strike the photoplate to produce a barely detectable image. The actual number of ions required will vary with the emulsion and the mass and energy of the ions.

3.2. Analytical Procedure

In order to obtain a meaningful analysis of a material by rf spark-source mass spectrography, one has to ensure that the resolved ion beam is truly representative of the sample. Biases that may occur in the analytical process have to be evaluated and corrected.

It is imperative that the electrodes be representative of the sample to be characterized, and that they are compositionally homogeneous at the scale of volume that is consumed during an analysis. In a trace-element determination about 0.1 mg of the electrodes are removed and only a fraction of this material is actually ionized. Once the ions are formed only part of them are drawn from the ionization region, and only a fraction of these ions is permitted to pass through the energy-defining slit of the spectrograph. Additional ions are lost when the beam passes through the rest of the mass-analyzing system. In the final analysis the analyst relies on about 10^{13} ions to represent the electrode material, which, in turn, one would like to interpret as being representative of a rock sample of several lb, for example. In order that such inferences can reliably be made, one has to ensure that any one of the sampling steps in the analytical procedure yields a representative sample, and that wherever biases are introduced these are known and can be taken into account. Attempts have been made to evaluate separately some of the biases involved in mass-spectrographic analysis. Since most of these biases may not be dealt with directly, however, overall correction factors have been determined by the use of samples of known composition. The correction factors depend on the operating characteristics of the spectrograph and may be applied only to analyses that are carried out under the same conditions under which these factors were determined.

The relative amounts of charge deposited on the photoplate in different mass spectra may be determined by intercepting a fraction of the total ion beam entering the magnetic field. This is accomplished with a beam monitor located behind the energy stop of the instrument, as shown in Fig. 4. The ion currents striking the monitor are measured and integrated with a dc amplifier.

For the analysis of a sample a series of exposures is produced on the photoplate covering, normally, an exposure range from 10^{-13} to 10^{-6} C. After an exposure has been made the plate is moved vertically, with the aid of the photoplate elevator, so that an unexposed portion of the plate is located behind the magnet gap, and the next exposure is laid down. Normally, 15–20 exposures may be recorded on a plate. The lowest exposures are chosen so that some of the isotopes of the major constituent(s) are recorded in the linear portion of the response curve of the photographic emulsion; with an increased exposure length mass lines of lower and lower concentrations will appear. The exposure is normally increased by a factor of two to three depending on the concentration range to be covered on one plate. In most instruments practically the whole elemental mass range may be covered in a single spectrum.

The first step in the evaluation of a recorded mass spectrum is the determination of the masses of the different lines. The orientation in a spectrum is usually not very difficult once a certain amount of experience has been gained, because in most cases the major elements in a given sample are known and the operator soon learns to recognize the typical isotopic composition pattern of different elements. A mass scale along the photoplate may be determined by using the fact that the displacement of mass lines along the line of focus follows a defined relationship. In the case of the Mattauch–Herzog geometry a quadratic relationship holds. If d_x is the displacement of a line of unknown mass M_x from an arbitrary reference point in the spectrum, the mass of this line is given by

$$M_x = (d_x + d_0)^2 h^2$$

where d_0 and h are constants which can be evaluated from the knowledge of the masses of two lines and their respective displacement from the arbitrary reference point.

After the masses of all lines in the spectrum have been established the ion species producing them may be identified. In view of the large variety of ion types formed in a rf discharge the unambiguous identification of lines, especially of weak ones, may be difficult. The spectrum may be further complicated by the occurrence of charge-exchange lines, i.e., lines which are due to ions, such as A^{n+}, which have changed their charge during passage through the instrument by collision with rest-gas atoms or molecules, B, to yield the following:

$$A^{n+} + B \rightarrow A^{m+} + B^{(n-m)+}$$

The effect on the mass spectrum depends on where in the spectrograph the collision occurred. Charge exchange within the electrostatic analyzer or magnetic analyzer will lead to easily recognizable continua. Charge exchange between the electrostatic analyzer and the magnetic analyzer will lead to well-defined mass lines at mass positions $(n/m^2) \times M$, where n is the charge of the ion while passing through the electric field, m is the charge while passing through the magnetic field, and M is its mass.

The following possible contributions to a mass have to be considered: (a) singly charged ions, all unresolved isobars; (b) multiply charged ions; (c) singly charged polymer monoelemental ions; (d) singly charged polymer polyelemental ions; (e) multiply charged polymer ions; and (f) charge-exchange ions. In general the abundance of polymer ions, mono or poly-elemental, and of charge-exchange ions is relatively low and restricted to the major elements in the sample.

The presence of an element in the sample is recognized by the occurrence of its characteristic isotope-abundance pattern. In order that a mass line may be definitely assigned to an element one should be able to detect at least one other isotope of the element and the intensity ratio of the two lines should give the correct isotope ratio within 5% or better.

In the quantitative evaluation of a set of mass spectra absolute concentrations are not derived directly, but different constituents are referred to an isotope of the matrix or of a trace element, or, alternatively, to an isotope of an added internal standard. In order to arrive at absolute concentrations the concentration of the reference element either has to be known or has to be determined by an alternate method. The initial concentrations evaluated from the mass spectra are always in terms of atom fractions (at. %, ppm atomic, or ppb atomic), since they are based on isotopic ratios.

A large number of methods are being used to convert the recorded mass spectra into absolute concentrations, varying in the number of correction factors involved, the manner in which they are established, and the types of biases which are removed by their application.

In the simplest (semiquantitative) evaluation method the ratio of the impurity to that of a reference element is visually estimated from the ratio of the minimum exposures in which impurity element and reference element are barely detectable, with appropriate corrections for the relative isotopic abundances of the isotope lines used. If in a series of graded exposures an isotope of the reference element with relative abundance I_r becomes just detectable in an exposure of E_r relative units, where E is obtained from the integrated charge collected with the total beam monitor, and if an isotope of the trace element with relative abundance I_t is first detected in an exposure of E_t relative units, then if C_r is the concentration of the reference element in the sample, in at. %, the concentration of the trace element C_t is given by

$$C_t = C_r \frac{E_r}{E_t} \frac{I_r}{I_t} \text{ (at. \%)}$$

This method will naturally yield only very rough estimates, since it is based on a visual estimate of the optical density of the lines, neglects background corrections and line-width variations, and assumes that all elements are treated and recorded identically in the analytical process. The results obtained in this manner normally agree within a factor of two to three with those obtained by other analytical techniques.

The method will become considerably more refined if the line densities are measured with a densitometer and are adjusted for line-width variations and background and if a correction factor for different elemental sensitivies R_t is incorporated in the computation. In this case the ratio E_r/E_t is obtained in the following manner. For both lines the density is measured with a microdensitometer in several exposures of increasing length and plotted *versus* the logarithm of the exposure. The exposures required to achieve a line of the same density for the reference element and the unknown are read from the plot; the density for which the comparison is made is chosen so that it will fall on the linear portion of the response curve of the photographic emulsion.

The relative-sensitivity coefficient R_t is a measure of the sensitivity of the total measuring procedure for the line of the trace-element isotope compared with the sensitivity for the line of the reference-element isotope; to the latter an arbitrary value of unity is assigned. R_t is determined by the use of a standard which contains the trace element and the reference element in known proportions. The concentration of the trace element is then calculated from

$$C_t = C_r \frac{E_r}{E_t} \frac{I_r}{I_t} \frac{1}{R_t}$$

R_t varies from element to element and the charge state of the ions, as well as with the operating conditions of the mass spectrograph. A particular value of R_t may, therefore, only be applied to analyses carried out under the same experimental conditions under which the correction factor was determined. The relative sensitivity coefficients may be thought of as a composite of: (a) an ionization correction; (b) a transmission correction; and (c) a detection correction due to ion mass and energy dependence of the emulsion sensitivity. In some cases line-width-variation corrections are also incorporated in R_t.

The acquisition of densitometer data and their reduction to concentrations is time consuming and prone to the introduction of human error, especially if a large number of elements have to be determined; consequently, in many cases only a fraction of the available information is extracted. The plate evaluation has been made more efficient, however, by automatization of the densitometer measurements and the inclusion of computer programs for the determination of emulsion calibration and concentration computations from densitometer data.

3.3. Sample-Handling Techniques

Conducting and semiconducting materials may be used after appropriate shaping directly as electrodes. Care has to be taken, however, that in the shaping process none of the impurities that are being studied are introduced into the sample. The analysis of such electrodes characterizes, naturally, only the electrode material consumed in the analytical process. Whether this is a representative sample of the bulk of the material from which the electrodes were formed is a different question. If the composition of the material is not uniform, individual analyses by spark-source mass spectrography may deviate considerably from those obtained by other analytical methods which utilize comparatively larger amounts of sample. Metal powders may be formed into electrode rods by compacting them under pressure. In cases where the physical properties of the metal prohibit this the powder may be mixed with graphite as a binder. Analyses of low-boiling metals such as mercury or gallium can be accomplished by the use of liquid-nitrogen-cooled electrodes.

Nonconducting materials cannot be ionized directly in a rf spark source, since for the initiation and maintenance of the discharge at least one of the electrodes must be conducting. A number of electrode types have been described which will permit the maintenance of a spark and the ionization of insulating material in it. Some investigators have packed hollow, high-purity metal tubes with insulator powder. Others have backed electrodes of the nonconductor with a high-purity metal strip, or used cup electrodes packed with the sample and faced by a pointed, conducting-probe electrode. All of these electrode types have the disadvantage that the relative contribution of conductor and insulator to the total ion beam is not very reproducible. This makes the quantitative evaluation of the mass spectra obtained with these electrodes difficult. The contribution of sample to the ion beam is more reproducible, if the powdered insulator is well mixed with a high-purity conducting material and pressed into electrode rods. Elements that have been used for this purpose include gold, silver, copper, titanium, tin, nickel, aluminum, and carbon. The selection of a particular conductor will vary with the analytical problem, since the mass spectrum of the added conductor is superposed on that of the sample. Considerable effort may have to be spent to obtain a truly homogeneous mixture of conductor and insulator.

Impurities of liquids have been determined successfully by drying or electroplating them on high-purity metal electrodes; water impurities in the ppb range have been determined on 30-mg samples.

The analysis of traces of carbon, nitrogen, oxygen, and hydrogen is difficult by rf spark-source mass spectrography. This is due to the fact that these elements are present in the residual gas of the mass spectrograph and are ionized in the rf discharge along with the sample. Reduction of the analytical background has been achieved by: (a) flushing the instrument parts which are opened to atmospheric pressures during the changing of a sample with dry argon; (b) removing surface contaminations by cathode etching through the initiation of a glow discharge in the ion source at about 25 μ pressure; (c) the use of cryogenic pumping in the ion source; and (d) the presparking of the electrodes. The ion-source pressure has to be kept in the 10^{-9} torr range during the analysis for these elements. Although concentrations of oxygen, nitrogen, carbon, and hydrogen of a few ppm and somewhat lower may be detected, quantitative determinations are further complicated by the unavailability of certified gas-free samples, lack of standards containing gases in the low-ppm range, and the heterogeneity of the gas distribution in most samples.

Trace-element determinations of organic materials have also been carried out by mass spectrography. Here again the addition of a conductor to the sample is generally required. In some cases the organic sample was mixed directly with graphite and pressed into electrode rods. This results, however, in excessive background spectra, and it is preferable to first carry out a low-temperature ashing of the sample; the ash is then mixed with graphite and formed into electrodes.

As the sample consumption in the spark-source method is rather small the technique may be readily applied to the analysis of microsamples. Small, rigid conducting particles have been handled by constructing special electrode holders of high-purity materials. Small powdered samples have been analyzed by pressing electrodes from metal powders in which the sample proper is confined to the very tip of the electrode rod. With this "tipped-electrode" type trace-element concentrations at the ppm level were determined in 0.01 mg of sample.

Other analytical problems involving small samples are the analysis of small compositional irregularities, such as inclusions, segregations, or layering. In conducting samples such compositional inhomogeneities have been studied by sparking a pointed-probe electrode against a flat sample electrode. With careful probe control impurities of a few ppm may be detected in zones as narrow as 10 mils. The shape of the probe electrode is critical for obtaining sufficiently well-defined spark areas. The contribution of the probe electrode to the total ion beam reduces the sensitivity of the method, and appropriate corrections have to be applied for its presence if quantitative determinations are to be carried out.

The isotope-dilution technique, generally used only for elements which can be ionized in a gas source or thermal source, has been applied to a limited extent to trace analysis by rf spark-source mass spectrography. Concentration determinations of copper in nickel oxide, antimony in tin, and lead in cadmium have been described. The samples are dissolved, the spike solutions are added, and part of the solution is either dried or electroplated on high-purity electrodes.

3.4. Detection Limit, Precision, and Accuracy

The minimum number of atoms required to produce a measurable signal at the detector is defined as the absolute detection limit. If no disturbing influences are present, such as line overlap or excessive background, the absolute detection limit for an element is determined by the sensitivity of the detector for the element, its isotopic composition and ionization efficiency, as well as the transmission efficiency of the instrument. Based on the following observations the absolute detection limit may be estimated to be of the order of 5×10^{11} to 5×10^{12} atoms: (a) one out of 5×10^{7} atoms removed from the electrodes reaches the detector; (b) 10^{4}–10^{5} ions/mm^{2} are required to produce a detectable mass line; and (c) an average line covers about 0.1 mm^{2}. It was assumed in the calculation that all ions from the element are combined in one line, i.e., a monoisotopic element, and the formation of multiply charged and polyatomic ions was neglected. Measured absolute detection limits have been quoted as 6×10^{11} and 10^{13} atoms. The modifier "absolute" in conjunction with detection limit should, however, not be construed to indicate that the given figures present the absolute limit of detection of this method since the number of atoms required to produce a detectable line may be lowered by: (a) improved ion-source design, permitting higher ionization and extraction efficiencies; (b) improved transmission of the spectrograph; (c) increased detector sensitivity; and (d) improved signal-to-noise ratio of the analytical process.

The detection limit is defined as the minimum concentration of an element that may be determined in an analysis. The detectable concentration is reduced by increasing the exposure length. However, limitations exist as to the length of time over which an exposure can reasonably be extended and the size of the sample that is available. A maximum exposure of 10^{-6} C can be obtained within a reasonable length of time. Measurements of samples containing elements with isotopic concentrations at the ppb level as well as extrapolations of concentration determinations of elements in the ppm range show that in a 10^{-6}-C exposure detection limits of a few ppb can be obtained for almost all elements; in some cases detection limits even below 1 ppb have been reported. Exceptions to these low detection limits are the gases hydrogen, nitrogen, and oxygen, and carbon where it lies in the low-ppm range for reasons discussed above.

Apart from sample-size limitations and reasonable exposure length a number of other factors may prohibit the detection of the lowest possible concentrations: (a) spectral overlap due to background lines, charge-exchange lines, multiply charged ions, molecular ions, and isobars may occur; (b) the continuous plate background may become excessive. The general plate background increases with the exposure length, and continua and halation around major lines occurs. These phenomena are related to ion scatter in the instrument, charge-exchange processes, surface charging of the photoplate, and secondary ion impact.

Little can usually be done if the detection of an element is limited by line overlap. An increase in the resolution of the mass spectrograph would help; limits are, however, set by the fact that with a given instrument only a certain maximum resolution can be practically achieved, and that increased resolution will result in a loss of sensitivity since narrower collimating slits have to be used in order to achieve it.

The possibilities of improving detection limits which are imposed by excessive plate background have been and are still being investigated very actively. Obviously a decrease of the residual pressure throughout the mass spectrograph would be beneficial. Background reductions have been achieved by preventing the very intense ion beam of the major component to strike the plate, by breaking it at the point where this ion species would be recorded, and leaving a gap. The halation around intense lines has been reduced also by placing grounded, conduction strips on the plate at the point of impact of the strong ion beams. Special developers and developing conditions can suppress the development of the background and produce clearer photoplates; very considerable improvements in this respect have been achieved by the use of gelatin-free photoplates.

The smallest concentration difference that may be detected is determined by the precision and sensitivity, i.e., the slope of the analytical curve, of the analytical method. The precisions quoted for mass-spectrographic analyses vary considerably. Visual concentration estimates can be carried out within a factor of two to three. Generally, precisions of 20–30%, and in favorable cases of 10%, are obtained, if densitometry is used for plate evaluation and appropriate correction factors are applied. In view of the fact that the early electrical concentration determinations with rf spark-source mass spectrometry showed precisions of 2.5%, i.e., about a factor of 10 better than the results obtained with photographic detection, it has been widely believed that the photoplate recording technique is responsible for the poor reproducibility. This was, however, at variance with the observation that isotopic composition measurements could be made on photoplates with a precision of 5% and better. It has been recently shown that with proper experimental methods the photoplate is inherently capable of yielding results with a precision of 1–2 standard deviations.

Major sources of poor reproducibility are: (a) variability of spark parameters, (b) sample inhomogeneities, and (c) densitometer errors. The recognition that fluctuations in the ionization process of the rf spark source are a major cause for poor reproducibility has led some investigators to abandon the rf spark source completely. Others have attempted to obtain better control over the spark parameters, such as rf pulse length, repetition frequency, spark voltage, spark-gap width, and spark location. In this second approach the spark parameters are fixed as closely as possible throughout the calibration and during the recording of the analytical mass spectra. As the average ion current flowing through the instrument has to be varied in order to be able to record exposures of different intensity, an ion-beam chopper, shown in Fig. 4, is set into the beam path. This device interrupts the constant ion beam with a variable frequency, thus providing a control over the average ion current flowing through the instrument without changes in the sparking conditions. The use of the ion-beam chopper has led to analytical precisions of 4–5%. Similar reproducibilities have been achieved by using a low-voltage discharge source in place of the rf spark source. The use of this type of ion source has, however, so far been very limited.

For accurate measurements a comparative standard of sufficient homogeneity is essential in order to correct for differences in elemental sensitivities. The accuracy of the method will critically depend on the constancy of these relative-sensitivity coefficients and on the availability of standards which match the sample in chemical composition and physical form. In general one can expect that the achievable accuracy is slightly less than the precision obtained by the same analyst.

3.5. Geochemical Applications

Although rf spark-source mass spectrographs have been used extensively in solving a wide variety of analytical problems in industrial research their application in geochemical studies has been very limited.

Leipziger[263,264] was the first to determine ^{207}Pb–^{206}Pb ages of uraninite and pitchblende samples by rf spark-source mass spectrography. After suitable shaping the minerals may be used directly as electrodes. The analytical procedure, therefore, requires no chemical treatment, which presents a certain advantage over the more conventional methods. At the time the determinations were carried out the limited precision with which the isotope ratios could be measured photographically (5%) presented a drawback. With the present better understanding of correct plate evaluation a considerable part of the uncertainty can, however, be removed.

The early applications of the technique to the determination of trace elements in geologic materials were of a preliminary nature, aimed largely at the demonstration of the usefulness of the method.[265–268] The standard rock samples G-1, W-1, and S-1, and a number of native metals and sulfides, were used as samples. Natural graphite crystals were also analyzed.[269]

The quantitative aspect of the method for geologic samples was developed through the work of Taylor[270,271] and Nicholls and his associates.[272] The technique is generally used to supplement concentration determinations carried out by wet chemical, emission spectrographic, x-ray-fluorescence, and other methods. Examples of its application include an investigation concerning the composition of meteorite-impact glass[273] and trace-element studies of andesites[274] and rhyolitic volcanic rocks.[275]

The main advantageous features of the method, such as small sample size, the low detection limits, the wide range of elements that can be determined simultaneously in a single analysis, and the wide concentration range that may be covered in a single analysis, will no doubt lead to further applications in geochemical studies. Complete analysis of many natural waters by conventional methods is difficult without preconcentration. With rf spark-source mass spectrography impurities in the ppb range have been detected in 30-mg-water samples and the whole element range was covered in the analysis. The technique will be especially useful where it is difficult to obtain large sample volumes as in the study of connate waters, fluid inclusions, or liquids equilibrated with solid phases in hydrothermal experiments. Further applications making use of the small sample techniques will include the analysis of small single crystals and of solid phases which have been precipitated in hydrothermal-equilibrium studies. If the rf-probe technique can be extended to insulators it will become possible to study major-, minor-, and trace-element-concentration variations simultaneously in zoned crystals. In view of the low detection limits and the possibility of carrying out simultaneously isotopic and elemental abundance studies, the method may also be applicable to the study of spallation products in meteorites.

REFERENCES

1. Inghram, M. G., and Hayden, R. J., A Handbook on Mass Spectroscopy, Nuclear Science Series, Report No. 14, National Academy of Sciences–National Research Council, Washington, 1954.
2. Duckworth, H. E., *Mass Spectroscopy*, Cambridge University Press, Cambridge, 1958.
3. McDowell, C. A., ed., *Mass Spectrometry*, McGraw-Hill, New York, 1963.
4. Herzog, L. F., Marshall, D. J., Kendall, B. R., and Cambey, L. A., in *Advances in Analytical Chemistry and Instrumentation*, C. N. Reilley, ed., Interscience, New York, 1964, Vol. 3, pp. 143–183.
5. Roboz, J., in *Trace Analysis, Physical Methods*, G. H. Morrison, ed., Interscience, New York, pp. 435–509.
6. Chupakhin, M. S., *Zh. Anal. Khim.* **14**, 331 (1959).
7. Dattner, J., and Fischler, J., *Brit. J. Appl. Phys.* **14**, 728 (1963).
8. Nier, A. O., *Phys. Rev.* **77**, 789 (1950).
9. Kistemaker, J., Natl. Bur. Standards (US) Circ. 522, 1953, p. 243.
10. Boerboom, A. J. H., and Tasman, H. A., *Physica* **24**, 683 (1958).
11. Shackleton, N. J., *J. Sci. Instr.* **42**, 689 (1965).
12. Han, I., and Fritz, G. J., *Anal. Chem.* **37**, 1442 (1965).
13. McKinney, C. R., McCrea, J. M., Epstein, S., Allen, H. A., and Urey, H. C., *Rev. Sci. Instr.* **21**, 724 (1950).
14. Epstein, S., Natl. Bur. Standards (US) Circ. 522, 1953, p. 133.
15. Akers, R. J., *J. Sci. Instr.* **44**, 871 (1967).
16. Chupakhin, M. S., *Zh. Anal. Khim.* **15**, 155 (1960).

17. Craig, H., *Geochim. Cosmochim. Acta* **12**, 133 (1957).
18. Galimov, E. M., Grinenko, V. A., and Ustinov, V. I., *Zh. Anal. Khim.* **20**, 547 (1965).
19. Ustinov, V. I., Galimov, E. M., and Grinenko, V. A., *ibid.* **20**, 1180 (1965).
20. Thode, H. G., Graham, R. L., and Ziegler, J. A., *Can. J. Res.* **23**, 40 (1945).
21. Nier, A. O., Eckelmann, W. R., and Lupton, R. A., *Anal. Chem.* **34**, 1358 (1962).
22. Letolle, B., Marce, A., and Fontes, J. C., *Bull. Soc. Franc. Miner. Crist.* **88**, 417 (1965).
23. Matus, L., Opauszky, I., and Kiss, I., *Pribory i Tekhn. Eksperim.* **11**, 158 (1966).
24. Vetshtein, V. E., *Metrod. Vopr. Izotopnoi Geol., Akad. Nauk SSSR* **68** (1965).
25. Dibeler, V. H., Natl. Acad. Sci.–Natl. Res. Council Publ. 400, 1956, p. 55.
26. Mohler, F. L., *U.S. Natl. Bur. Standards Techn. Notes* **51**, 1 (1960).
27. Craig, H., *Science* **133**, 1833 (1961).
28. Epstein, S., and Mayeda, T., *Geochim. Cosmochim. Acta* **4**, 213 (1953).
29. Dansgaard, W., *ibid.* **3**, 253 (1953).
30. Teis, R. V., Naidin, D. P., Zadorzhnyi, N. K., and Stolyarova, S. S., *Geochem. Intern.* **1**, 44 (1964).
31. Halevy, E., and Payne, B. R., *Science* **156**, 669 (1967).
32. Washburn, H. W., Berry, C. E., and Hall, L. G., U.S. Natl. Bur. Standards Circ. 522, 1953, p. 141.
33. Boyer, P. D., Graves, D. J., Suelter, C. H., and Dempsey, M. E., *Anal. Chem.* **33**, 1906 (1961).
34. Il'in, L. A., *Vestn. Leningr. Univ.* **21**, 85 (1966).
35. Majzoub, M., *J. Chem. Phys.* **63**, 563 (1966).
36. Compston, W., and Epstein, S., *Trans. Amer. Geophys. Union* **39**, 511 (1958).
37. Vetshtein, V. E., *Khim. Prom. Ukraini* **1**, 38 (1966).
38. O'Neil, J. R., and Epstein, S., *J. Geophys. Res.* **71**, 4955 (1966).
39. Trofimov, A. V., *Zh. Anal. Khim.* **8**, 353 (1953).
40. Vol'nov, L. I., Chamova, V. N., and Salimova, K. M., *ibid.* **2**, 1262 (1966).
41. Staschewski, D., *Ber. Bunsenges. Physik. Chem.* **68**, 454 (1964).
42. Staschewski, D., *ibid.* **69**, 426 (1965).
43. Bottinga, Y., *J. Chem. Phys.* **72**, 800 (1968).
44. Dostrovsky, I., and Klein, F. S., *Anal. Chem.* **24**, 414 (1952).
45. Il'in, L. A., Maksimov, N. A., and Panteleeva, N. S., *Vopr. Med. Khim.* **12**, 211 (1966).
46. Tamas, J., and Opauszky, I., *Magy. Kem. Folyoirat* **71**, 352 (1965).
47. McCrea, J. M., *J. Chem. Phys.* **18**, 849 (1950).
48. Bolotnikov, A. A., and Finkel'shtein, Y. B., *Pribory i Tekhn. Eksperim.* **9**, 172 (1964).
49. Vedder, R., *Kernenergie* **3**, 890 (1960).
50. Deines, P., Ph.D. thesis, Pennsylvania State University, 1967, 230 pp.
51. Sharma, T., and Clayton, R. N., *Geochim. Cosmochim. Acta* **29**, 1347 (1965).
52. Epstein, S., Graf, D. L., and Degens, E. T., in *Cosmic and Isotopic Chemistry*, H. Craig, S. L. Miller, and G. J. Wasserburg, eds., North Holland, Amsterdam, 1964, p. 169.
53. Epstein, S., Buchsbaum, R., Lowenstam, H., and Urey, H. C., *Bull. Geol. Soc. Amer.* **62**, 417 (1951).
54. Epstein, S., Buchsbaum, R., Lowenstam, H., and Urey, H. C., *ibid.* **64**, 1315 (1953).
55. Galimov, E. M., *Tr. Mosk. Inst. Neftekhim. i Gaz. Prom.* **56**, 92 (1966).
56. Clayton, R. N., *J. Chem. Phys.* **34**, 724 (1961).

57. Anbar, M., and Guttman, S., *Intern. J. Appl. Radiation Isotopes* **5**, 233 (1959).
58. Shakhashiri, B. Z., and Gordon, G., *Talanta* **13**, 142 (1966).
59. Baertschi, P., and Schwander, H., *Helv. Chim. Acta* **35**, 1748 (1952).
60. Schwander, H., *Geochim. Cosmochim. Acta* **4**, 261 (1953).
61. Dontsova, E. I., *Geokhimiya*, 824 (1959), in translation.
62. Longinelli, A., and Craig, H., *Science* **156**, 56 (1967).
63. Rafter, T. A., *New Zealand J. Sci.* **10**, 493 (1967).
64. Baertschi, P., and Silverman, S. R., *Geochim. Cosmochim. Acta* **1**, 317 (1951).
65. Clayton, R. N., and Mayeda, T. K., *ibid.* **27**, 43 (1963).
66. Tudge, A. P., *ibid.* **18**, 81 (1960).
67. Maass, I., *Chem. Tech.* **10**, 17 (1958).
68. Maass, I., *Kernenergie* **3**, 883 (1960).
69. Craig, H., *Geochim. Cosmochim. Acta* **3**, 53 (1953).
70. Eckelmann, W. R., Broecker, W. S., Whitlock, D. W., and Allsup, J. R., *Bull. Amer. Assoc. Petrol. Geol.* **46**, 699 (1962).
71. Nesmelova, Z. N., and Dubrova, N. V., *Tr. Vses. Neft. Nauchn.-Issled. Geologo-razved. Inst.* **227**, 250 (1964).
72. Zhurov, Y. A., and Karpov, A. K., *Nauchn.-Tekhn. Sb. Po Geol. Razrabotke i Transp. Prirodn. Gaza* **71** (1963).
73. Keith, M. L., Anderson, G. M., and Eichler, R., *Geochim. Cosmochim. Acta* **28**, 1757 (1964).
74. Van Slyke, D. D., Folch, J., and Plazin, J., *J. Biol. Chem.* **136**, 509 (1940).
75. Grosse, A. V., and Kirshenbaum, A. D., *Anal. Chem.* **24**, 584 (1952).
76. Swain, C. G., Tsuchihashi, G., and Taylor, L. J., *ibid.* **35**, 1415 (1963)
77. Rittenberg, D., and Ponticorvo, L., *Intern. J. Appl. Radiation Isotopes* **1**, 208 (1956).
78. Jordan, R. B., and Odell, A. L., *Anal. Chem.* **39**, 681 (1967).
79. Keeling, C. D., *Geochim. Cosmochim. Acta* **13**, 322 (1958).
80. Keeling, C. D., *ibid.* **24**, 277 (1961).
81. Craig, H., and Keeling, C. D., *ibid.* **27**, 549 (1963).
82. Naughton, J. J., and Terada, K., *Science* **120**, 580 (1954).
83. Thode, H. G., *Can. Petrol. Geol. Bull.* **12**, 242 (1965).
84. Bigeleisen, J., *Science* **147**, 463 (1965).
85. Weber, J. N., *Bibliography: Geochemistry of the Stable Isotopes of Carbon and Oxygen*, Materials Research Lab. Monogr. 1, Pennsylvania State Univ., 125 pp.
86. Thode, H. G., Wanless, R. K., and Wallouch, R., *Geochim. Cosmochim. Acta* **5**, 286 (1954).
87. Engel, A. E. J., Clayton, R. N., and Epstein, S., *Intern. Geol. Congr. Sympos. Geochem. Prosp.*, 1956, p. 3.
88. Cheney, E. S., and Jensen, M. L., *Econ. Geol.* **61**, 44 (1966).
89. Perry, E. C., Jr., and Bonnichsen, B., *Science* **153**, 528 (1966).
90. James, H. L., *Bull. Geol. Soc. Amer.* **70**, 1623 (1959).
91. Petrovskaya, N. V., and Grinenko, L. N., *Geol. Rudn. Mestorozhd.* **2**, 3 (1962).
92. Belevtsev, Y. N., Lugovaya, I. P., and Mel'nik, Y. P., *Dokl. Akad. Nauk SSSR* **173**, 678 (1967).
93. Clayton, R. N., Natl. Acad. Sci.–Natl. Res. Council Publ. 1075, 1963, p. 185.
94. Clayton, R. N., in *Studies in Analytical Geochemistry*, D. M. Shaw, ed., Roy. Soc. Can. Spec. Publ. 6, 1963, p. 42.
95. Clayton, R. N., and Epstein, S., *J. Geol.* **66**, 352 (1958).

96. Clayton, R. N., and Epstein, S., *ibid.* **69**, 447 (1961).
97. O'Neil, J. R., and Clayton, R. N., in *Cosmic and Isotopic Chemistry*, H. Craig, S. L. Miller, and G. J. Wasserburg, eds., North Holland, Amsterdam, 1964, p. 157.
98. Sharma, T., Mueller, R. F., and Clayton, R. N., *J. Geol.* **73**, 664 (1965).
99. Silverman, S. R., *Amer. Assoc. Petrol. Geol. Bull.* **44**, 1256 (1960).
100. Silverman, S. R., in *Cosmic and Isotopic Chemistry*, H. Craig, S. L. Miller, and G. J. Wasserburg, eds., North Holland, Amsterdam, 1964, p. 92.
101. Silverman, S. R., and Epstein, S., *Amer. Assoc. Petrol. Geol. Bull.* **42**, 998 (1958).
102. Dakhnov, V. N., and Galimov, E. M., *Sov. Geol.* **9**, 117 (1966).
103. Dakhnov, V. N., and Galimov, E. M., *Geol. Nefti i Gaza* **10**, 59 (1966).
104. Colombo, U., Gazzarrini, F., Gonfiantini, R., Sironi, G., and Tongiorgi, E., in *Advances in Organic Geochemistry*, Pergamon Press, London, 1966, p. 279.
105. Colombo, U., Gazzarrini, F., Sironi, G., Gonfiantini, R., and Tongiorgi, E., *Nature* **205**, 1303 (1965).
106. Compston, W., *Geochim. Cosmochim. Acta* **18**, 1 (1960).
107. Grinberg, I. V., and Petrikovskaya, M. E., *Naukova Dumka Kiev*, 148 (1965).
108. Jeffery, P. M., Compston, W., Greenhalgh, D., and de Laeter, J., *Geochim. Cosmochim. Acta* **7**, 255 (1955).
109. Nakai, N., *J. Earth Sci. Nagoya Univ.* **8**, 174 (1960).
110. Sackett, W. M., *Marine Geol.* **2**, 173 (1964).
111. Sackett, W. M., Eckelmann, W. R., Bender, M. L., and Be, A. W. H., *Science* **148**, 235 (1965).
112. Sackett, W. M., Eckelmann, W. R., Bender, M. L., and Be, A. W. H., *Erdoel Kohle* **19**, 562 (1966).
113. Sackett, W. M., and Thompson, R. R., *Amer. Assoc. Petrol. Geol. Bull.* **47**, 525 (1963).
114. Abelson, P. H., and Hoering, T. C., *Geol. Soc. Amer. Bull.* **71**, 1811 (1960).
115. Abelson, P. H., and Hoering, T. C., *Proc. Natl. Acad. Sci. U.S.* **47**, 623 (1961).
116. Rankama, K., *Geol. Soc. Amer. Bull.* **59**, 389 (1948).
117. Rankama, K., *J. Geol.* **56**, 199 (1948).
118. Rankama, K., *ibid.* **58**, 75 (1950).
119. Rankama, K., *Science* **119**, 506 (1954).
120. Rankama, K., *Geochim. Cosmochim. Acta* **5**, 142 (1954).
121. Rankama, K., *Bull. Comm. Geol. Finlande* **166**, 5 (1954).
122. Rankama, K., *Econ. Geol.* **49**, 541 (1954).
123. Hoefs, J., and Schidlowski, M., *Science* **155**, 1096 (1967).
124. Hoering, T. C., Geol. Soc. Amer. Spec. Paper 68, 1961, p. 199.
125. Hoering, T. C., Carnegie Inst. Washington Year Book 61, 1962, p. 190.
126. Hoering, T. C., Natl. Acad. Sci.–Natl. Res. Council Publ. 1075, 1963, p. 196.
127. Bondesen, E., Pedersen, K. R., and Jorgensen, O., *Medd. Gronland* **164**, 1 (1967).
128. Prashnowsky, A. A., and Schidlowskii, M., *Nature* **216**, 560 (1967).
129. Marmo, V., *Geol. Foren. i Stockholm Forh.* **75**, 89 (1953).
130. Mars, K. E., *J. Geol.* **59**, 131 (1951).
131. Meinschein, W. G., *Science* **150**, 601 (1965).
132. Welin, E., *Geol. Foren. i Stockholm Forh.* **87**, 509 (1965).
133. Craig, H., *Econ. Geol.* **48**, 600 (1953).
134. Craig, H., *Geochim. Cosmochim. Acta* **6**, 186 (1955).
135. Schutz, D. F., *Isotopics* **2**, 1 (1965).

136. Craig, H., *Science* **154**, 1544 (1966).
137. Craig, H., and Gordon, L. I., Univ. Rhode Is. Occ. Publ. 3, 1965, p. 277.
138. Dansgaard, W., *Deep-Sea Res.* **6**, 346 (1960).
139. Eriksson, E., and Bolin, B., U.S. Atom. Energy Comm. Sympos. Ser. 5, 1965, p. 675.
140. Lloyd, R. M., *Geochim. Cosmochim. Acta* **30**, 801 (1966).
141. Weber, J. N., *Geokhimiya* p. 674 (1965).
142. Deevey, E. S., Jr., and Stuiver, M., *Limnol. Oceanogr.* **9**, 1 (1964).
143. Deevey, E. S., Jr., Stuiver, M., and Nakai, N., *Intern. Assoc. Theor. Appl. Limnol. Congr. Proc.* **15**, 284 (1964).
144. Oana, S., and Deevey, E. S., Jr., *Amer. J. Sci.* **258**, 253 (1960).
145. Aldaz, L., and Deutsch, S., *Earth Planet. Sci. Letters* **3**, 267 (1967).
146. Dansgaard, W., *Fysisk Tidsskrift*, 49 (1958).
147. Deutsch, S., Ambach, W., and Eisner, H., *Earth Planet. Sci. Letters* **1**, 197 (1966).
148. Epstein, S., Natl. Acad. Sci.–Natl. Res. Council Publ. 845, 1960, p. 102.
149. Epstein, S., and Sharp, R. P., *J. Geol.* **67**, 88 (1959).
150. Epstein, S., and Sharp, R. P., *J. Geophys. Res.* **72**, 5595 (1967).
151. Epstein, S., Sharp, R. P., and Goddard, I., *J. Geol.* **71**, 698 (1963).
152. Epstein, S., Sharp, R. P., and Gow, A. J., *J. Geophys. Res.* **70**, 1809 (1965).
153. Epstein, S., Sharp, R. P., and Vidziunas, I., *ibid.* **65**, 4043 (1960).
154. Gonfiantini, R., *ibid.* **70**, 1815 (1965).
155. Gonfiantini, R., and Picciotto, E., *Nature* **184**, 1557 (1959).
156. Gonfiantini, R., Togliatti, V., Tongiorgi, E., de Breuck, W., and Picciotto, E., *ibid.* **197**, 1096 (1963).
157. Gonfiantini, R., Togliatti, V., Tongiorgi, E., de Breuck, W., and Picciotto, E., *J. Geophys. Res.* **68**, 3791 (1963).
158. Lorius, C., *Polar Record* **12**, 211 (1964).
159. Macpherson, D. S., and Krouse, H. R., Natl. Acad. Sci.–Natl. Res. Council Publ. 1488, 1967, p. 180.
160. Picciotto, E., Deutsch, S., and Aldaz, L., *Earth Planet. Sci. Letters* **1**, 202 (1966).
161. Bodvarsson, G., *Jokull* **12**, 49 (1962).
162. Clayton, R. N., Friedman, I., Graf, D. L., Mayeda, T. K., Meents, W. F., and Shimp, N. F., *J. Geophys. Res.* **71**, 3869 (1966).
163. Craig, H., *Science* **133**, 1703 (1961).
164. Degens, E. T., *Geol. Rundschau* **52**, 625 (1962).
165. Epstein, S., in *Researches in Geochemistry*, P. H. Abelson, ed., John Wiley & Sons, New York, 1959, p. 217.
166. Graf, D. L., Friedman, I., and Meents, W. F., Ill. State Geol. Surv. Circ. 393, 1965.
167. Banwell, C. J., in *Nuclear Geology on Geothermal Areas*, E. Tongiorgi, ed., Lab. Geol. Nucl. Pisa, Italy, 1963, p. 95.
168. Ferrara, G. C., Ferrara G., and Gonfiantini, R., *ibid.*, p. 277.
169. Ferrara, G., and Gonfiantini, R., *Trans. Amer. Geophys. Union* **46**, 170 (1965).
170. Hulston, J. R., and McCabe, W. J., *Geochim. Cosmochim. Acta* **26**, 383 (1962).
171. Craig, H., in *Nuclear Geology on Geothermal Areas*, E. Tongiorgi, ed., Lab. Geol. Nucl. Pisa, Italy, 1963, p. 17.
172. Craig, H., Boato, G., and White, D. E., Natl. Acad. Sci.–Natl. Res. Council Publ. 400, 1956, p. 29.

173. Fontes, J. C., Glangeaud, L., Gonfiantini, R., and Tongiorgi, E., *Acad. Sci. Paris C.R.* **256**, 472 (1963).
174. White, D. E., Craig, H., and Begemann, F., in *Nuclear Geology on Geothermal Areas*, E. Tongiorgi, ed., Lab. Geol. Nucl. Pisa, Italy, 1963, p. 9.
175. Allen, P., and Keith, M. L., *Nature* **208**, 1278 (1965).
176. Baertschi, P., *Schweiz. Mineral. Petrog. Mitt.* **37**, 73 (1957).
177. Clayton, R. N., and Degens, E. T., *Amer. Assoc. Petrol. Geol. Bull.* **43**, 890 (1959).
178. Keith, M. L., and Parker, R. H., *Marine Geol.* **3**, 115 (1965).
179. Keith, M. L., and Weber, J. N., *Geochim. Cosmochim. Acta* **28**, 1787 (1964).
180. Keith, M. L., and Weber, J. N., *Science* **150**, 498 (1965).
181. Letolle, R., and Tivollier, J., *Acad. Sci. Paris C.R.* **263**, 1824 (1966).
182. Lloyd, R. M., *J. Geol.* **72**, 84 (1964).
183. Taft, W. H., and Harbaugh, J. W., *Stanford Univ. Publ. Geol. Sci.* **8**, 1 (1964).
184. Weber, J. N., *Geochim. Cosmochim. Acta* **31**, 2342 (1967).
185. Weber, J. N., *Amer. J. Sci.* **265**, 586 (1967).
186. Weber, J. N., *Geochim. Cosmochim. Acta* **32**, **33** (1968).
187. Weber, J. N., Bergenback, R. E., Williams, E. G., and Keith, M. L., *J. Sed. Petrol.* **35**, 36 (1965).
188. Weber, J. N., and Raup, D. M., *Geochim. Cosmochim. Acta* **30**, 681 (1966).
189. Weber, J. N., Williams, E. G., and Keith, M. L., *J. Sed. Petrol.* **34**, 814 (1965).
190. Lowenstam, H. A., and Epstein, S., *J. Geol.* **62**, 207 (1954).
191. Urey, H. C., Lowenstam, H. A., Epstein, S., and McKinney, C. R., *Geol. Soc. Amer. Bull.* **62**, 399 (1951).
192. Bowen, R., *Paleotemperature Analysis*, Elsevier, Amsterdam, 1966, 265 pp.
193. Bowen, R., *Earth Sci.* **2**, 199 (1966).
194. Emiliani, C., *Science* **154**, 851 (1966).
195. Emiliani, C., *J. Geol.* **74**, 109 (1966).
196. Donn, W. L., and Shaw, D. M., *ibid.* **75**, 497 (1967).
197. Donn, W. L., and Shaw, D. M., *Science* **157**, 722 (1967).
198. Shackleton, N. J., *Nature* **215**, 15 (1967).
199. Shackleton, N. J., *ibid.* **218**, 79 (1968).
200. Longinelli, A., *ibid.* **207**, 716 (1965).
201. Longinelli, A., *ibid.* **211**, 923 (1966).
202. Clayton, R. N., Jones, B. F., and Berner, R. A., *Geochim. Cosmochim. Acta* **32**, 415 (1968).
203. Degens, E. T., and Epstein, S., *ibid.* **28**, 23 (1964).
204. Russell, K. L., Deffeyes, K. S., Fowler, G. A., and Lloyd, R. M., *Science* **155**, 189 (1967).
205. Spotts, J. H., and Silverman, S. R., *Amer. Min.* **51**, 1144 (1966).
206. Weber, J. N., *Science* **145**, 1303 (1964).
207. Weber, J. N., *Geochim. Cosmochim. Acta* **28**, 1257 (1964).
208. Bonatti, E., *Science* **153**, 534 (1966).
209. Friedman, I., and Hall, W. E., *J. Geol.* **71**, 238 (1963).
210. Northrop, D. A., and Clayton, R. N., *ibid.* **74**, 174 (1966).
211. O'Neil, J. R., and Epstein, S., *Science* **152**, 198 (1966).
212. Schwarcz, H. P., *J. Geol.* **74**, 38 (1966).
213. O'Neil, J. R., and Epstein, S., *Trans. Amer. Geophys. Union* **45**, 113 (1964).
214. Borshchevskiy, Y. A., and Khristianov, V. K., *Geokhimiya*, 844 (1965).

215. Fontes, J. C., *Geol. Rundschau* **55**, 172 (1965).
216. Fontes, J. C., and Nielson, H., *Acad. Sci. Paris C.R.* **262**, 2685 (1966).
217. Gonfiantini, R., and Fontes, J. C., *Nature* **200**, 644 (1963).
218. Uklonskii, A. S., Tsapenko, M. N., and Gluschchenko, V. M., *Dokl. Akad. Nauk UzSSR* **23**, 44 (1966).
219. Berner, R. A., *Science* **159**, 195 (1967).
220. Gross, M. G., *J. Geol.* **72**, 170 (1964).
221. Gross, M. G., and Tracey, J. I., Jr., *Science* **151**, 1082 (1966).
222. Hodgson, W. A., *Geochim. Cosmochim. Acta* **30**, 1223 (1966).
223. Murata, K. J., Friedman, I. I., and Madsen, B. M., *Science* **156**, 1484 (1967).
224. Degens, E. T., *Neues Jahrb. Geol. Pal. Monatsh.*, 72 (1959).
225. Dontsova, E. I., *Geokhimiya*, 430 (1966).
226. Dontsova, E. I., and Milovskii, A. V., *Geokhimiya*, 655 (1967).
227. Garlick, G. D., *Earth Planet. Sci. Letters* **1**, 361 (1966).
228. Taylor, H. P., Jr., and Epstein, S., *Geol. Soc. Amer. Bull.* **73**, 461 (1962).
229. Taylor, H. P., Jr., and Epstein, S., *J. Petrol.* **4**, 51 (1963).
230. Deines, P., *Geochim. Cosmochim. Acta* **32**, 613 (1968).
231. Kukharenko, A. A., and Dontsova, E. I., *Econ. Geol. USSR* **1**, 31 (1964).
232. Taylor, H. P., Jr., Frechen, J., and Degens, E. T., *Geochim. Cosmochim. Acta* **31**, 407 (1967).
233. Vinogradov, A. P., Kropotova, O. I., Epstein, E. M., and Grinenko, V. A., *Geokhimiya*, 499 (1967).
234. Vinogradov, A. P., Kropotova, O. I., Orlov, Y. L., and Grinenko, V. A., *Geokhimiya*, 1395 (1966).
235. Anderson, A. T., *J. Geol.* **75**, 323 (1967).
236. Hahn-Weinheimer, P., *Geol. Rundschau* **55**, 197 (1965).
237. Paulitsch, P., and Hahn-Weinheimer, P., *Naturwiss.* **48**, 597 (1961).
238. Garlick, G. D., and Epstein, S., *Geochim. Cosmochim. Acta* **31**, 181 (1967).
239. Schwarcz, H. P., *Geol. Soc. Amer. Bull.* **77**, 879 (1966).
240. Schwarcz, H. P., and Clayton, R. N., *Can. J. Earth Sci.* **2**, 72 (1965).
241. Taylor, H. P., Jr., Geol. Soc. Amer. Spec. Paper 76, 1963, p. 163.
242. Taylor, H. P., Jr., Albee, A. L., and Epstein, S., *J. Geol.* **71**, 513 (1963).
243. Boató, G., *Geochim. Cosmochim. Acta* **6**, 209 (1954).
244. Clayton, R. N., *Science* **140**, 192 (1963).
245. Trofimov, A. V., *Dokl. Akad. Nauk SSSR* **72**, 663 (1950).
246. Reuter, J. H., Epstein, S., and Taylor, H. P., Jr., *Geochim. Cosmochim. Acta* **29**, 481 (1965).
247. Taylor, H. P., Jr., Duke, M. B., Silver, L. T., and Epstein, S., *ibid.* **29**, 489 (1965).
248. Taylor, H. P., Jr., and Epstein, S., in *Cosmic and Isotopic Chemistry*, H. Craig, S. L. Miller, and G. J. Wasserburg, eds., North Holland, Amsterdam, 1964, p. 181.
249. Taylor, H. P., Jr., and Epstein, S., *Science* **153**, 173 (1966).
250. Vinogradov, A. P., Dontsova, E. I., and Chupakhin, M. S., *Geochim. Cosmochim. Acta* **18**, 278 (1960).
251. Vinogradov, A. P., Kropotova, O. I., Vdovykin, G. P., and Grinenko, V. A., *Geokhimiya*, 267 (1967).
252. Walter, L. S., and Clayton, R. N., *Science* **156**, 1357 (1967).
253. Keith, M. L., and Anderson, G. M., *Science* **141**, 634 (1963).
254. Pearson, F. J., Jr., U.S. Atom. Energy Comm. Rept. CONF-650652, 1966, p. 357.

255. Hoering, T. C., and Bufalini, M., *Carnegie Inst. Washington Papers Geophys. Lab.* **7**, 205 (1961).
256. Cassie, G. E., Cuthbert, J., and Silk, C., *J. Sci. Instr.* **43**, 283 (1966).
257. McCullough, H., and Krouse, H. R., *Rev. Sci. Instr.* **36**, 1132 (1965).
258. Hagan, P. J., and de Laeter, J. R., *J. Sci. Instr.* **43**, 662 (1966).
259. Moreland, P. E., Jr., Stevens, C. M., and Waling, D. B., *Rev. Sci. Instr.* **38**, 760 (1967).
260. Ridley, R. G., and Young, W. A. P., U.K. Atom. Energy Auth. Rept. AWRE 0-37, 1965.
261. Ahearn, A. J., ed., *Mass Spectrometric Analysis of Solids*, Elsevier, New York, 1966.
262. Deines, P., The Principles of Spark Source Mass Spectrography and its Application to the Determination of Trace Element Concentrations, Experiment Station Circular 78, The College of Earth and Mineral Science, The Pennsylvania State University, 1970.
263. Leipziger, F. D., *Appl. Spectry.* **17**, 158 (1963).
264. Leipziger, F. D., and Croft, W. J., *Geochim. Cosmochim. Acta* **28**, 268 (1964).
265. Brown, R., and Wolstenholme, W. A., Paper 75, presented at the 11th Annual Conference on Mass Spectrometry and Allied Topics, San Francisco, 1963.
266. Brown, R., and Wolstenholme, W. A., *Nature* **201**, 598 (1964).
267. Herzog, L. F., Eskew, T. J., and Deines, P., *Trans. Am. Geophys. Union* **45**, 112 (1964).
268. Herzog, L. F., Eskew, T. J., and Deines, P., The Pennsylvania State University Mineral Industries Experiment Station Special Report No. 1–65, 1965.
269. Cookson, G. G., Fletcher, W., and Tushingham, R., Conf. Ind. Carbon and Graphite, 1965, London, 1966, p. 77.
270. Taylor, S. R., *Geochim. Cosmochim. Acta* **29**, 1243 (1965).
271. Taylor, S. R., *Nature* **205**, 34 (1965).
272. Nicholls, G. D., Graham, A. L., Williams, E., and Wood, M., *Anal. Chem.* **39**, 584 (1967).
273. Taylor, S. R., *Geochim. Cosmochim. Acta* **30**, 1121 (1966).
274. Taylor, S. R., and White, A. R. J., *Bull. Volcanol.* **29**, 177 (1966).
275. Ewart, A., Taylor, S. R., and Capp, A. C., *Contr. Mineral. and Petrol.* **18**, 76 (1968).

Cromite fusion, p. 61.